Rhinovirus Infections

Rhinovirus Infections

Rethinking the Impact on Human Health and Disease

Edited by

NATHAN BARTLETT
Viral Immunology and Respiratory Disease Group, Priority Centre for Healthy Lungs, School of Biomedical Sciences and Pharmacy, Faculty of Health and Medicine, University of Newcastle, Callaghan, NSW, Australia

PETER WARK
Priority Research Centre for Healthy Lungs, HMRI, University of Newcastle, Callaghan, NSW, Australia
Department of Respiratory and Sleep Medicine, John Hunter Hospital, New Lambton, NSW, Australia

DARRYL KNIGHT
School of Biomedical Sciences and Pharmacy, Faculty of Health and Medicine, University of Newcastle, Callaghan, NSW, Australia

ELSEVIER

ACADEMIC PRESS
An imprint of Elsevier

Academic Press is an imprint of Elsevier
125 London Wall, London EC2Y 5AS, United Kingdom
525 B Street, Suite 1650, San Diego, CA 92101, United States
50 Hampshire Street, 5th Floor, Cambridge, MA 02139, United States
The Boulevard, Langford Lane, Kidlington, Oxford OX5 1GB, United Kingdom

British Library Cataloguing-in-Publication Data
A catalogue record for this book is available from the British Library

Library of Congress Cataloging-in-Publication Data
A catalog record for this book is available from the Library of Congress

ISBN: 978-0-12-816417-4

For Information on all Academic Press publications
visit our website at https://www.elsevier.com/books-and-journals

Publisher: Stacy Masucci
Acquisition Editor: Katie Chan
Editorial Project Manager: Kristi Anderson
Production Project Manager: Stalin Viswanathan
Cover Designer: Limbert Matthew

Typeset by MPS Limited, Chennai, India

CONTENTS

LIST OF CONTRIBUTORS

AusREC
Priority Research Centre for Healthy Lungs, Hunter Medical Research Institute, The University of Newcastle, Callaghan, NSW, Australia; Telethon Kids Institute, The University of Western Australia, Nedlands, WA, Australia; Robinson Research Institute, University of Adelaide, Adelaide, SA, Australia

Nathan Bartlett
Priority Research Center for Healthy Lungs, Faculty of Health and Medicine, University of Newcastle, Newcastle, Australia

Yury A. Bochkov
Department of Pediatrics, University of Wisconsin School of Medicine and Public Health, Madison, WI, United States

Punnam Chander-Veerati
School of Medicine and Public Health, University of Newcastle, Callaghan, NSW, Australia; Priority Research Centre for Healthy Lungs, Hunter Medical Research Institute, The University of Newcastle, Callaghan, NSW, Australia

Sarah Croft
Faculty of Science and Technology, Centre for Research in Therapeutic Solutions, University of Canberra, Canberra, ACT, Australia

Michael R. Edwards
Airway Disease Infection Section, National Heart and Lung Institute, Imperial College, London, United Kingdom; MRC & Asthma UK Centre in Allergic Mechanisms of Asthma, London, United Kingdom

Camille Esneau
University of Newcastle, Callaghan, NSW, Australia; Priority Research Centre for Healthy Lungs, Faculty of Health and Medicine, University of Newcastle, Newcastle, Australia

Luke W. Garratt
Telethon Kids Institute, The University of Western Australia, Nedlands, WA, Australia

James E. Gern
Departments of Pediatrics and Medicine, University of Wisconsin-Madison, Madison, WI, United States

Reena Ghildyal
Faculty of Science and Technology, Centre for Research in Therapeutic Solutions, University of Canberra, Canberra, ACT, Australia

Jason Girkin
University of Newcastle, Callaghan, NSW, Australia; Priority Research Center for Healthy Lungs, Faculty of Health and Medicine, University of Newcastle, Newcastle, Australia

Thomas Iosifidis
Telethon Kids Institute, The University of Western Australia, Nedlands, WA, Australia

Sebastian L. Johnston
Airway Disease Infection Section, National Heart and Lung Institute, Imperial College, London, United Kingdom; MRC & Asthma UK Centre in Allergic Mechanisms of Asthma, London, United Kingdom

Anthony Kicic
Telethon Kids Institute, The University of Western Australia, Nedlands, WA, Australia; Occupation and Environment, School of Public Health, Curtin University, Bentley, WA, Australia; Paediatrics, Medical School, Faculty of Healthy and Medical Science, The University of Western Australia, Nedlands, WA, Australia; Department of Respiratory and Sleep Medicine, Perth Children's Hospital, Nedlands, WA, Australia; Centre for Cell Therapy and Regenerative Medicine, School of Medicine and Pharmacology, The University of Western Australia, Nedlands, WA, Australia

Ngan Fung Li
Priority Research Centre for Healthy Lungs, Hunter Medical Research Institute, The University of Newcastle, Callaghan, NSW, Australia; School of Biomedical Sciences and Pharmacy, University of Newcastle, Callaghan, NSW, Australia

Su-Ling Loo
Priority Research Centre for Healthy Lungs, Hunter Medical Research Institute, The University of Newcastle, Callaghan, NSW, Australia; School of Biomedical Sciences and Pharmacy, University of Newcastle, Callaghan, NSW, Australia

Kevin Looi
Telethon Kids Institute, The University of Western Australia, Nedlands, WA, Australia

Patrick Mallia
Faculty of Medicine, National Heart & Lung Institute, Imperial College London, London, United Kingdom

Steven Maltby
Priority Research Center for Healthy Lungs, Faculty of Health and Medicine, University of Newcastle, Newcastle, Australia

Gary McLean
Molecular Immunology, Cellular Molecular and Immunology Research Centre, London Metropolitan University, London, United Kingdom; National Heart and Lung Institute, Imperial College London, London, United Kingdom

Sai P. Narla
Diamantina Institute, The University of Queensland, Brisbane, QLD, Australia; Department of Respiratory Medicine, Princess Alexandra Hospital, Brisbane, QLD, Australia

Brian Gregory George Oliver
Woolcock Institute of Medical Research, The University of Sydney, Glebe, NSW, Australia; School of Life Sciences, University of Technology Sydney, Ultimo, NSW, Australia

Prabuddha Pathinayake
School of Medicine and Public Health, University of Newcastle, Callaghan, NSW, Australia; Priority Research Centre for Healthy Lungs, Hunter Medical Research Institute, The University of Newcastle, Callaghan, NSW, Australia

Andrew T. Reid
School of Medicine and Public Health, University of Newcastle, Callaghan, NSW, Australia; Priority Research Centre for Healthy Lungs, Hunter Medical Research Institute, The University of Newcastle, Callaghan, NSW, Australia

Andrew I. Ritchie
Airway Disease Infection Section, National Heart and Lung Institute, Imperial College, London, United Kingdom; Royal Brompton and Harefield NHS Trust, London, United Kingdom

Aran Singanayagam
Faculty of Medicine, National Heart & Lung Institute, Imperial College London, London, United Kingdom

Roberto Solari
National Heart and Lung Institute, Imperial College London, London, United Kingdom

Erika N. Sutanto
Telethon Kids Institute, The University of Western Australia, Nedlands, WA, Australia

John W. Upham
Diamantina Institute, The University of Queensland, Brisbane, QLD, Australia; Department of Respiratory Medicine, Princess Alexandra Hospital, Brisbane, QLD, Australia

Peter Wark
Centre for Healthy Lungs, HMRI, University of Newcastle, New Lambton, NSW, Australia

Teresa Williams
Priority Research Centre for Healthy Lungs, Hunter Medical Research Institute, The University of Newcastle, Callaghan, NSW, Australia; School of Biomedical Sciences and Pharmacy, University of Newcastle, Callaghan, NSW, Australia

FOREWORD

In 1963, the name *rhinovirus* was given to a series of recently identified enterovirus-like isolates from individuals with common cold symptoms. These viruses were distinguished from the enteroviruses by their sensitivity to acid pH and preferential replication at 33°C. Over the next 30 years the rhinoviruses were recognized as one of the most common infections of mankind. These infections were typically associated with common cold symptoms and approximately 50% of all common cold illnesses were caused by rhinoviruses. Although lower respiratory illness in infants with rhinovirus was described, the strong association of rhinovirus infection with the common cold together with a widely held view that "restricted" replication at 37°C limited the virus to the upper respiratory tract led to a common perception that rhinovirus infections were of little medical importance.

The development of polymerase chain reaction as a sensitive diagnostic tool in the 1990s led to a remarkable reassessment of the epidemiology, clinical manifestations, and approach to treatment of rhinovirus infections. The use of PCR led to the discovery of a new species of rhinovirus, rhinovirus C, that unlike the known RV-A and RV-B species cannot be detected by standard cell culture methods. In contrast to early studies using cell culture detection that had described rhinovirus occurring predominately in spring and fall epidemics, population surveys using PCR found that infections occur year-round and revealed that 30%–40% of rhinovirus infections are asymptomatic.

The availability of PCR also changed our perception of the medical importance of the rhinoviruses. It is now clear that the rhinoviruses are an important cause of lower respiratory disease in some populations. Exacerbations of asthma, chronic obstructive pulmonary disease, and cystic fibrosis are frequently associated with rhinovirus infection and prevention or treatment of the infection has become a target of efforts to prevent these exacerbations. It is also clear that the early observations of bronchiolitis and pneumonia associated with rhinovirus infection in young infants were correct and rhinovirus is now recognized as an important cause of these syndromes. The role of rhinovirus as a cause of pneumonia in immunocompromised hosts and in older children and adults, however, remains less clear. The frequent occurrence of coinfections with other known lower respiratory pathogens together with the high prevalence of rhinovirus infection in the general population suggests caution in the extrapolation of rhinovirus detection in the nose to causation of these lower respiratory tract syndromes. The potential that rhinovirus infections in young infants may, together with certain host and environmental factors, be a cause of asthma has emerged as another area of active investigation. Progress in this area of investigation may provide the impetus for a more vigorous exploration of approaches to treatment and prevention of these infections.

There have been extensive efforts to develop effective interventions for rhinovirus infections beginning with investigation of vaccines in the 1960s and interferon in the 1970s and 1980s. These early studies were done in the context of our perception that rhinovirus was a common cold virus. The limited medical importance of the common cold suggested that any treatment must be extraordinarily effective, completely safe, and inexpensive. Despite repeated efforts, no antiviral therapy was ever able to achieve these high standards. The recognition of the role of rhinovirus in exacerbations of lower respiratory symptoms in hosts with underlying lung disease has changed the assumptions around treatment approaches. It is clear that interventions that even modestly reduce the incidence of exacerbations of asthma or COPD will be medically important and cost effective. Given this new understanding there has been a renewed interest in previously studied drugs as well as in the development of new interventions for treatment or prevention of rhinovirus infection and illness. The discovery of the RV-C species, which attaches to cells via a unique receptor and is constitutively resistant to the capsid-binding antivirals, has complicated the development of antiviral approaches but the recognition of the role of the innate immune response in pathogenesis and host response to rhinovirus infection has provided a new avenue of exploration for prevention of infection and illness. Interventions designed to manipulate the innate response either directly, such as type I interferon and Toll-like receptor agonists or antagonists, or indirectly by attempts to alter the airway microbiome with probiotics may prove to be a fruitful approach.

The story of the rhinoviruses describes a remarkable evolution in our perception of these viruses from a consensus that these infections had little medical importance to a recognition that rhinovirus infections are an important cause of respiratory illness in certain patient populations and potentially play a role as a cause of chronic lung disease. This dramatic shift in perception has occurred, and continues to evolve, largely due to the availability of ever more sophisticated molecular diagnostic techniques. This timely book provides a review of our current understanding as well as a view of the road ahead for this familiar and yet misunderstood group of viruses.

Ronald B. Turner

CHAPTER 1

Rhinovirus structure, replication, and classification

Camille Esneau[1], Nathan Bartlett[1] and Yury A. Bochkov[2]
[1]Priority Research Centre for Healthy Lungs, Faculty of Health and Medicine, University of Newcastle, Newcastle, Australia
[2]Department of Pediatrics, University of Wisconsin School of Medicine and Public Health, Madison, WI, United States

1.1 INTRODUCTION

Non-influenza viral respiratory tract illnesses are estimated to result in 39.5 billion USD in total healthcare expenditures, including both direct (visit to the physician and drugs prescriptions) and indirect (missed work days) costs. This total ranks higher than the total expenses due to asthma and chronic obstructive pulmonary disease (COPD).[1] While the contribution of rhinovirus (RV) to respiratory tract illness was initially thought to be quite low, RV is now recognized as an important mediator of upper respiratory tract infection[2] responsible for up to 60% of yearly respiratory illnesses.[3,4] Further, RV infections are important causes of more severe illnesses including asthma and COPD exacerbations,[5] community-acquired pneumonia,[6] and bronchiolitis.[7] Currently, there is no commercialized therapy to prevent or treat RV infection. Although early clinical trials showed that protective immunity could be induced by formalin-inactivated RVs,[8,9] vaccine development has been hampered by the diversity of co-circulating RV types and limited cross-reactivity of induced adaptive immune responses between different RVs. Several classes of antiviral drugs (capsid-binding compounds, protease inhibitors, soluble receptors, etc.) were not sufficiently safe and efficient either.[10–12] RVs are classified into three species, RV-A, -B, and -C, consisting of over 160 types based on genetic homology, more than all other *Enterovirus* genus members with human tropism combined.[13] Strategies to categorize RV types have been shaped by increased understanding of RV biology and have passed through several iterations since the initial discovery of RV in the 1950s. An understanding of RV structure, replication, genetics and diversity is necessary to inform novel treatment strategies to reduce the

Rhinovirus Infections
DOI: https://doi.org/10.1016/B978-0-12-816417-4.00001-9

burden of RV-induced disease. This chapter will provide an overview of the structural and genetic features that distinguish RV species, as well as historical and current methods used for RV classification (e.g., serotype, receptor usage, and genetic type).

1.2 BIOLOGY OF HUMAN RHINOVIRUSES

1.2.1 Rhinovirus genome organization

RV is a positive-sense, single-stranded RNA virus belonging to the *Enterovirus* genus in the family *Picornaviridae*.[13] The RV genome can vary between 7079 bases for RV-C1 to 7233 bases for RV-B92 and contains a single open reading frame that functions as a messenger RNA.[14] The RV genome contains a high percentage of adenine (31%–34%) and uracil bases (25%–30%) and is lower in guanine and cytosine content (19%–22% and 18%–22%, respectively).[15] The RV coding region is approximately 2150 codons in length, consisting of 11 genes organized in the same general structure as in other Enteroviruses (Fig. 1.1).[16,17] The 5′-proximal coding region, termed P1, includes the structural genes VP1 (1D), VP2 (1B), VP3 (1C) and VP4 (1A), while non-structural genes are

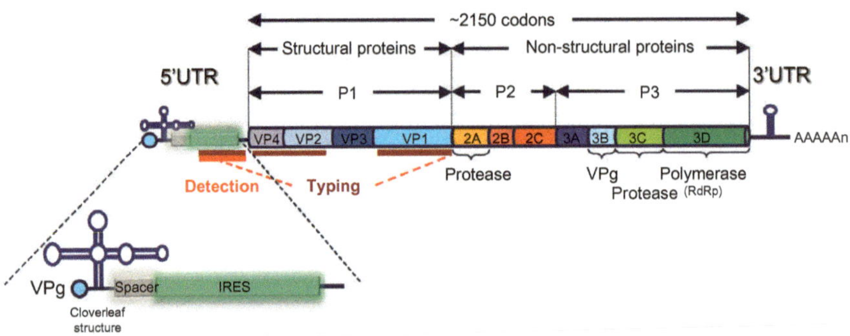

Figure 1.1 ***Schematic organization of rhinovirus genome.*** The RV single-stranded positive-sense RNA genome is about 7.1–7.2 kb long and composed of 5′ and 3′-UTRs and a coding region of approximately 2150 codons divided into the P1 (coding for structural proteins), P2 and P3 (coding for non-structural proteins) regions. The coding region is flanked by the 5′UTR and 3′UTR. The 5′UTR is important for viral RNA translation and replication and contains the IRES sequence and a cloverleaf structure, 3′UTR forms a conserved stem-loop structure preceding a poly(A) tail. A VPg protein is covalently bound to the 5′-end of the genome. Genomic regions used for RV detection and typing are indicated by yellow and red lines, respectively. *IRES*, internal ribosome entry site; *RV*, Rhinovirus; *UTRs*, untranslated regions; *VPg*, viral protein genome-linked.

located within the P2 (2A, 2B, 2C) and P3 (3A, 3B, 3C, 3D) regions towards the 3′ end. Non-structural genes encode the viral polymerase (3D) and viral proteases (2A and 3C), which are essential for virus replication and polyprotein processing, respectively.[18] In addition, the 3B gene codes for a small VPg (*V*iral *P*rotein *g*enome-linked) protein, which is essential for replication initiation.[19] The remaining non-structural proteins contribute to a number of functions facilitating viral replication. Although limited data are available specifically on RV, 2B and 2C proteins have been associated with alterations of the host cell membranes for related poliovirus and coxsackievirus.[20,21] Moreover, the 3A protein has been reported to cause Golgi apparatus disruption for RV-A16 and RV-A1, but not for RV-B14 or RV-A2.[22]

The coding region of the genome is flanked by two untranslated regions (UTRs) at both the 5′ and 3′ termini that contain several structural elements regulating viral RNA translation and replication. The 5′UTR is approximately 600 nucleotides long and begins with a cloverleaf secondary structure of 80–84 bases, which supports protein binding and is important for viral replication.[23] This structure is followed by a short pyrimidine-rich spacer sequence and an internal ribosome entry site (IRES), a complex structure with multiple stem-loops that allows direct entry of the 40S host cell ribosomal subunit during cap-independent translation.[15] The 3′UTR is shorter in size (40–50 bases) and forms a highly conserved stem followed by a poly-adenine tail.[24] In addition to the structural motifs within the 5′- and 3′UTRs, replication of picornaviruses depends on the *cis*-acting replication element (*cre*) located within the polyprotein coding sequence. *Cre* is required for uridylylation of the tyrosine residue in VPg that functions as a primer for initiation of RNA synthesis during genome replication. Notably, *cre* is positioned in different regions of the genome in each RV species. In RV-A, the *cre* is located within the 2A protease sequence,[25] while in RV-B and RV-C it is located in the structural proteins VP1 and VP2, respectively.[26]

RVs exhibit a higher degree of genetic diversity compared with other picornaviruses, leading to their classification into three species comprising over 160 types based on phylogenetic analysis.[27] Both intra- and interspecies recombination events combined with the low fidelity of the viral RNA polymerase[28] are thought to result in the broad diversity of RV species/types.[15] Overall, the most sequence variability is observed between capsid proteins (VP1, VP2, VP3) of different RV types,[15] especially in the regions that are mapped to the external surface of the virus and are the potential sites of immune recognition. Thus this variability is thought to contribute to immune evasion. Within a single species, RV types can have up to 30% (RV-A and RV-B) or 36% (RV-C) genome variability.[27] The RV-A and RV-C species are overall more genetically related to each other (42% between species mean distance across the genome) than to RV-B (43%–45% mean distance).[29] The RV-B genome is the longest among the three species, whereas RV-C genome is the shortest, primarily because of some long deletions in the VP1 protein region.[30]

1.2.2 Rhinovirus structural organization

RV is a non-enveloped virus and its virion is composed of four capsid proteins, surrounding the positive-sense single-stranded RNA genome with the VPg protein covalently bound to its 5′end. The first three-dimensional RV structure of B14 strain was published in 1985 and then refined at a higher resolution in 1990 by Rossmann and colleagues.[31,32] The crystal structure identified a 30-nanometer indiameter capsid made up of repeating protomers, each containing one copy of the viral proteins VP1, VP2, VP3, and VP4. A total of 60 protomers per virion are organized into twelve 5-unit ensembles termed pentamers, forming an icosahedral structure (Fig. 1.2). VP1, VP2, and VP3 each form 8-stranded antiparallel β-barrel structures and are present on the capsid surface.[31] VP1 is the most external protein and the main site of interaction with cellular receptors. VP4 is smaller than the other structural proteins and localized on the internal surface of the capsid, interacting with the viral genome.[32] The crystal structures of additional RV-A (A1, A2, A16) and RV-B3 strains were also determined at atomic resolution and shown to be overall similar to that of B14.[34–37]

The surface-exposed parts of capsid proteins VP1, VP2, and VP3 are the main site of antibody recognition and contain the most highly immunogenic regions, termed neutralizing immunogens (NIms). Four NIms were initially identified for RV-A16 and RV-B14 types,[31,38] but these regions appear to vary between RV types and are different for RV-A2.[39] The three surface capsid proteins also form a canyon structure (~2.5-nm depression) that surrounds each five-fold axis of the capsid and is the receptor binding site for many RVs. The floor of the canyon is composed of the conserved amino acid residues that are not accessible to neutralizing antibodies.[40] A hydrophobic pocket, formed by VP1 residues, is situated underneath the canyon and thought to be occupied by a "pocket factor" of cellular origin (a fatty acid molecule) in RV-A.[35]

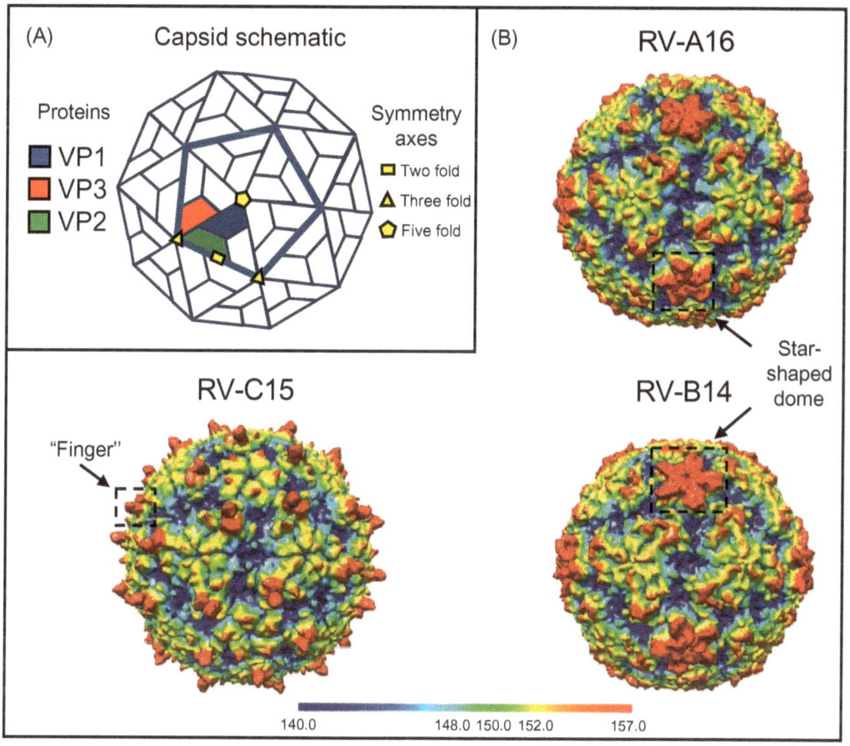

Figure 1.2 ***Rhinovirus structure.*** (A) Schematic representation of a RV capsid showing the arrangement of three surface proteins (biological unit) and symmetry axes. Five protomers consisting of one copy each of four viral capsid proteins (internal VP4 is not shown) are organized in pentamers (outlined in *blue*) forming an icosahedral structure. (B) Comparison of RV-A, -B, and -C capsid structures highlighting differences between species. Colors indicate radial distances to the center of the virion. Structure images were made using Chimera software.[33] RCSB Protein Databank entries: RV-A16 (1AYM), RV-B14 (4RHV), and RV-C15 (5K0U). *RV*, Rhinovirus. *(Courtesy of Dr. Jean-Yves Sgro (University of Wisconsin-Madison)).*

The first cryo-EM atomic structure of RV-C was only recently published by Liu et al. and some regions where it differs from other RVs were identified.[41] In contrast to RV-A and B (and many other picornaviruses), the RV-C15a capsid forms spiky "fingers" on the outer surface of the virion that are predicted to function as dominant immunogenic sites. The "finger" structure is formed by VP1 and VP2 residues that are highly variable among different RV-C types. The RV-C15a capsid lacks a protruding star-shaped "plateau" around each of the fivefold vertices (due to several large deletions in VP1 loops) and has narrower, noncontinuous canyons compared with RV-A and RV-B.

1.2.3 Rhinovirus replication cycle

RV primarily infects the upper layer of epithelial cells in the host respiratory tract. The RV replication cycle takes 8–12 hours to complete[42] and consists of four general steps: (1) binding of the viral particle to its receptor and internalization into the host cell by endocytosis, (2) uncoating and release of viral RNA into the cytoplasm, (3) viral RNA translation and replication in the cytoplasm, and (4) assembly and release of the new infectious particles (Fig. 1.3).

1.2.3.1 Binding and entry

Different RV types use one of three known cellular receptors to mediate host cell binding and internalization. The "major receptor group" of RV, which consists of all RV-B and most RV-A types, binds to the

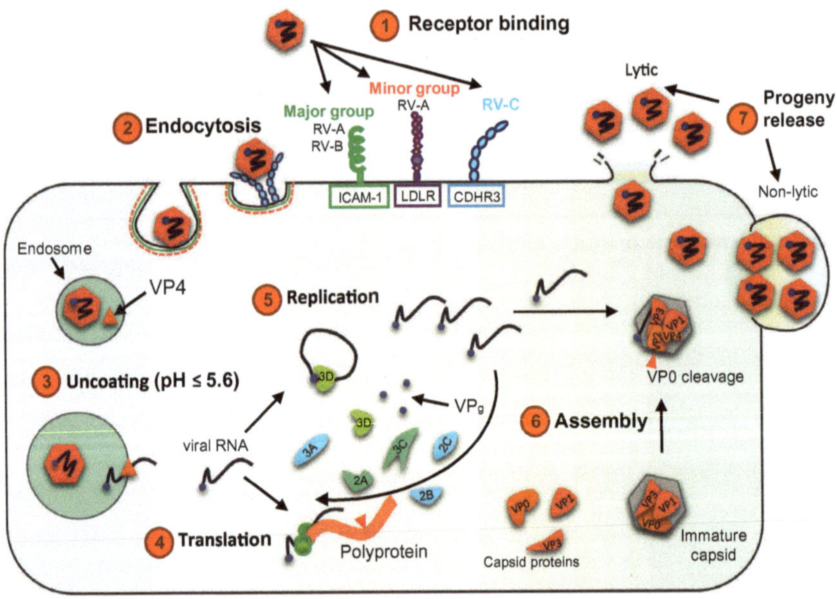

Figure 1.3 ***Schematic representation of rhinovirus infectious cycle.*** Rhinovirus enters the cell via endocytosis. Following uncoating at pH ≤ 5.6, the viral RNA acts as a mRNA allowing translation of the genome to a large polyprotein, which subsequently yields mature viral proteins. Viral RNA also acts as a template for the viral polymerase allowing the production of new viral genomes. Structural proteins VP1 and VP3 and the VP0 precursor assemble into empty capsids. Mature viral particles (complete with the viral RNA) are formed upon the final cleavage of VP0 into VP2 and VP4. Finally, cell lysis or non-lytic exocytosis involving autophagy components allow the release of the new infectious viral particles.

intercellular adhesion molecule-1 (ICAM-1).[43–46] Interestingly, major group RV-A89 and RV-A54 strains have been shown to also bind to heparan sulfate as an alternative receptor in cells lacking ICAM-1 expression.[47,48] In contrast, the RV "minor receptor group" consisting of a small number of RV-A types ($n = 10$) binds to the low-density lipoprotein receptor (LDLR).[43,49] Major group RVs bind to ICAM-1 at the base of the narrow canyon,[50] while minor group RVs bind to LDLR outside of the canyon region, on the star-shaped dome structure present at the vertex.[51] RV-C viruses were recently demonstrated to bind the cadherin-related family member 3 (CDHR3), a membrane protein that is highly expressed in airway epithelium and lung tissue.[52] Neither the CDHR3 receptor binding site on the RV-C surface nor molecular mechanisms of virus entry to host cell have been determined yet. Similarly to some RV-A types, RV-C15 strain was readily adapted to use heparan sulfate as an additional receptor by serial passaging in HeLa-E8 cells.[53]

Following receptor binding, the mechanism of viral particle internalization varies depending on the receptor and the host cell type. Most of the studies on RV cell entry were done in HeLa and some other established cell lines. Upon binding to HeLa cells in vitro, major group RV-B14 enter the cell via clathrin/dynamin-dependent endocytosis.[54] This mechanism has also been demonstrated in undifferentiated primary bronchial epithelial cells infected with RV-A16.[55] However, in contrast to HeLa cells, RV-B14 endocytosis in rhabdomyosarcoma cells was clathrin-independent.[56] For minor group RV-A2, clathrin/dynamin-mediated endocytosis has also been confirmed.[57] However, in a different study, RV-A2 also effectively replicated when clathrin and dynamin were inhibited,[58,59] suggesting alternate cellular internalization mechanisms. Further studies are needed to fully understand RV binding and entry mechanisms in primary airway epithelial cells.

1.2.3.2 Uncoating

Following cellular internalization, RV uncoating is more (minor group) or less (major group) pH dependent.[60] Acidification of the endosome triggers conformational changes, such as the release of the N-terminal regions of VP4 and VP1 capsid proteins. These changes in the viral particle can be assessed by measuring their sedimentation coefficients. Loss of VP4 from the intact viral particle (150S) leads to the formation of the intermediate subviral particle called subviral particle A (135S), prior to the formation of the subviral particle B (80S), which is RNA-free.[61]

For minor group RVs, these conformational changes are thought to result from pH changes directly, leading to the dissociation of the viral particle from its receptor LDLR.[62] VP4 then forms an aqueous pore in the endosomal membrane, allowing the release of viral RNA into the cytoplasm as early as 10-minute post-infection, followed by the subsequent degradation of the capsid proteins in the lysosomes.[63] In contrast, uncoating of major group viruses is receptor-mediated: ICAM-1 provides an uncoating signal in addition to the changes in pH, which may alternatively result in disruption of the endosomal membrane in physiological conditions.[64] More recent results have demonstrated additional variability within major group RV types. For example, RV-A89 uncoated at 20°C, but other tested RV types that bind ICAM-1 (RV-A16, RV-B3, RV-B14) did not.[65] Further, RV-A16 and RV-A89 were directed intracellularly to the recycling endosomes, while RV-B3 an RV-B14 were directed to the late endosomes, suggesting differing mechanisms of viral uncoating even between closely related RVs.[65] The physiological relevance of these differences remains unclear.

1.2.3.3 Translation of the polyprotein and replication

Following release of the viral genome into the host cell cytoplasm, host ribosomes bind within the highly structured 5′UTR IRES and the viral RNA is translated into a large single polyprotein of approximately 250 kDa in size.[15] Co- and posttranslationally, viral proteases $2A^{pro}$ and $3C^{pro}$ cleave the polyprotein, releasing smaller polypeptides. The first autocatalytic cleavage by $2A^{pro}$ occurs between the P1 and P2 regions of the polyprotein. Subsequently, cleavage by the viral protease $3C^{pro}$ (or its precursor $3CD^{pro}$) releases mature functional viral proteins.[24] Some viral proteins are initially cleaved to form stable precursors (2BC, 3AB, 3CD) that might perform alternative functions compared with their mature forms. VP4 and VP2 remain bound in the VP0 precursor until the viral particle assembly and maturation, and 3C protease and 3D polymerase initially form the $3CD^{pro}$ precursor. $3CD^{pro}$ exhibits protease activity,[66] but could lack polymerase activity by analogy with poliovirus.[67]

Most of the current knowledge on replication of picornaviruses comes from research based on poliovirus. Although there is a lack of information on RV replication, the fact that chimeric polio/RV viruses could replicate efficiently with minimal changes suggests that replication of these two viruses is overall similar.[68,69] A ribonucleoprotein complex formed around the 5′-end cloverleaf RNA structure and consisting of the viral 3CD precursor and the cellular poly(rC) binding protein interacts with the cellular poly(A) binding protein (PABP1) bound to the 3′-end poly(A) tail, thus linking the ends and circularizing the viral RNA.[19,70,71] Once separated from the 3CD^{pro} precursor, the RNA-dependent-RNA polymerase 3D (3D^{pol}) is the main actor of RV replication, however, other non-structural viral proteins are also involved. First, 3D^{pol} performs a *cre*-mediated uridylylation of the VPg protein.[25,72,73] Uridylylated VPg (VPg-pUpU$_{OH}$) is then translocated to the 3′-end of the plus-strand poly (A) tail to prime the initiation of viral RNA replication. Second, the 3D^{pol} initiates synthesis of a negative-strand RNA, starting from the 3′ end of the viral RNA genome. Then, based on this negative-sense template, the polymerase synthesizes a large number of new, complementary RNAs with positive-sense polarity. The new positive-sense RNA serves as a messenger RNA template for the production of viral proteins and is also packaged into newly formed viral particles.

1.2.3.4 Assembly and release of infectious viral particles

Limited data exist on the assembly and release of infectious RV virions. However, it is believed that the mechanism is similar to that described for poliovirus. Structural proteins VP1 and VP3 interact with the VP0 precursors (VP2 + VP4) to form protomers, which then assemble into pentamers and form a capsid. The VP0 precursor remains intact until virus assembly. The final autocatalytic cleavage of VP0 into mature VP2 and VP4 takes place at the same time as the viral RNA binds to the capsid, in a process termed "maturation cleavage".[24] This final cleavage event allows the formation of infectious viral particles.[74,75] The specific mechanism underlying this cleavage event remains unclear, however, a cotranslational N-terminal myristoylation of VP4 was shown to play an important role in the viral maturation cleavage of related enteroviruses.[76,77]

For RV as a non-enveloped virus, progeny release was thought to occur via lysis of the host cell plasma membrane. In agreement with this hypothesis, strong cytopathic effect (CPE) is usually observed in established cell lines (HeLa, WI-38) and undifferentiated primary airway cells infected with RV strains that have been adapted for efficient growth in these cells. However, RV replicates in foci and cause little visible damage in differentiated airway epithelium both in vivo and in vitro suggesting a potential non-lytic release pathway.[78–80] Visualization of virus shedding in differentiated bronchial epithelial cells infected with "reporter" RV-C15 strain, genetically engineered to express green fluorescent protein

(GFP) during viral replication, revealed occasional, rounded, GFP-expressing cells that become detached from the epithelial layer, leaving gaps in the epithelium that subsequently contracted over the course of the experiment.[81]

Recently, non-lytic release through vesicle formation and stimulation by autophagic components has been described for poliovirus and coxsackievirus.[82,83] In 2011, Klein et al. showed that the minor group rhinovirus RV-A2 induced the autophagic pathway in cultured human embryonic kidney cells and that viral titer in culture increased upon treatment with autophagy inducer, while blocking autophagy had the opposite effect.[84] Interestingly, another minor group strain, RV-A1, used in that study, neither induced autophagy nor responded to autophagic stimuli indicating the use of a different virus release pathway. Formation of autophagosome-like structures has been also reported both for minor group (RV-A2) and major group (RV-B14) strains in another study.[85] In agreement with previous findings, Chen et al. demonstrated that RV-A2 and poliovirus could be located in phosphatidyl-serine vesicles, allowing a non-lytic transmission of multiple viral particles at the same time, ultimately resulting in a higher infectivity than upon infection with individual virions.[86]

1.3 CLASSIFICATION OF HUMAN RHINOVIRUSES

1.3.1 Early methods of classification

Following the first isolation of RVs in 1953, the number of recognized RVs has increased rapidly.[87,88] Identification and characterization of RVs was performed through two general steps: (1) detection of RV in clinical samples and (2) serotyping of the isolated virus. In earlier studies, RV detection from a clinical sample was based on certain physical and biochemical properties, including size and shape of virion, buoyant density, acid lability, chloroform/ether resistance, and growth (or not) in specific cell lines (e.g., Hep-2, HeLa). This approach was suitable to separate RV from other respiratory viruses before performing their antigenic characterization.[87] In the 1990s, polymerase chain reaction (PCR) became the favored method for detection of RVs in clinical samples due to its higher sensitivity and shorter turnaround time.[89–91]

1.3.1.1 Serotyping

Initial attempts to classify RVs were based on "serotyping," which relies on host neutralizing antibody response to identify distinct virus types through antigenic cross-reactivity. Assessment of cross-reactivity required the propagation (several passages) of clinical isolates in cell lines (WI-38, MRC-5, and HeLa) in vitro and titration by plaque assay.[92–95] Susceptible cell cultures were then infected at a constant multiplicity of infection with the unknown RV and the CPE was documented in the presence or absence of reference sera directed against known RV serotypes. Reference sera were produced by injection of virus into an experimental animal (e.g., guinea pig or rabbit). After an incubation period, the absence of CPE in the infected culture indicated that a specific reference serum could neutralize viral replication allowing assignment of the RV isolate to a particular serotype. In some cases, serotyping results were ambiguous, as animal serum could (1) have non-specific neutralization effects, and (2) cross-react with multiple antigenically related RVs.[96,97] Antigenic characterization of a single clinical isolate using this approach could take up to 6-7 days.[92] By 1987, cross-neutralization assays identified 100 distinct RV serotypes,[88,98] which were subsequently differentiated into two species, RV-A and B, with 75 and 25 serotypes, respectively.

1.3.1.2 Additional characterizations of rhinoviruses

In conjunction with serotyping, RV isolates can be described according to their characteristics, such as cellular receptor usage or drug sensitivity.

Cell receptor tropism

Early characterization of RVs distinguished between M and H strains, based on their ability (or not) to grow in monkey (M) kidney cell lines in addition to human (H) cell lines, however, it was of a limited epidemiologic importance.[99,100] This distinction was later abandoned in favor of the minor/major group distinction, which focuses on receptor usage. The major group RVs bind to ICAM-1 receptor while minor group RVs use LDLR.[43,44,49] Most M strains belong to the minor group, although the overlap between the M and H strain designations and minor/major groups is incomplete. However, this observation is likely due to the significant differences in amino acid composition and structure of the amino-terminal Ig-like domain (D1) of ICAM-1, which serves as the virus binding site, between humans and other mammalian species,[101] resulting in an inability of H strains to grow in monkey cell lines. Given the functional relevance of the receptor use for RV replication, the minor and major group appellations are still in use today.

Drug sensitivity

A further characterization approach is based on viral sensitivity to capsid-binding drug panels. *Enterovirus* capsids contain surface pockets that vary in depth and shape. Exposure to molecules that intercalate into these surface pockets (e.g., pleconaril) can inhibit RV replication.[102,103] RVs were divided into antiviral groups A and B (not to be confused with species A and B) based on their sensitivity to compounds that inhibit their replication. Antiviral group A RVs ($n = 33$ serotypes) are susceptible to elongated antiviral compounds, whereas the group B serotypes ($n = 67$) are inhibited by structurally shorter antiviral agents.[104]

1.3.2 Current rhinovirus classification

1.3.2.1 Technical improvements, rhinovirus C discovery, and new classification proposals

Current classification approaches are based on genome sequencing of RV isolates and analysis based on sequence homology. Initial characterization using reverse transcription (RT)−PCR assays was first reported in the 1980s, as a more efficient classification method than previous tissue culture isolation.[90,105−107] The first full-genome sequence was reported for RV-B14 in 1984 and revealed a similar genomic organization between RV and other members of the *Enterovirus* genus.[108]

In 2006, while aiming to improve detection of respiratory pathogens in clinical samples, Arden et al. used an extended PCR primer panel on respiratory samples, which resulted in a consequent improvement of microbial pathogens detection, 44.4% of which were RVs. Partial sequencing of the detected RVs identified a distinct genetic lineage termed human rhinovirus (HRV)-A2, which clustered separately from other RV-A strains.[109] Similar reports were made identifying a novel RV species by different names (e.g., HRV-X[110] and HRV-C[30,111,112]). These newly recognized RV types had not been previously noticed, as they were unable to replicate in vitro in standard cell lines used for virus isolation and serotyping, later attributed to the lack of their cognate receptor on undifferentiated cell lines.[29,52,109,113] Characterization of the coding sequence of one of these isolates, HRV-QPM, revealed that this isolate had only 51.9% sequence homology with the reference strain RV-A1.[29] Following this discovery, a genetic homology classification similar to other *enterovirus* classification systems was proposed in 2010, introducing the third RV species, RV-C, classified to 33 types using a threshold of 13% nucleotide divergence in VP1 gene, and, additionally, to 28 "provisionally assigned types" (pat) with >10% divergence in VP2/VP4 (when VP1 sequences were lacking).[114] This proposal for classification based on genome homology rather than serotyping was later extended to all three RV species[27] and accepted by International Committee on Taxonomy of Viruses[13] in 2012.

1.3.2.2 Genotyping

Under the current nomenclature, RVs can be abbreviated RV, followed by the species letter (A, B, or C) and the type number. As previous taxonomies have used the term "human rhinovirus," the abbreviation "HRV" is also frequently found in the literature. Detection and typing of RV isolates are performed using RT-PCR techniques on specific viral genes, followed by sequencing and comparison with previously characterized reference sequences.[14,115]

Initial identification of a RV type is generally based on the 5′UTR sequence amplification with universal diagnostic primers that anneal to highly conserved motifs providing the most sensitive detection assay in both conventional and real-time PCR formats (Fig. 1.1). Generation of evolutionary trees based on comparison of different genomic regions (e.g., VP1, VP2/VP4, or 5′UTR) does not always yield the same results.[15] Sequencing of the 5′UTR can still be used as an indicator of RV type in most cases;[105] however, this region is sometimes the site of recombination between RV-A and -C species, and therefore neither ideal nor sufficient for accurate typing.[116] VP1 and VP2/VP4 genes exhibit the most intra- and intertypic sequence variability, and therefore, are the most appropriate for unequivocal RV species and type assignments;[15] however, the high sequence variability in these regions impedes design of universal diagnostic primers. Although VP2/VP4 is usually predictive of the type, VP1 sequence comparison is required for a definitive assignment of a new RV type when partial VP2/VP4 (and/or 5′UTR) analysis indicates low sequence identity with all known reference types.[27]

A new RV isolate is assigned to a species when exhibiting >70% sequence identity with other members of that species. If there is >90% nucleotide sequence identity from the reference sequence of VP2/VP4 and 87% of VP1 with an existing RV type, the new isolate is classified to the same type. The Picornaviridae Study Group Subcommittee website lists all known reference sequences for RV species. Through August 2018, this website reports 80 RV-A, 32 RV-B, and 56 RV-C types based on the current classification method (Fig. 1.4).[118]

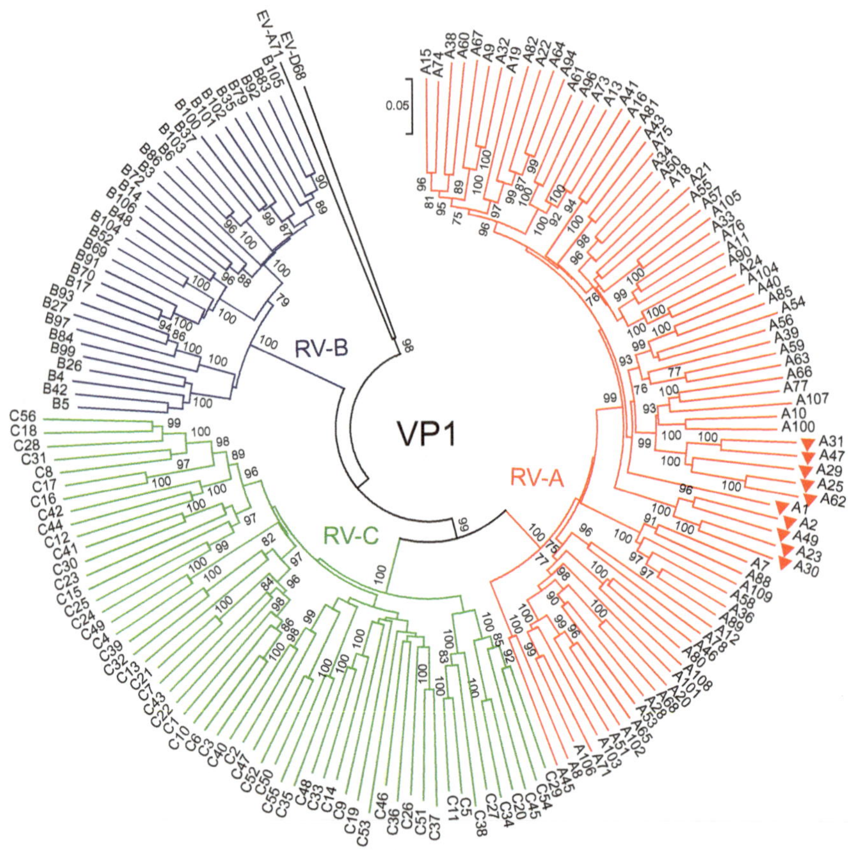

Figure 1.4 ***Neighbor-joining phylogenetic tree of currently recognized rhinovirus types within species A, B, and C based on VP1 sequence.*** The evolutionary distances were computed using the *p*-distance method in MEGA 6.06 software[117] and are in the units of the number of base differences per site. All major nodes are labeled with bootstrap values (500 replicates, with its value more than 75%). Branch lengths are proportional to nucleotide similarity (*p*-distance). Enteroviruses (EV-A71 and EV-D68) were included as an outgroup. RV-A types belonging to the minor receptor group are indicated by a red triangle. *RV*, Rhinovirus.

1.3.3 Challenges in rhinovirus classification and diversity

RV classification approaches have been revised several times. Improvement in detection and typing methods have considerably increased the knowledge on RVs and allowed the identification of RV-C species more than 50 years after the initial RV discovery. However, the existing diagnostic techniques do not always yield unequivocal RV typing results.

The use of differing classification systems sometimes makes comparisons of RV types across studies difficult. Moreover, early studies obviously did not include information on RV types that were not identified or assigned at the time (e.g., RV-C). When studies do report findings using updated classification methods, sequencing costs and detection failure rates can affect the ability to assess accurately which types and strains are present and prevalent. Individual primers used for detection and molecular typing can also affect the spectrum of types detected. These technical limitations are being addressed through improved detection and typing techniques as well as the increasing use of full-genome sequencing data to identify novel RV types.[109,115,116,119,120]

Antigenic variability between genetically related RV types provides additional challenges to classification. For example, when comparing an A101 prototype strain to a circulating strain isolated 8 years later, Rathe et al. found that while the two isolates retained sufficient sequence identity to be assigned to the same type, they no longer exhibited antigenic cross-reactivity in serotyping assays. This result suggests an antigenic divergence between the isolates despite being assigned to the same type by genetic analysis.[121]

In many clinical studies, molecular typing of RV isolates is based on sequencing of only one genomic region (e.g., 5′UTR, VP2/VP4, or VP1). However, partial sequences obtained from different regions can lead to varying type assignments in some cases.[14] While studying type circulation over a 3-year period in Cambodia, Naughtin et al. compared typing using VP1 or VP2/VP4 nucleotide sequence homology, as well as protein sequence homology in these two regions.[122] While most clinical isolates could be attributed to the same genotype when using these two regions, typing results differed between VP1 and VP2/VP4 for some RV-C isolates.

These findings highlight that current classification approaches do not always correctly assign genetic type or specifically assess antigenic differences between virus isolates. Therefore, additional information, such as serological characterization of clinical isolates using virus neutralization tests, inclusion of amino acid identity thresholds for typing,[123] or new tools such as protein-based typing microarray could help answering this need.[124]

1.4 SUMMARY

The RVs are currently classified into three species (RV-A, B, and C) according to phylogenetic sequence criteria and distinct genomic features. They have genome organizations and capsid structures overall similar to those of other Enteroviruses. With the discovery of the third RV species in 2006, some unique biological features of RV-C were identified such as the use of a novel cellular receptor CDHR3, expressed only on fully-differentiated airway epithelial cells; "finger" structures on the outer surface of the virion that were predicted to function as dominant immunogenic sites; and the *cre* localized in VP2 gene. Effective identification and classification of RV represent an important foundation for the development of efficient treatment strategies against RVs. Attempts to develop vaccines against RV have been successful at inducing adaptive immune responses, but these responses have so far failed to provide protection across multiple RV types.[125] As up to 20–30 RV types circulate in a population at any given time and the spectrum of types changes from season to season,[126–128] this presents a challenge for vaccine design. It has been recently shown for the first time that serum neutralizing antibodies against multiple RV types (up to 49) can be generated using a straightforward polyvalent inactivated RV vaccine approach.[129] Identification of the most clinically relevant virus types has also been proposed, as targets for new therapies. As RVs are among the most diverse species of non-influenza respiratory viruses, the development of high-throughput, cost-effective, and systematic detection and typing methods would be highly beneficial to fully understand RV diversity, seasonality, and circulation patterns.[130]

REFERENCES

1. Fendrick AM, Monto AS, Nightengale B, Sarnes M. The economic burden of non-influenza-related viral respiratory tract infection in the United States. *Arch Intern Med.* 2003;163(4):487–494.
2. Makela MJ, Puhakka T, Ruuskanen O, et al. Viruses and bacteria in the etiology of the common cold. *J Clin Microbiol.* 1998;36(2):539–542.

3. Morikawa S, Kohdera U, Hosaka T, et al. Seasonal variations of respiratory viruses and etiology of human rhinovirus infection in children. *J Clin Virol.* 2015;73:14–19.
4. Visseaux B, Burdet C, Voiriot G, et al. Prevalence of respiratory viruses among adults, by season, age, respiratory tract region and type of medical unit in Paris, France, from 2011 to 2016. *PLoS One.* 2017;12(7):e0180888.
5. Edwards MR, Bartlett NW, Hussell T, Openshaw P, Johnston SL. The microbiology of asthma. *Nat Rev Microbiol.* 2012;10(7):459–471.
6. Lieberman D, Shimoni A, Shemer-Avni Y, Keren-Naos A, Shtainberg R, Lieberman D. Respiratory viruses in adults with community-acquired pneumonia. *Chest.* 2010;138(4):811–816.
7. Van Leeuwen JC, Goossens LK, Hendrix RM, Van Der Palen J, Lusthusz A, Thio BJ. Equal virulence of rhinovirus and respiratory syncytial virus in infants hospitalized for lower respiratory tract infection. *Pediatr Infect Dis J.* 2012;31(1):84–86.
8. Mitchison DA. Prevention of colds by vaccination against a rhinovirus: a report by the Scientific Committee on Common Cold Vaccines. *Br Med J.* 1965;1 (5446):1344–1349.
9. Perkins JC, Tucker DN, Knope HL, et al. Evidence for protective effect of an inactivated rhinovirus vaccine administered by the nasal route. *Am J Epidemiol.* 1969;90 (4):319–326.
10. Turner RB, Wecker MT, Pohl G, et al. Efficacy of tremacamra, a soluble intercellular adhesion molecule 1, for experimental rhinovirus infection: a randomized clinical trial. *JAMA.* 1999;281(19):1797–1804.
11. Hayden FG, Turner RB, Gwaltney JM, et al. Phase II, randomized, double-blind, placebo-controlled studies of ruprintrivir nasal spray 2-percent suspension for prevention and treatment of experimentally induced rhinovirus colds in healthy volunteers. *Antimicrob Agents Chemother.* 2003;47(12):3907–3916.
12. Hayden FG, Herrington DT, Coats TL, et al. Efficacy and safety of oral pleconaril for treatment of colds due to picornaviruses in adults: results of 2 double-blind, randomized, placebo-controlled trials. *Clin Infect Dis.* 2003;36(12):1523–1532.
13. Taxonomy, International Committee on Taxonomy of Viruses. <https://talk.ictvonline.org/taxonomy/>; 2017 Accessed 17.08.18.
14. Palmenberg AC, Gern JE. Classification and evolution of human rhinoviruses. *Methods Mol Biol.* 2015;1221:1–10.
15. Palmenberg AC, Spiro D, Kuzmickas R, et al. Sequencing and analyses of all known human rhinovirus genomes reveal structure and evolution. *Science.* 2009;324 (5923):55–59.
16. Rueckert RR, Wimmer E. Systematic nomenclature of picornavirus proteins. *J Virol.* 1984;50(3):957–959.
17. Lee WM, Wang W, Rueckert RR. Complete sequence of the RNA genome of human rhinovirus 16, a clinically useful common cold virus belonging to the ICAM-1 receptor group. *Virus Genes.* 1995;9(2):177–181.
18. Kelly JT, Busse WW. Host immune responses to rhinovirus: mechanisms in asthma. *J Allergy Clin Immunol.* 2008;122(4):671–682.
19. Paul AV, van Boom JH, Filippov D, Wimmer E. Protein-primed RNA synthesis by purified poliovirus RNA polymerase. *Nature.* 1998;393(6682):280–284.
20. Aldabe R, Barco A, Carrasco L. Membrane permeabilization by poliovirus proteins 2B and 2BC. *J Biol Chem.* 1996;271(38):23134–23137.
21. van Kuppeveld FJ, Hoenderop JG, Smeets RL, et al. Coxsackievirus protein 2B modifies endoplasmic reticulum membrane and plasma membrane permeability and facilitates virus release. *EMBO J.* 1997;16(12):3519–3532.
22. Mousnier A, Swieboda D, Pinto A, et al. Human rhinovirus 16 causes Golgi apparatus fragmentation without blocking protein secretion. *J Virol.* 2014;88(20):11671–11685.

23. Xiang W, Harris KS, Alexander L, Wimmer E. Interaction between the 5′-terminal cloverleaf and 3AB/3CDpro of poliovirus is essential for RNA replication. *J Virol.* 1995;69(6):3658–3667.
24. Palmenberg AC. Proteolytic processing of picornaviral polyprotein. *Annu Rev Microbiol.* 1990;44:603–623.
25. Gerber K, Wimmer E, Paul AV. Biochemical and genetic studies of the initiation of human rhinovirus 2 RNA replication: identification of a cis-replicating element in the coding sequence of 2A(pro). *J Virol.* 2001;75(22):10979–10990.
26. Cordey S, Gerlach D, Junier T, Zdobnov EM, Kaiser L, Tapparel C. The cis-acting replication elements define human enterovirus and rhinovirus species. *RNA.* 2008;14 (8):1568–1578.
27. McIntyre CL, Knowles NJ, Simmonds P. Proposals for the classification of human rhinovirus species A, B and C into genotypically assigned types. *J Gen Virol.* 2013;94 (Pt 8):1791–1806.
28. Ferrer-Orta C, Ferrero D, Verdaguer N. RNA-Dependent RNA polymerases of picornaviruses: from the structure to regulatory mechanisms. *Viruses.* 2015;7 (8):4438–4460.
29. McErlean P, Shackelton LA, Lambert SB, Nissen MD, Sloots TP, Mackay IM. Characterisation of a newly identified human rhinovirus, HRV-QPM, discovered in infants with bronchiolitis. *J Clin Virol.* 2007;39(2):67–75.
30. Lau SK, Yip CC, Tsoi HW, et al. Clinical features and complete genome characterization of a distinct human rhinovirus (HRV) genetic cluster, probably representing a previously undetected HRV species, HRV-C, associated with acute respiratory illness in children. *J Clin Microbiol.* 2007;45(11):3655–3664.
31. Rossmann MG, Arnold E, Erickson JW, et al. Structure of a human common cold virus and functional relationship to other picornaviruses. *Nature.* 1985;317 (6033):145–153.
32. Arnold E, Rossmann MG. Analysis of the structure of a common cold virus, human rhinovirus 14, refined at a resolution of 3.0 A. *J Mol Biol.* 1990;211(4):763–801.
33. Pettersen EF, Goddard TD, Huang CC, et al. UCSF Chimera—a visualization system for exploratory research and analysis. *J Comput Chem.* 2004;25(13):1605–1612.
34. Kim SS, Smith TJ, Chapman MS, et al. Crystal structure of human rhinovirus serotype 1A (HRV1A). *J Mol Biol.* 1989;210(1):91–111.
35. Oliveira MA, Zhao R, Lee WM, et al. The structure of human rhinovirus 16. *Structure.* 1993;1(1):51–68.
36. Zhao R, Pevear DC, Kremer MJ, et al. Human rhinovirus 3 at 3.0 A resolution. *Structure.* 1996;4(10):1205–1220.
37. Verdaguer N, Blaas D, Fita I. Structure of human rhinovirus serotype 2 (HRV2). *J Mol Biol.* 2000;300(5):1179–1194.
38. Sherry B, Mosser AG, Colonno RJ, Rueckert RR. Use of monoclonal antibodies to identify four neutralization immunogens on a common cold picornavirus, human rhinovirus 14. *J Virol.* 1986;57(1):246–257.
39. Appleyard G, Russell SM, Clarke BE, Speller SA, Trowbridge M, Vadolas J. Neutralization epitopes of human rhinovirus type 2. *J Gen Virol.* 1990;71(Pt 6):1275–1282.
40. Rossmann MG, He Y, Kuhn RJ. Picornavirus-receptor interactions. *Trends Microbiol.* 2002;10(7):324–331.
41. Liu Y, Hill MG, Klose T, et al. Atomic structure of a rhinovirus C, a virus species linked to severe childhood asthma. *Proc Natl Acad Sci USA.* 2016;113 (32):8997–9002.
42. Harris 2nd JM, Gwaltney Jr. JM. Incubation periods of experimental rhinovirus infection and illness. *Clin Infect Dis.* 1996;23(6):1287–1290.

43. Abraham G, Colonno RJ. Many rhinovirus serotypes share the same cellular receptor. *J Virol.* 1984;51(2):340–345.
44. Greve JM, Davis G, Meyer AM, et al. The major human rhinovirus receptor is ICAM-1. *Cell.* 1989;56(5):839–847.
45. Tomassini JE, Graham D, DeWitt CM, Lineberger DW, Rodkey JA, Colonno RJ. cDNA cloning reveals that the major group rhinovirus receptor on HeLa cells is intercellular adhesion molecule 1. *Proc Natl Acad Sci USA.* 1989;86(13):4907–4911.
46. Staunton DE, Merluzzi VJ, Rothlein R, Barton R, Marlin SD, Springer TA. A cell adhesion molecule, ICAM-1, is the major surface receptor for rhinoviruses. *Cell.* 1989;56(5):849–853.
47. Vlasak M, Goesler I, Blaas D. Human rhinovirus type 89 variants use heparan sulfate proteoglycan for cell attachment. *J Virol.* 2005;79(10):5963–5970.
48. Khan AG, Pichler J, Rosemann A, Blaas D. Human rhinovirus type 54 infection via heparan sulfate is less efficient and strictly dependent on low endosomal pH. *J Virol.* 2007;81(9):4625–4632.
49. Hofer F, Gruenberger M, Kowalski H, et al. Members of the low density lipoprotein receptor family mediate cell entry of a minor-group common cold virus. *Proc Natl Acad Sci USA.* 1994;91(5):1839–1842.
50. Colonno RJ, Condra JH, Mizutani S, Callahan PL, Davies ME, Murcko MA. Evidence for the direct involvement of the rhinovirus canyon in receptor binding. *Proc Natl Acad Sci USA.* 1988;85(15):5449–5453.
51. Hewat EA, Neumann E, Conway JF, et al. The cellular receptor to human rhinovirus 2 binds around the 5-fold axis and not in the canyon: a structural view. *EMBO J.* 2000;19(23):6317–6325.
52. Bochkov YA, Watters K, Ashraf S, et al. Cadherin-related family member 3, a childhood asthma susceptibility gene product, mediates rhinovirus C binding and replication. *Proc Natl Acad Sci USA.* 2015;112(17):5485–5490.
53. Bochkov YA, Watters K, Basnet S, et al. Mutations in VP1 and 3A proteins improve binding and replication of rhinovirus C15 in HeLa-E8 cells. *Virology.* 2016;499:350–360.
54. Grunert HP, Wolf KU, Langner KD, Sawitzky D, Habermehl KO, Zeichhardt H. Internalization of human rhinovirus 14 into HeLa and ICAM-1-transfected BHK cells. *Med Microbiol Immunol.* 1997;186(1):1–9.
55. Lau C, Wang X, Song L, et al. Syk associates with clathrin and mediates phosphatidylinositol 3-kinase activation during human rhinovirus internalization. *J Immunol.* 2008;180(2):870–880.
56. Khan AG, Pickl-Herk A, Gajdzik L, Marlovits TC, Fuchs R, Blaas D. Human rhinovirus 14 enters rhabdomyosarcoma cells expressing icam-1 by a clathrin-, caveolin-, and flotillin-independent pathway. *J Virol.* 2010;84(8):3984–3992.
57. Snyers L, Zwickl H, Blaas D. Human rhinovirus type 2 is internalized by clathrin-mediated endocytosis. *J Virol.* 2003;77(9):5360–5369.
58. Huber M, Brabec M, Bayer N, Blaas D, Fuchs R. Elevated endosomal pH in HeLa cells overexpressing mutant dynamin can affect infection by pH-sensitive viruses. *Traffic.* 2001;2(10):727–736.
59. Bayer N, Schober D, Huttinger M, Blaas D, Fuchs R. Inhibition of clathrin-dependent endocytosis has multiple effects on human rhinovirus serotype 2 cell entry. *J Biol Chem.* 2001;276(6):3952–3962.
60. Jurgeit A, McDowell R, Moese S, Meldrum E, Schwendener R, Greber UF. Niclosamide is a proton carrier and targets acidic endosomes with broad antiviral effects. *PLoS Pathog.* 2012;8(10):e1002976.

61. Weiss VU, Subirats X, Pickl-Herk A, et al. Characterization of rhinovirus subviral A particles via capillary electrophoresis, electron microscopy and gas-phase electrophoretic mobility molecular analysis: Part I. *Electrophoresis*. 2012;33(12):1833–1841.
62. Brabec M, Baravalle G, Blaas D, Fuchs R. Conformational changes, plasma membrane penetration, and infection by human rhinovirus type 2: role of receptors and low pH. *J Virol*. 2003;77(9):5370–5377.
63. Brabec-Zaruba M, Pfanzagl B, Blaas D, Fuchs R. Site of human rhinovirus RNA uncoating revealed by fluorescent in situ hybridization. *J Virol*. 2009;83(8):3770–3777.
64. Schober D, Kronenberger P, Prchla E, Blaas D, Fuchs R. Major and minor receptor group human rhinoviruses penetrate from endosomes by different mechanisms. *J Virol*. 1998;72(2):1354–1364.
65. Ganjian H, Zietz C, Mechtcheriakova D, Blaas D, Fuchs R. ICAM-1 binding rhinoviruses enter HeLa cells via multiple pathways and travel to distinct intracellular compartments for uncoating. *Viruses*. 2017;9:4.
66. Davis GJ, Wang QM, Cox GA, et al. Expression and purification of recombinant rhinovirus 14 3CD proteinase and its comparison to the 3C proteinase. *Arch Biochem Biophys*. 1997;346(1):125–130.
67. Harris KS, Reddigari SR, Nicklin MJ, Hammerle T, Wimmer E. Purification and characterization of poliovirus polypeptide 3CD, a proteinase and a precursor for RNA polymerase. *J Virol*. 1992;66(12):7481–7489.
68. Cheney IW, Naim S, Shim JH, et al. Viability of poliovirus/rhinovirus VPg chimeric viruses and identification of an amino acid residue in the VPg gene critical for viral RNA replication. *J Virol*. 2003;77(13):7434–7443.
69. Todd S, Towner JS, Semler BL. Translation and replication properties of the human rhinovirus genome in vivo and in vitro. *Virology*. 1997;229(1):90–97.
70. Herold J, Andino R. Poliovirus RNA replication requires genome circularization through a protein-protein bridge. *Mol Cell*. 2001;7(3):581–591.
71. Todd S, Nguyen JH, Semler BL. RNA-protein interactions directed by the 3′ end of human rhinovirus genomic RNA. *J Virol*. 1995;69(6):3605–3614.
72. Gerber K, Wimmer E, Paul AV. Biochemical and genetic studies of the initiation of human rhinovirus 2 RNA replication: purification and enzymatic analysis of the RNA-dependent RNA polymerase 3D(pol). *J Virol*. 2001;75(22):10969–10978.
73. Yang Y, Rijnbrand R, McKnight KL, et al. Sequence requirements for viral RNA replication and VPg uridylylation directed by the internal cis-acting replication element (cre) of human rhinovirus type 14. *J Virol*. 2002;76(15):7485–7494.
74. Basavappa R, Syed R, Flore O, Icenogle JP, Filman DJ, Hogle JM. Role and mechanism of the maturation cleavage of VP0 in poliovirus assembly: structure of the empty capsid assembly intermediate at 2.9 A resolution. *Protein Sci*. 1994;3(10):1651–1669.
75. Lee WM, Monroe SS, Rueckert RR. Role of maturation cleavage in infectivity of picornaviruses: activation of an infectosome. *J Virol*. 1993;67(4):2110–2122.
76. Chow M, Newman JF, Filman D, Hogle JM, Rowlands DJ, Brown F. Myristylation of picornavirus capsid protein VP4 and its structural significance. *Nature*. 1987;327(6122):482–486.
77. Corbic Ramljak I, Stanger J, Real-Hohn A, et al. Cellular *N*-myristoyltransferases play a crucial picornavirus genus-specific role in viral assembly, virion maturation, and infectivity. *PLoS Pathog*. 2018;14(8):e1007203.
78. Lopez-Souza N, Dolganov G, Dubin R, et al. Resistance of differentiated human airway epithelium to infection by rhinovirus. *Am J Physiol Lung Cell Mol Physiol*. 2004;286(2):L373–381.

79. Winther B, Farr B, Turner RB, Hendley JO, Gwaltney Jr JM, Mygind N. Histopathologic examination and enumeration of polymorphonuclear leukocytes in the nasal mucosa during experimental rhinovirus colds. *Acta Otolaryngol Suppl.* 1984;413:19–24.
80. Mosser AG, Brockman-Schneider R, Amineva S, et al. Similar frequency of rhinovirus-infectible cells in upper and lower airway epithelium. *J Infect Dis.* 2002;185(6):734–743.
81. Griggs TF, Bochkov YA, Basnet S, et al. Rhinovirus C targets ciliated airway epithelial cells. *Respir Res.* 2017;18(1):84.
82. Bird SW, Maynard ND, Covert MW, Kirkegaard K. Nonlytic viral spread enhanced by autophagy components. *Proc Natl Acad Sci USA.* 2014;111(36):13081–13086.
83. Robinson SM, Tsueng G, Sin J, et al. Coxsackievirus B exits the host cell in shed microvesicles displaying autophagosomal markers. *PLoS Pathog.* 2014;10(4): e1004045.
84. Klein KA, Jackson WT. Human rhinovirus 2 induces the autophagic pathway and replicates more efficiently in autophagic cells. *J Virol.* 2011;85(18):9651–9654.
85. Jackson WT, Giddings Jr TH, Taylor MP, et al. Subversion of cellular autophagosomal machinery by RNA viruses. *PLoS Biol.* 2005;3(5):e156.
86. Chen YH, Du W, Hagemeijer MC, et al. Phosphatidylserine vesicles enable efficient en bloc transmission of enteroviruses. *Cell.* 2015;160(4):619–630.
87. Gwaltney Jr. JM, Jordan Jr. WS. Rhinoviruses and respiratory disease. *Bacteriol Rev.* 1964;28:409–422.
88. Hamparian VV, Colonno RJ, Cooney MK, et al. A collaborative report: rhinoviruses—extension of the numbering system from 89 to 100. *Virology.* 1987;159 (1):191–192.
89. Johnston SL, Pattemore PK, Sanderson G, et al. Community study of role of viral infections in exacerbations of asthma in 9-11 year old children. *BMJ.* 1995;310 (6989):1225–1229.
90. Gama RE, Horsnell PR, Hughes PJ, et al. Amplification of rhinovirus specific nucleic acids from clinical samples using the polymerase chain reaction. *J Med Virol.* 1989;28(2):73–77.
91. Ireland DC, Kent J, Nicholson KG. Improved detection of rhinoviruses in nasal and throat swabs by seminested RT-PCR. *J Med Virol.* 1993;40(2):96–101.
92. Arruda E, Crump CE, Rollins BS, Ohlin A, Hayden FG. Comparative susceptibilities of human embryonic fibroblasts and HeLa cells for isolation of human rhinoviruses. *J Clin Microbiol.* 1996;34(5):1277–1279.
93. Porterfield JS. Titration of some common cold viruses (rhinoviruses) and their antisera by a plaque method. *Nature.* 1962;194:1044–1047.
94. Fiala M. Plaque formation by 55 rhinovirus serotypes. *Appl Microbiol.* 1968;16 (10):1445–1450.
95. Dolan TM, Fenters JD, Fordyce PA, Holper JC. Rhinovirus plaque formation in WI-38 cells with methylcellulose overlay. *Appl Microbiol.* 1968;16(9):1331–1336.
96. Conant RM, Hamparian VV. Rhinoviruses: basis for a numbering system. 1. HeLa cells for propagation and serologic procedures. *J Immunol.* 1968;100(1):107–113.
97. Conant RM, Hamparian VV. Rhinoviruses: basis for a numbering system. II. Serologic characterization of prototype strains. *J Immunol.* 1968;100(1):114–119.
98. [No authors listed]. Rhinoviruses: a numbering system. *Nature.* 1967;213 (5078):761–762.
99. Taylor-Robinson D, Tyrrell DA. Serotypes of viruses (rhinoviruses) isolated from common colds. *Lancet.* 1962;1(7227):452–454.
100. Hilleman MR. Present knowledge of the rhinovirus group of viruses. *Curr Top Microbiol Immunol.* 1967;41:1–22.

101. Bella J, Kolatkar PR, Marlor CW, Greve JM, Rossmann MG. The structure of the two amino-terminal domains of human ICAM-1 suggests how it functions as a rhinovirus receptor and as an LFA-1 integrin ligand. *Proc Natl Acad Sci USA*. 1998;95 (8):4140–4145.
102. Kaiser L, Crump CE, Hayden FG. In vitro activity of pleconaril and AG7088 against selected serotypes and clinical isolates of human rhinoviruses. *Antiviral Res*. 2000;47 (3):215–220.
103. Ledford RM, Patel NR, Demenczuk TM, et al. VP1 sequencing of all human rhinovirus serotypes: insights into genus phylogeny and susceptibility to antiviral capsid-binding compounds. *J Virol*. 2004;78(7):3663–3674.
104. Andries K, Dewindt B, Snoeks J, et al. Two groups of rhinoviruses revealed by a panel of antiviral compounds present sequence divergence and differential pathogenicity. *J Virol*. 1990;64(3):1117–1123.
105. Torgersen H, Skern T, Blaas D. Typing of human rhinoviruses based on sequence variations in the 5′ non-coding region. *J Gen Virol*. 1989;70(Pt 11):3111–3116.
106. Hyypia T, Auvinen P, Maaronen M. Polymerase chain reaction for human picornaviruses. *J Gen Virol*. 1989;70(Pt 12):3261–3268.
107. Gama RE, Hughes PJ, Bruce CB, Stanway G. Polymerase chain reaction amplification of rhinovirus nucleic acids from clinical material. *Nucleic Acids Res*. 1988;16 (19):9346.
108. Stanway G, Hughes PJ, Mountford RC, Minor PD, Almond JW. The complete nucleotide sequence of a common cold virus: human rhinovirus 14. *Nucleic Acids Res*. 1984;12(20):7859–7875.
109. Arden KE, McErlean P, Nissen MD, Sloots TP, Mackay IM. Frequent detection of human rhinoviruses, paramyxoviruses, coronaviruses, and bocavirus during acute respiratory tract infections. *J Med Virol*. 2006;78(9):1232–1240.
110. Kistler A, Avila PC, Rouskin S, et al. Pan-viral screening of respiratory tract infections in adults with and without asthma reveals unexpected human coronavirus and human rhinovirus diversity. *J Infect Dis*. 2007;196(6):817–825.
111. Lee WM, Kiesner C, Pappas T, et al. A diverse group of previously unrecognized human rhinoviruses are common causes of respiratory illnesses in infants. *PLoS One*. 2007;2(10):e966.
112. Lamson D, Renwick N, Kapoor V, et al. MassTag polymerase-chain-reaction detection of respiratory pathogens, including a new rhinovirus genotype, that caused influenza-like illness in New York State during 2004-2005. *J Infect Dis*. 2006;194 (10):1398–1402.
113. Lau SK, Yip CC, Woo PC, Yuen KY. Human rhinovirus C: a newly discovered human rhinovirus species. *Emerg Health Threats J*. 2010;3:e2.
114. Simmonds P, McIntyre C, Savolainen-Kopra C, Tapparel C, Mackay IM, Hovi T. Proposals for the classification of human rhinovirus species C into genotypically assigned types. *J Gen Virol*. 2010;91(Pt 10):2409–2419.
115. Faux CE, Arden KE, Lambert SB, et al. Usefulness of published PCR primers in detecting human rhinovirus infection. *Emerg Infect Dis*. 2011;17(2):296–298.
116. Bochkov YA, Grindle K, Vang F, Evans MD, Gern JE. Improved molecular typing assay for rhinovirus species A, B, and C. *J Clin Microbiol*. 2014;52(7):2461–2471.
117. Tamura K, Stecher G, Peterson D, Filipski A, Kumar S. MEGA6: molecular evolutionary genetics analysis version 6.0. *Mol Biol Evol*. 2013;30(12):2725–2729.
118. *The Picornavirus Pages*. Available from: <http://www.picornaviridae.com>.
119. Westerhuis BM, Wiewel MA, Ende AA, et al. A chip-based rapid genotyping assay to discriminate between rhinovirus species A, B and C. *J Clin Virol*. 2018;99-100:10–14.

120. Ottolini MG, Malloy AMW. Rhinovirus—new insights into a complex epidemiology. *J Infect Dis.* 2018;218(6):845–847.
121. Rathe JA, Liu X, Tallon LJ, Gern JE, Liggett SB. Full-genome sequence and analysis of a novel human rhinovirus strain within a divergent HRV-A clade. *Arch Virol.* 2010;155(1):83–87.
122. Naughtin M, Sareth R, Sentilhes AC, et al. Genetic diversity of human rhinoviruses in Cambodia during a three-year period reveals novel genetic types. *Infect Genet Evol.* 2015;35:42–49.
123. Arden KE, Faux CE, O'Neill NT, et al. Molecular characterization and distinguishing features of a novel human rhinovirus (HRV) C, HRVC-QCE, detected in children with fever, cough and wheeze during 2003. *J Clin Virol.* 2010;47(3):219–223.
124. Niespodziana K, Stenberg-Hammar K, Megremis S, et al. PreDicta chip-based high resolution diagnosis of rhinovirus-induced wheeze. *Nat Commun.* 2018;9(1):2382.
125. Glanville N, Johnston SL. Challenges in developing a cross-serotype rhinovirus vaccine. *Curr Opin Virol.* 2015;11:83–88.
126. Monto AS, Bryan ER, Ohmit S. Rhinovirus infections in Tecumseh, Michigan: frequency of illness and number of serotypes. *J Infect Dis.* 1987;156(1):43–49.
127. Olenec JP, Kim WK, Lee WM, et al. Weekly monitoring of children with asthma for infections and illness during common cold seasons. *J Allergy Clin Immunol.* 2010;125(5):1001–1006.e1.
128. Lee WM, Lemanske Jr RF, Evans MD, et al. Human rhinovirus species and season of infection determine illness severity. *Am J Respir Crit Care Med.* 2012;186(9):886–891.
129. Lee S, Nguyen MT, Currier MG, et al. A polyvalent inactivated rhinovirus vaccine is broadly immunogenic in rhesus macaques. *Nat Commun.* 2016;7:12838.
130. Tang JW, Lam TT, Zaraket H, et al. Global epidemiology of non-influenza RNA respiratory viruses: data gaps and a growing need for surveillance. *Lancet Infect Dis.* 2017;17(10):e320–e326.

CHAPTER 2

Rhinovirus diversity and virulence factors

Camille Esneau[1,*], Sarah Croft[2,*], Su-Ling Loo[3] and Reena Ghildyal[2]
[1]University of Newcastle, Callaghan, NSW, Australia
[2]Faculty of Science and Technology, Centre for Research in Therapeutic Solutions, University of Canberra, Canberra, ACT, Australia
[3]School of Biomedical Sciences and Pharmacy, University of Newcastle, Callaghan, NSW, Australia

2.1 INTRODUCTION

The antigenic diversity of rhinoviruses (RVs) has been an obstacle to the development of effective cross-reactive antiviral treatments,[1] with the RV genus consisting of more than 167 subtypes[2–4] within three species, utilizing three different receptors for cell entry.[5–7] There are no current vaccines or licensed antivirals that specifically target RV-induced respiratory disease. Several candidate antivirals have been developed but none have been translated into clinical use. These have included the broadly active nucleoside analogue ribavirin, as well as RV-specific capsid binders, and rupintrivir, a small molecule inhibitor of RV 3C protease.[8] Pleconaril, a broadly reactive capsid binder, has shown promise in clinical trials in reducing the number of RV-induced asthma exacerbations but is yet to be approved for clinical use.[9] Rupintrivir did not significantly decrease viral load or illness severity in natural infection due primarily to a requirement for the drug to be taken either prior to or immediately after infection, and further trials were terminated.[10] None of the candidates have so far been effective against all the RV subtypes tested. A feasible strategy could be to identify relevant subtypes contributing to an increased illness severity and then specifically targeting these. To enable such a strategy, a deep understanding of RV subtype association with disease, and the molecular mechanisms thereof, is required.

RV infection can cause a wide range of symptoms from a mild cold to fatal asthma exacerbation or be asymptomatic.[11–13] Could this variety of clinical outcomes be a manifestation of the large number of subtypes,

* Authors contributed equally to the chapter.

Rhinovirus Infections
DOI: https://doi.org/10.1016/B978-0-12-816417-4.00002-0

with each or group of subtypes giving rise to a narrow range of disease effects? Recent literature clearly shows an effect of RV species on illness severity, with two species (RV-A and RV-C) being often found in illness requiring hospitalization while the third (RV-B) is mostly found in asymptomatic infections.[14–19] Emerging, but currently limited literature links specific subtypes with illness severity while work from the Palmenberg group has clearly shown a subtype specific host directed function of the RV proteases.[20]

RV encodes two proteases, 2A (2Apro) and 3C (3Cpro), which are the main virulence factors being directly responsible for the damage to host cells that results in cell death and probably contributes to disease symptoms.[21] RV proteases are essential for RV replication, being responsible for polyprotein processing, and cause the characteristic host cell gene expression shutoff by cleaving host factors required for eukaryotic transcription and translation.[22] RV proteases also cleave intermediates in several signaling pathways that lead to a change in the cellular milieu that favors RV replication, while stalling cellular processes. RV proteases are well conserved across all RV subtypes[2] yet show species-specific variation in the host factors cleaved[20] and may contribute to the species/subtype-specific illness.

In this chapter, we bring together the current clinical and experimental knowledge on RV diversity, which may be associated with disease severity, and discuss the underlying factors affecting RV virulence with a focus on the role of RV proteases.

2.2 SPECIES AND SUBTYPES IN CLINICAL PRESENTATION

2.2.1 Species and illness severity

RV infections can be asymptomatic or result in one or more of a wide range of symptoms including mild upper respiratory tract manifestations, severe lower respiratory tract illnesses (LRTI), and exacerbation of chronic diseases such as asthma.[11–13] This variation in or lack of symptoms suggests that the severity of the illness depends on the infecting virus subtype as well as other factors such as the host response[23,24] and environmental factors.[25] While host immune and inflammatory signaling pathways have been extensively studied, particularly in the context of asthma exacerbations induced by RV infection, the contribution of particular RV species or subtypes in the development of a severe illness remains relatively unexplored. With more than 167 subtypes classed into three species, it is likely that RV-subtype specific factors are major contributors to symptom severity in respiratory illnesses.

Determination of RV species in numerous clinical studies shows almost unequivocally that RV-A and RV-C alternate in prevalence, while RV-B is underrepresented.[14–18] In addition to being less common, RV-B is generally associated with less severe symptoms and more likely to be found in asymptomatic infections.[19] RV-B isolates clustered in a low virulence group in RV-positive samples from children enrolled into the Childhood Origin of Asthma birth cohort, causing a higher proportion of asymptomatic infections and being up to 10 times less likely to cause a severe illness than RV-A. RV-B was unlikely to cause an illness even in children with a high-risk profile.[25] The lower association of RV-B with illness is consistent with known phylogenetic relationships; RV-A and RV-C are somewhat more closely related (58% mean genome identity between species) while the RV-B species is genetically (slightly) more distant (55%–57% mean genome identity)[2,26,27] suggesting that some species-specific factors may be responsible for virulence.

2.2.1.1 Association of rhinovirus species C with severe illnesses and asthma

The inability to grow RV-C viruses using traditional lab-based cell culture systems delayed their discovery by 50 years, until sequencing and phylogenetic analysis of genome sequences from clinical isolates revealed a RV species that formed a distinct sublineage from RV-A and RV-B species.[28] When characterizing RV-QPM isolates, which are now classified as the RV-C3 subtype,[29] McErlean et al. found that 50% of isolates originated from samples of patients with lower respiratory tract symptoms (bronchiolitis, asthma, and chronic obstructive pulmonary disease). This finding led to the hypothesis that this group of RV could be associated with more severe LRTI disease than that caused by infections with RV-A or RV-B.[30] This hypothesis is supported by several studies published in the following years including hospitalized and nonhospitalized subjects. In samples from patients (mostly children under 10 years of age) referred for testing for a virus infection, Wisdom et al. found that RV-C was almost two times more likely to be associated with acute respiratory tract illnesses than with asymptomatic infections, matching the rates of respiratory syncytial virus (RSV), and more frequently associated with cough in comparison to RV-A.[31] In hospitalized children under 5 years of age, Miller et al. found that RV-C was more likely to cause severe illness requiring oxygen supplementation. Children infected with RV-C were more likely to wheeze when compared with RV-A.[32] Lau et al. found that RV-C was more frequently associated with pneumonia than RV-A in hospitalized adults.[17] An Australian study found that RV-C, but not RV-A, was significantly associated with upper respiratory tract symptoms.[33] Finally, a 21-year prospective study in children under 5 years enrolled when presenting with an upper respiratory tract illness (URTI) or LRTI found a stronger association of RV-C with LRTI.[34]

Wheeze, a high-pitched whistling sound whilst breathing, is a symptom of airway obstruction. It is highly prevalent in preschool children aged 5 years or younger. RV was detected in almost 70% of preschool children presenting at a hospital with acute wheeze.[35] RV-infected children were found to be significantly younger, had a shorter fever duration than the RV-negative group but significantly higher frequency of chest retraction and wheezing in a recent study of children <18 years old hospitalized with acute LRTI.[36]

Wheezing in preschool children with RV, especially in the first 3 years of life, is associated with a risk of developing asthma at a later age, more so than wheezing associated with RSV.[37] Both RV-A and RV-C subtypes are significantly associated with asthma exacerbations,[18,25,38] but RV-C seems to have a stronger link to development of asthma highlighting its role in disease in early life. In a retrospective case—control study comparing hospitalized children with RV-positive asthma exacerbations with control non-allergic children with a RV-positive respiratory illness, Mak et al. found that RV-C rates were significantly higher in children with asthma than in children with other respiratory illnesses. Moreover, more subjects with RV-C infection required oxygen supplementation.[39] In children presenting to the emergency department with an asthma exacerbation, 98% of whom were subsequently hospitalized, Bizzintino et al. reported that children with RV-C infection presented with a higher asthma exacerbation severity score and more severe asthma and RV-C was the main virus identified in severe asthma attacks.[40] Children with RV-C infection and wheezing or asthma exacerbation were found to have significantly lower forced expiratory volume in 1 second and required oxygen supplementation more frequently.[32,39] This suggests that RV-C may cause more severe disease, resulting in a reduced lung capacity and increased respiratory problems. Opinion is divided as to whether children show differences in response to RV infection or whether wheezing with RV reveals a preexisting impairment that promotes asthma mainly in predisposed children.

2.2.1.2 More than just diversity in the species

Many studies have suggested that, particularly in the very young, RV-C is associated with more severe illnesses, more so than RV-A. However, studies are not unanimous on this conclusion, and differences that are observed between species could be caused by more than just virulence associated with RV species. Cox et al. found that children infected with RV-C not only had an increased risk of future hospitalization but were also more likely to have been hospitalized in the past.[41] This suggests that infection with RV-C might not only be more severe, but also an indicator of at-risk populations who may be more prone to a severe outcome. Moreover, a greater incidence of RV-C in severe illness might not necessarily mean that this species has an increased virulence. Despite finding a greater association with LRTI and bronchiolitis, Renwick et al. found that LRTI associated with RV-C were, in general, milder than LRTI associated with RV-A, contradicting previous observations of RV-C in severe illnesses.[42] Similar results were reported by Arden et al. when examining RV species in a cohort of children with asthma exacerbation without hospitalization. They found that RV-A was associated with worse asthma symptoms and a longer-lasting cough.[43]

Age might also be a factor contributing to higher prevalence of one RV species over another in illness. In the previously mentioned study, Wisdom et al. found that RV-C tends to be more frequent in young children while the older age groups included in the study mainly showed cases of RV-A species. This indicates that RV-C might be more prevalent in early childhood, which is when RV infections produce more severe symptoms.[19] This observation is also in agreement with the observation by Cox et al. that children with RV-C infection are more likely to have had respiratory illness in the past, suggesting that RV-C causes severe disease in the context of lowered, weakened, or immature immune response.[41]

In summary, symptoms associated with RV infection involve many environmental and immunological factors, which need to be considered when studying RV-induced illness severity. While the interplay between these factors remains to be completely understood, it is apparent that RV species are not equal in their contribution to disease. Considering RV-A,

RV-B, and RV-C species individually allows us to find association with illness that might not necessarily be obvious when examining RV genus as a whole. Following the same logic, each RV species contains dozens of subtypes, which might not contribute equally to the development of severe illnesses.

Data from laboratory-based basic research and some clinical studies suggest that there may be a case for subtype association with disease; these are discussed below.

2.2.2 Toward the identification of more pathogenic subtypes

Early vaccination trials showed that protective immunity against one RV subtype could be achieved. Cross-subtype protection was poor due to the high antigenic diversity that exists between RV subtypes.[1] While new approaches allowed the development of multivalent inactivated RV vaccines able to target 25 subtypes in mice and 50 subtypes in rhesus macaques, this still represented less than half of identified RV subtypes.[44] Because of these limitations in targeting all RV subtypes, identification of virulent subtypes could be an asset for drug development, leading to narrowing of the antigenic focus for a vaccine or small molecule antivirals.

2.2.2.1 Cases of rhinovirus subtypes with increased pathogenicity

In 2003, a Santa Cruz RV variant (SC-RV) of RV-A82 was responsible for unusually high morbidity and mortality rates in an aged care facility.[45] A more recent study found a predominance of RV-A21 in adult inpatients with community-acquired pneumonia (CAP) and in outpatients with URTI. The RV-A21 subtype represented 21% of the RV-positive infection and was more often associated with severe cases of CAP.[46] Characterization of subtypes associated with increased severity demonstrates the genetic fluidity of RVs. Indeed, SC-RV had 90% identity with RV-A82 but showed divergence in the neutralization epitopes on VP2 and VP4, which could have led to immune evasion and increased pathogenicity.[47] In support of this, RV-A21 isolates displayed significant divergence from the prototype subtype and could not be neutralized with prototype subtype antiserum.[46]

The low fidelity of the viral polymerase and recombination events are important factors for genetic divergence of RV.[48,49] This is exemplified by RVA-101: full genome sequencing showed a significant divergence of the circulating subtype when compared with the original reference subtype sequenced 8 years earlier. The majority of the nonconservative variations were present on VP1-VP3 genes coding for the capsid proteins—responsible for antigenic variability and receptor specificity—and on the 3D gene coding for the viral polymerase.[50] While this particular study did not compare virulence between the two isolates, these genetic variations within a same subtype could result in changes in viral proteins and favor the emergence of a more virulent variant. Such emergence was observed in the closely related enterovirus EV-D68 (first recognized as RV-87). EV-D68 is an atypical enterovirus that shares properties with RV (growth at lower temperature and acid lability). Because of these properties, this virus was first classified as RV-87 but is now commonly used to define the outgroup (closest relative not belonging to RV) to establish RV phylogenetic relationships.[30,32,47,51,52] Before 2014, EV-D68 was infrequently reported to be associated with illness, but recent years have seen an increased incidence of EV-D68 infections associated with severe respiratory outcomes.[53] In addition to severe respiratory illnesses, the emergence of EV-D68 was accompanied by expanded tropism to the nervous system and association with neurological disease such as acute flaccid myelitis.[54] Following this increased prevalence, genomic characterization of the virus showed that the more severe outcomes were accompanied by an increased genetic diversity in the VP1 capsid-coding gene as well as amino acid substitutions in the viral proteases 3Cpro and 2Apro.[55,56] Given the close relationship of EV-D68 with RV and the similarities in genetic divergence mechanisms, the emergence of pathogenic RV subtypes could occur in a similar way.

2.2.2.2 An increased need for surveillance to identify circulating subtypes

Noninfluenza respiratory viruses (e.g., RV), although an important burden in LRTI, lack surveillance data.[28,57] Increasing surveillance of RV subtypes would allow us to understand the underlying genetic changes associated with clinical consequences, and perhaps allow the identification of emerging RV subtypes. Many subtypes of RV coexist at the same time in a population,[58] and RV circulation patterns are still unclear.[25] Surveillance of RVs has historically been hindered by their genetic diversity. For example, early PCR primers failed to detect all subtypes, or led to confusion between RV-A and RV-C species, requiring frequent improvement of the assays.[28,59–61] Due to the late discovery of RV-C species[30,62] the presence of this species was not included and/or detected in earlier studies.[34] As described earlier, although RV-C was highlighted as a more virulent species at the time of its discovery,[17,32] more recent studies suggest that there might be no difference in severity between RV-A and RV-C species.[51,63] Variations in study design, season of study, age, and patient selection may contribute to the discrepancies between older and more recent studies. Similarly, as viruses are opportunistic and can cause more severe illnesses in vulnerable populations, such as those with chronic respiratory diseases, it is difficult to determine if these variants are more virulent or if the virus is taking advantage of the compromised host for its spread and survival. These limitations highlight the challenges to understand RV virulence and the need to establish a consistent surveillance protocol to understand RV circulation and identify pathogenic RV species and subtypes. It is clear, however, that RV-A and C cause more frequent symptomatic disease than RV-B, the latter being rarely identified in the hospitalized cases of RV disease.

Better availability of typing methods has allowed recent study cohorts to identify specific RV subtypes that occur regularly. In a 1-year cohort study in Vietnam, Tran et al. found that RV-A12, -A78, and -C40 were recurring often and across several months.[63] Van der Linden et al. reported similar results when analyzing RV-positive samples collected from hospitalized and nonhospitalized subjects in an Amsterdam clinic. Subtypes RV-A12 and -A78 occurred in a high number of the samples, which were collected from 2007 to 2014. In addition, RV-A12 was identified in patients with respiratory illness during winter, 3 years in a row, suggesting that RV-A12 could be linked to a high number of illnesses, over a long time period.[64] These findings suggest that some RV subtypes may have a higher ability to spread in a population and thus occur more frequently. However, the data remain inadequate, and further studies are needed to determine if these subtypes are truly more pathogenic.

In an effort to assess subtype association with clinical illness, we conducted a meta-analysis of clinical studies where RV was genotyped. Studies were identified by a PubMed search using the following keywords: *rhinovirus*, *subtype*, *genetic diversity*, *molecular typing*, *subtype*, and *phylogeny*. To be included in the analysis, studies had to utilize genome-based typing and provide a description of the illness caused by infection. When multiple gene sequences were used for typing, the VP4/VP2 sequence was preferred. When subtypes were presented in a phylogenetic tree, the closest subtype to the clinical isolate was used. The design for included studies is available in Table 2.1. We listed the RV subtypes identified in each study to characterize the occurrence of more frequent subtypes. For

Table 2.1 Details of clinical studies used in the meta-analysis.

Study reference	Study location	Sample collection year	Subject selection criteria	Total samples	HRV-positive samples	Sequencing method	HRV + samples selected for sequencing	No. of sequences used in meta-analysis (A/B/C)	% Species distribution (A/B/C)
[31]	United Kingdom	Sep 2006–Feb 2007	Samples sent to diagnosis for ARI	456 (345 individuals)	111/456 (24.3%)	VP4/VP2	110/111 (99%)	64 (56/8/0)	62.5/7.6/29.8
[65]	Italy	Oct 2008–Sep 2009	Children and adults hospitalized ARI	1500 (985 individuals)	405/1500 (27%)	VP4/VP2	195/405 (48%)	128 (110/18/0)	56.4/9.2/34.3
[66]	South Africa	May 2004–Nov 2005	Children wheezing	220	128/220 (58.2%)	5'UTR and VP4/VP2	71/128 (55.5%)	31 (23/8/0)	36.6/11.3/52.1
[14]	Unites States	Oct 2003–Apr 2005	Children control and hospitalized	2737	438/2737 (16%)	VP1	326/438 (74.4%)	201 (179/22/0)	59/8.9/32
[67]	Japan	Jan–Dec 2010	Children and adults ARI symptoms	501	92/501 (18.4%)	VP4/VP2	92/92 (100%)	49 (47/2/0)	51/2.1/46.7
[68]	Panama	Aug 2010–Jun 2011	Children hospitalized ARI	118 (420 individuals)	62/118 (52.5%)	VP4/VP2	62/62 (100%)	40 (35/5/0)	59.6/8/32.2
[69]	France	Oct 2009–May 2010	Children and adults influenza/RSV negative	232 (209 individuals)	63/232 (27%)	VP4/VP2	63/63 (100%)	37 (33/4/0)	52/6/40
[70]	Spain	Dec 2003–Jun 2010	Children SARI	3987	949/3987 (23.8%)	VP4/VP2	397/949 random selection (40%)	39 (0/39/0)	51.4/9.8/38.8
[71]	Paraguay	May 2010–Dec 2011	Children hospitalized acute LTRI	101	34/101 (33.7%)	VP4/VP2	34/34 (100%)	18 (13/1/4)	72/5.5/22

(*Continued*)

Table 2.1 (Continued)

Study reference	Study location	Sample collection year	Subject selection criteria	Total samples	HRV-positive samples	Sequencing method	HRV + samples selected for sequencing	No. of sequences used in meta-analysis (A/B/C)	% Species distribution (A/B/C)
[72]	Italy	2010	Children hospitalized	Not specified	90	5′UTR or VP4/VP2	73/90 (81%)	72 (40/5/27)	55/7/38
[73]	Sweden	Nov 2006–Sep 2010	Children and adults ARI	11,498	1840/11498 (16%)	VP4/VP2	170/1840 (9%)	106 (64/11/31)	56/9.6/32.4
[74]	United Kingdom	Nov 2010–Feb 2011	Patients or staff resp. symptoms	23	23/23 (100%)	VP4/VP2	21/23 (91%)	19 (12/3/4)	50/7/28.5
[75]	United Kingdom	Dec 2010–Mar 2011	Adults cystic fibrosis	626 (98 individuals)	138/626 (22%)	5′UTR	61/138 (44%)	54 (26/15/0)	69/26.2/4.8
[18]	China	2009–12	Children hospitalized ARI	1567	223/1567 (14%)	VP4/VP2	89/223 (39.9%)	84 (0/0/84)	54.7/5.4/39.4
[76]	Argentina	Jun 2008–May 2010	Children ARI, LRTI and/or URTI	620	252/620	5′UTR and VP4/VP2	45/252 (17.8%)	41 (21/1/19)	46.6/2/42.2
[77]	12 EU countries	2007–10	Adults prolonged RV shedding	4970 (2485 individuals)	554/2485	VP3/VP1 VP4/VP2 or 5′UTR	42/554 (7.5%)	24 (18/4/2)	56/28/13
[78]	The Netherlands	Nov 2009–Dec 2012	Children RV positive	120	120/120 (100%)	VP4/VP2	120/120 (100%)	107 (63/4/40)	55/9/35
[79]	Tanzania	Apr–Dec 2008	Children fever ≥38°	1005	244/1005 (24.2%)	5′UTR, VP4/VP2, VP1	244/244 (100%)	225 (119/38/68)	52/17/31

[51]	Cambodia	Jun 2007–Dec 2009	Children and adults ILI/ARTI	4170	455/4170 (10.9%)	VP4/VP2, VP1	88/455 (19.3%)	60 (36/10/14)	60/16.6/28.3
[80]	Indonesia	Mar 2010–Apr 2011	Adults ILI, ARI, SARI	106	42/106 (39.6%)	5′UTR	42/42 (100%)	41 (14/27/0)	33/64/2.3
[81]	Cyprus	Nov 2010–Oct 2013	Children ARI	485	116/485 (23.9%)	VP4/VP2	116/116 (100%)	68 (36/5/27)	52.9/7.4/39.7
[82]	Kenya	2008	Children ILI	517	131/517 (25%)	VP4/VP2	37/131 (28%)	26 (14/3/9)	54/12/35
[63]	Vietnam	Apr 2010–May 2011	Children hospitalized ARI	1082	325/1082 (30%)	VP4/VP2	58/325 (17.8%)	58 (44/0/14)	75.9/0/24.1
[64]	The Netherlands	2007–12	Children samples for diagnostic	6258	1102/6258 (17.6%)	VP4/VP2	745/1102 (67.9%)	587 (309/56/222)	52.4/11.3/36.2
[83]	Australia	2006–09,	ILI, children and adults	555	116/555 (20.9%)	5′UTR or VP4/VP2	65/116 (56%)	44 (43/1/0)	68/1.5/30.7

ARI, acute respiratory illness; *EU*, European Union; *HRV*, human rhinovirus; *ILI*, Influenza like illness; *LRTI*, lower respiratory tract illness; *RSV*, respiratory syncytial virus; *RV*, rhinovirus; *SARI*, severe acute respiratory illness; *5′UTR*, 5′untranslated region; *URTI*, upper respiratory tract illness.

RV-A species, some subtypes were merged in the modern classification, and isolates typed as RV-A44, RV-A95, and RV-A98 in earlier studies were therefore changed to RV-A29, RV-A54, and RV-A8, respectively. Subtypes RV-A25 and RV-A62 are genetically closely related and were reported together in several studies. Therefore these two subtypes were counted together in the meta-analysis. Finally, subtypes identified as provisionally assigned types were excluded.

Twenty-three published studies for RV-A and RV-B species met the analysis criteria. Due to its more recent discovery, only fourteen studies were included for RV-C species, and 77/80, 28/32, and 50/56 subtypes were reported at least once for RV-A, RV-B, and RV-C respectively (Fig. 2.1). This finding is consistent with the broad circulation of RV subtypes. However, although nearly all RV subtypes were found overall, only a handful was reported in more than half of the studies. RV-A12 and -A78 were the most frequently occurring subtypes for A species, as they were reported in 70% and 65% of the studies, respectively. This result is consistent with the reoccurrence of these subtypes in longitudinal studies (discussed above).[63,64] RV-A49 and RV-A1, each reported in 61% and 56% of studies, were the only minor group RVs occurring in more than 50% of the studies analyzed. For RV-C species, the commonly studied RV-C subtypes RV-C15 and RV-C2 were the most frequently occurring subtypes (64% and 57% of studies respectively). RV-B6 and -B27 were the most frequent RV-B subtypes, but no RV-B subtypes occurred in more than 50% of the studies probably due to the fact that this subtype is underrepresented in clinical illnesses, and that 60% of the studies included only reported five or fewer RV-B subtypes.

These results support the idea that RV subtypes associated with respiratory disease do not occur equally. RV-A12, -A78, -C15, and -C2 appeared in a large number of studies, even though the studies included covered many countries, climates, years, and involved patients with different illnesses. However, this meta-analysis has some limitations. Some subtypes might be underrepresented as not 100% of the samples were typed in every study, and because provisionally assigned subtypes were excluded from the analysis. Recent studies generally utilize more samples and report a broader number of subtypes, and it is possible that typing a greater number of samples might reduce the differences in frequency of RV detection associated with respiratory disease. Finally, studies included in the analysis covered a broad range of illnesses (from illnesses requiring hospitalization to mild illnesses and asymptomatic RV detections) and sometimes

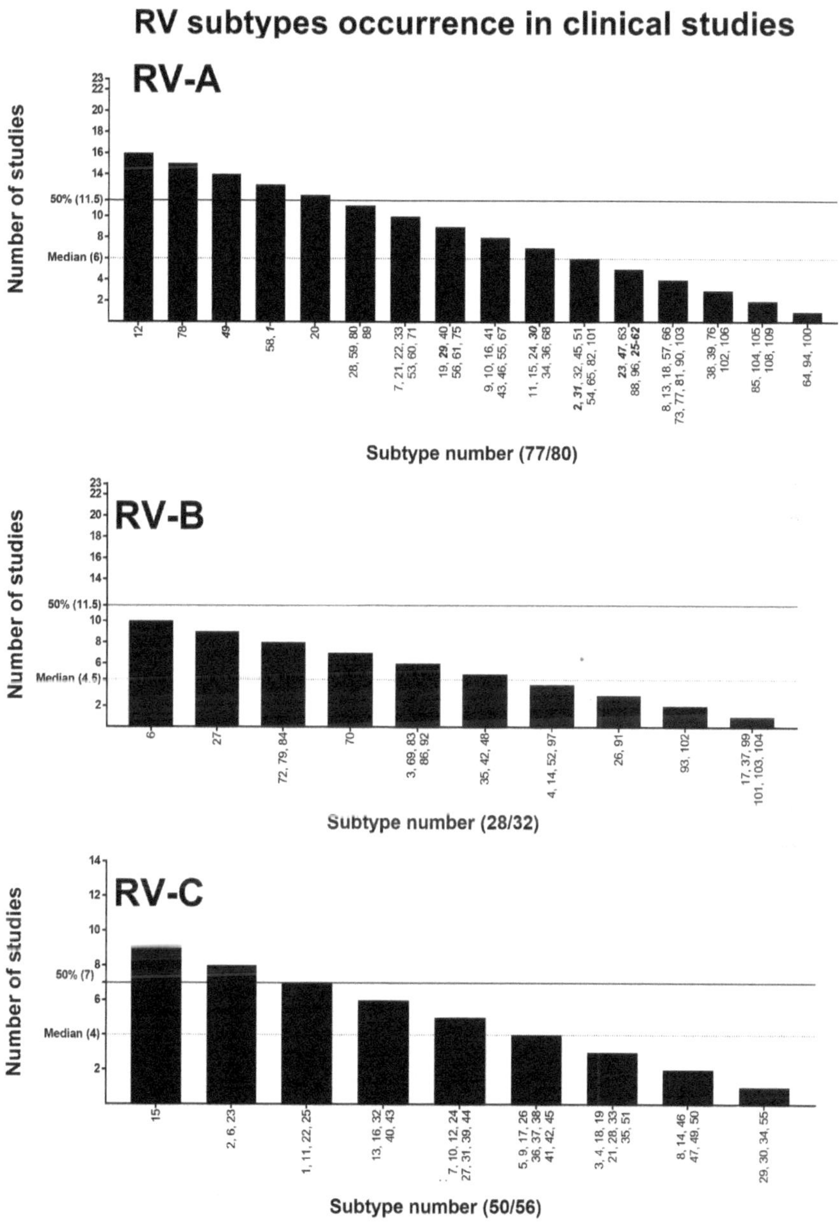

Figure 2.1 ***Rhinovirus (RV) subtypes occurrence in published studies.*** Each graph presents how often specific RV subtypes were found for A, B, and C species, respectively. The total number of studies used for the meta-analysis is shown on the *Y*-axis and subtype number is on the *X*-axis. Subtypes belonging to the minor group in RV-A are indicated in bold.

included subjects with a specific immune or inflammatory background (asthma, cystic fibrosis, and immunosuppression). Studies linking the subtype identified to symptom severity or host background remained rare. A sufficiently powered study that would allow stratification of RV occurrence with symptoms, geographical region or year of study would be necessary to confirm that individual subtypes are indeed more frequently associated with respiratory illnesses.

2.3 FACTORS CONTRIBUTING TO INCREASED SUBTYPE PATHOGENICITY

2.3.1 Rhinovirus diversity, receptor and cell tropism

Before the availability of genomic sequences, classification of RV subtypes was based on serology and receptor usage. The major group, so called as most (90%) RVs at the time were included in this group, used the intercellular adhesion molecule 1 (ICAM-I) to bind and enter cells. The remaining 10% used the low-density lipoprotein receptor (LDLR)[6,7] for cell attachment and entry. Bochkov et al. found that RV-C subtypes use a third receptor, Cadherin-related family member 3 (CDHR3).[84] Little is known about CDHR3 function in the lung, while ICAM-1 and LDLR are well-characterized and have been shown to be structurally and functionally distinct. ICAM-I is a transmembrane glycoprotein belonging to the immunoglobulin family and is involved in cellular communication,[85] while the LDLR family members are involved in transporting ligands involved in a variety of functions.[86] Interestingly, there have been three separate occasions where major group RVs have evolved to use the LDLR as a receptor.[87]

The use of different receptors has obvious consequences for RV tropism in the airway. Receptor usage also has consequences for the intracellular site of genome release with ramifications for the virus life cycle. When studying major group viruses, Ganjian et al. found that different subtypes were released at various subcellular compartments. RV-A89 and RV-A16 were directed to the endosomal recycling compartment for uncoating, whereas RV-B3 and RV-B14 were not. This observation could provide insight into why RV-Bs are less virulent. Differences were also observed within the species: RV-A89 could replicate even when the endosomal pH was neutralized, which was not the case for all the other RV-A subtypes studied.[88]

The choice of receptor by the RV subtype to enter cells may have consequences for the subsequent steps in viral replication processes. However, evidence from literature suggests that affinity to a particular receptor does not determine a fixed pathway of productive uncoating or intracellular trafficking. RV-B14 and RV-A89, despite binding ICAM-I, were routed into distinct endosomal compartments for release of their RNA into the cytosol.[89] RV-A89 was directed to the endosomal recycling compartment for uncoating, whereas RV-B14 was uncoated in endosomal compartments of the lysosomal route.[88]

In airway epithelial cell cultures differentiated at the air liquid interface (ALI), major and minor group RVs are able to replicate in both ciliated and unciliated cells,[90] while RV-C targets ciliated cells only.[91] A 2008 study using RV-A1a (LDLR-binding minor group) and RV-A16 (ICAM-1 binding major) showed that, when basal cells and cells immediately above the basal layer (suprabasal) were separated from ALI cultures, viral RNA copies were higher in basal cells compared with the suprabasal layer. The authors suggested that this susceptibility to viral replication was linked to an increased ICAM-1 expression ($\sim$4 fold) in basal cells, at least for the major group RV-A16.[92] However, the minor group RV-A1a also displayed more efficient replication in basal cells, indicating this may not be an ICAM-1 (receptor)-specific mechanism. Although LDLR distribution has not been as extensively studied, it is possible that it too is preferentially expressed in basal cells. Although RV subtypes have been classified on the basis of receptor affinity, recent evidence suggests that downstream processes of the viral replicative cycle are not straightforward or fixed, but different subtypes may deviate and direct uncoating and endocytic machinery. Cell surface expression of the RV receptors by various cell types present in differentiated epithelium is an important determinant in entry and replication impacting illness severity, the extent of which remains to be explored.

2.3.2 Viral load as an marker of illness severity

Viral load, or the amount of virus present, does not always correspond to the replicative rate of virus and may or may not be correlated with RV disease severity. One study developed a RV qPCR assay targeting a 210-bp region in the 5′untranslated region, to test three symptomatic patient populations analyzing nasopharyngeal swabs and midturbinate nasal flocked swabs. They were surprised to find that the mean viral loads of

RV were the same for populations of symptomatic children, university students, and the elderly, nor were these results affected by hospitalization.[93] This suggests that viral load in individuals may not be an indicator of the severity of RV illness. However, all RVs do not replicate at the same rate or have the same infectivity hence a low viral load of one subtype may cause similar infection and/or clinical disease as a significantly higher viral load of another subtype. In a 1972 study, 50 volunteers with a low titer (<2) of preexisting serum antibody to RV were infected with HRV14 (now RV-B14) or HRV39 (now RV-A39), both using ICAM-I receptor. The 50% human infectious dose was five times higher for HRV14 than that for HRV39, suggesting that the latter is able to produce successful infection with a lower number of viral particles, which could explain its higher detection in hospital admissions.[94] As discussed previously, RV-A species are more common in hospital admissions than RV-B and responsible for more severe symptoms. Studying the replicative ability of individual subtypes under different conditions might allow us to predict the likelihood of RV subtypes to cause severe illness.

Location of the infection in the respiratory tract, and corresponding temperature of the airways, may influence viral load of the infecting RV. RV infections were first thought to be limited to the upper respiratory tract due to the RV capsid being unstable at higher temperatures such as those encountered in the lower respiratory tract. In more recent studies, RV has been shown to replicate effectively at temperatures equivalent to that of the lower respiratory tract[95] and some RV-C subtypes have been shown to replicate equally efficiently at 33°C, 35°C, and 37°C.[96] RV-C virions might therefore be more stable at higher temperatures compared with RV-A or RV-B, which might confer a greater ability to colonize the lower respiratory tract and provoke a more severe illness as opposed to mild URTI. This result also impacts clinical studies, as the sampling location may influence the ability to detect specific virus subtypes.

Although it is difficult to conclude how viral load and replication correlate during respiratory illness, evidence suggests that viral load may have consequences for clinical outcomes of respiratory illness. Gerna et al. showed that when viral load was $\geq 10^6$ RNA copies/mL, single RV infections were associated with LRTI and wheezing in immune-competent children.[97] Another study demonstrated a significant positive correlation between viral load and disease severity among children aged 11 months or older.[98] Xiao et al. sought to identify correlations between RV subtypes and viral load in children, with their study suggesting that a high load of RV-A in the lower respiratory tract might be associated with disease severity in children younger than 2 years of age.[99] The importance of viral load has also been demonstrated in transplant recipients, who are prone to long periods of viral shedding.[100] Ogimi et al. found that high initial viral load was associated with prolonged shedding of RV in hematopoietic cell transplant recipients, and duration of shedding was not different acrossRV species.[101] The prolonged shedding is likely to be through the exaggerated induction of inflammatory cytokines, more acute illness, and a reduced ability to efficiently clear the pathogen. Slower replicating viruses that do not induce a robust immune response may peak in viral load later in the infectious cycle and persist for a longer period of time.

RV viremia (infectious virus in the blood) is very unusual. However cases of RV-C viremia associated with severe LRTI (pneumonia, repeated episodes of bronchiolitis) have been reported.[102] Studies detecting RVs in the blood during respiratory illness using classical culture-based diagnostic methods have been rare.[103,104] This may be because, as discussed below, RV-C is the predominant subtype detected in blood, which is difficult to culture in conventional cell lines. Recent studies using sensitive molecular methods have been able to characterize RV viral RNA in the blood of children presenting with respiratory illness.[105–107] However, these studies are approached with caution, because the detection of RV RNA in the blood may not represent true viremia with viable virus, and the mechanism by which the RNA enters the bloodstream remains unknown. Despite these limitations, RV RNA in the blood has been linked to respiratory disease and exacerbation. Lu et al. found that RV viral RNA could be detected in the blood of more than 1 out of 7 RV-infected children aged <10 years hospitalized with CAP. Interestingly, RV-A was the most common RV species detected from respiratory specimens in this group (48.8%), but RV-C RNA was detected in almost all positive blood samples (98.2%).[108]

Esposito et al. showed that RV nasopharyngeal load was significantly higher in the children with RV RNA detected in the blood and these children had higher respiratory rates, white blood cell counts, and C-reactive protein levels, as well as more often requiring oxygen therapy.[106] This suggests a link between nasopharyngeal viral load (though perhaps not subtype specific) and systemic RV RNA, as well as an association with more severe disease. In asthmatic children, Xatzipsalti et al. found that detection of RV RNA in blood was common in the early course of acute exacerbations, suggesting an association with asthma exacerbation pathogenesis.[105] However, this study did not draw any conclusion as to whether this was a consequence of asthma or due to a more virulent subtype. Studies characterizing the RV subtype/s involved in viremia during respiratory disease exacerbations are lacking. However, evidence points to a higher association of RV-C with severe illness leading to viremia in rare cases.

2.3.3 Rhinovirus diversity and immune response

Unlike influenza, RV causes minimal cytopathic effects on a small number of cells in the respiratory epithelium[109,110] suggesting that illness is not a result of acute damage to the airways but is caused by the host immune response and associated inflammation. Heinonen et al. performed transcriptome analysis on blood taken from healthy children as well as RV-positive children who either had symptoms or were asymptomatic. The authors found that symptomatic infections with RV created a specific transcriptomic signature, characterized by the overexpression of innate immunity and underexpression of adaptive immunity genes. In addition, asymptomatic infection displayed the same gene expression profile as RV-negative subjects.[12] This suggests that the intensity of the host immune response to infection is a major determinant of illness severity.

Several studies have characterized the differences between RV subtype replication and induction of host immune mediators in vitro. Wark et al. isolated RV-A43, RV-B48 (major group) and RV-A47 (minor group) from asthmatic patients and compared these with laboratory-adapted subtypes RV-A16, RV-B14 (major) and RV-A1b (minor). In undifferentiated, submerged bronchial epithelial cell monolayer cultures, more efficient infection and replication of major group RV subtypes was observed compared with minor group subtypes. Major group subtypes derived from clinical samples induced greater release of interleukin (IL)-6 and interferon (IFN) gamma-induced protein (IP)-10 compared with their laboratory cell culture—adapted counterparts. Infection with minor group RV subtypes induced significant production of IP-10, IL-6 and IFN-β in epithelial cells, causing apoptotic effects and resulting in lower viral load at 48 hours postinfection[111] compared with major group viruses. The robust proinflammatory response and antiviral IFN-β induced by minor group RV subtypes probably contributed to the less efficient viral replication. This suggests a relationship between the immune response elicited by viral infection and viral replication kinetics.

This has also been demonstrated in ALI-differentiated airway epithelial cell cultures, which more accurately model the human airway and have more recently been utilized to study RV infections. This is particularly relevant for RV-C viruses as they can only be propagated in differentiated epithelium that expresses CDHR3.[112] Nakagome et al. cloned the genome of RV-A16, A36, B52, B72, C2, C15, and C41 from respiratory samples and grew clinical isolates of RV-A7 and RV-B6 in cultured cells.[113] The study compared viral replication, cellular cytotoxicity, and cytokine secretion in ALI cultures of sinus epithelial cells. Virus binding and replication were lower in RV-B subtypes, compared with RV-A and RV-C subtypes. RV-B subtypes also induced significantly less cytokine and IFN expression compared with RV-A or RV-C. This links the rate or level of replication to the induction of inflammatory cytokines, supporting the idea that replication-induced inflammation underpins disease.

2.3.4 Host cell shutdown and disease

A contributing factor to subtype differences during RV-elicited antiviral responses may be the ability of RV subtypes to shut down host cell processes through the action of viral encoded proteases 2Apro and 3Cpro. These proteases are essential for polyprotein processing and host cell shut-off functions during infection, including cleavage of Phe/Gly-containing nucleoporin proteins (Nups) within nuclear pore complexes (NPC). Watters et al. found that RV subtypes displayed diverse 2Apro sequences that acted differentially on specific Nups, leading to distinct subtype-specific proteolytic efficiency and ability to alter the host nucleocytoplasmic trafficking.[114] In particular, RV-A and RV-C proteases showed slower kinetics than B04 and B52 in the disruption of nuclear import. The authors suggested that slower kinetics would allow antiviral signaling to occur before full nuclear transport shutoff, thus promoting the proinflammatory immune responses that have been observed with RV-A and RV-C infectious illness.[114] Enterovirus 71 has a similar genome organization as RV, with identified severe and mild subtypes. By swapping the 2Apro of Enterovirus 71 between severe and mild subtypes, it was shown that 2Apro was implicated in facilitating the replication and also modulating the virus induced cytotoxicity.[115] Further investigation is required to determine if 2Apro may be contributing to these processes during the replication of RV subtypes that cause more severe illness. 3Cpro may aid immune suppression of host cells in a more direct manner. RV-C has been shown to suppress type I IFN-β response in differentiated epithelium by disrupting RIG-I mediated IFN signaling via 3Cpro through a caspase-dependent mechanism.[116] These studies highlight the picornaviral protease as a major virulence factor. Their central role in RV disease is addressed in more detail in the next section.

2.4 RHINOVIRUS PROTEASES AND SUBTYPE-SPECIFIC DISEASE

RV proteases were initially described for their key role in RV replication as mediators of the *cis* and *trans* cleavage of RV polyprotein into individual proteins.[117] Later studies found that RV proteases also have proteolytic activity towards host proteins[118] and an important role in RV pathogenesis.

2.4.1 The proteases and proteolytic roles in viral replication

2Apro and 3Cpro both have a cysteine within the active site, however mutation of amino acids at loci that are characteristic of serine proteases resulted in a decrease in proteolytic activity.[119–121] Thus the RV proteases are closer to chymotrypsin-like serine proteases, rather than cysteine proteases. While 2Apro and the 3Cpro cleave different substrates, these substrates are often components of the same host cell pathway or process. For example, 2Apro and 3Cpro both target different intermediates in the host translation machinery.[122–125] Literature suggests that both proteases are required for full host cell shutoff, for example, Walker et al. have shown that translocation of 3Cpro into the nucleus requires 2Apro cleavage of nuclear pore proteins.[126]

2.4.1.1 2A protease

2Apro has a catalytic triad comprised of His18, Asp35, and Cys106 and a chymotrypsin-like fold.[127] A zinc motif within the zinc-binding domain of the N-terminus of 2Apro is essential to maintaining the structural stability of the protease.[127,128] The most effective 2A protease cleavage occurs at Tyrosine/Glycine bonds,[129,130] however the stereochemistry of the substrate also affects 2A proteolytic accessibility.[129] Importantly, the substrate processing kinetics of 2Apro often differs between RV-A, -B, or -C viruses. [20]

2.4.1.2 3C/3CD protease

RV translation results in one polyprotein that undergoes autoproteolysis by the viral proteases. The 3Cpro mediated cleavage of 3C from the adjacent 3D protein yields the protease and the 3D polymerase. Prior to this proteolysis, the 3Cpro domain of the polyprotein forms part of the 3CD protein (3D being the viral RNA dependent RNA polymerase) and is active. Using site-directed mutagenesis, studies have successfully mutated the cleavage site between the 3C and 3D domains, resulting in a 3CD (uncleavable) protein.[131] Interestingly, 3Cpro and 3CDpro cleave fluorogenic peptide substrates with similar kinetics.[131] Both 3C and 3CD proteases cleave most efficiently at Glutamate–Glycine bonds, with requirement for a hydrophobic amino acid five positions upstream of the cleavage site.[132] The conserved nature of RV 3Cpro qualifies it as a potential target of broad-spectrum antivirals.[133–135] Indeed, rupintrivir, an irreversible inhibitor of 3Cpro, has been the most likely candidate anti-RV drug to date[10,133] (formerly AG7088[134]). Rupintrivir inhibits 3Cpro by its interaction with a highly conserved amino acid sequence in 3Cpro.[136] While rupintrivir was active in clinical trials,[137] it was not effective in natural infection, and further trials were terminated.[138]

2.5 VARIANCES IN PROTEASES BETWEEN RHINOVIRUS SUBTYPES

The differences in activity of 2Apro from different RV subtypes have been reported.[20] The kinetics of the 2Apro activity was correlated with species; the proteolysis of eukaryotic initiation factor (eIF)4G was fastest with RV-A 2Apro, followed by RV-C 2Apro, and RV-B 2Apro was the slowest.[20] Similarly, the 2Apro mediated nuclear protein cleavage resulted in different cleavage profiles between RV-A16, -B04, -B14, -Cw12, and -Cw24.[20] Further work showed that 2Apro of different RV subtypes preferentially targeted either the nuclear import or export pathways.[114] Together, this indicates that the processing of 2Apro substrates may be specific to the species or subtypes of RV.[20] Importantly, as the host cell shutoff correlates with protease activity, and the latter has been shown to differ between subtypes, a difference in virulence of RV subtypes would be expected.

2.6 RHINOVIRUS PROTEASES AS VIRULENCE FACTORS

The RV proteases all confer virulence by altering the host cell in a manner that ultimately promotes RV replication. To promote viral replication, the protease cleaves cellular factors to either downregulate a host process, thereby allowing that cellular machinery to be utilized for RV replication, or to act as a countermeasure toward a restrictive host response.

2.6.1 Host cell shutoff

The proteolysis of cellular proteins by the RV 2Apro, 3Cpro, or 3CDpro includes, but is not limited to, nuclear pore protein alterations, translation and transcription machinery cleavage, and the cleavage of proteins associated with innate immunity.

2.6.1.1 Nuclear pore cleavage

The nuclear membrane provides a physical barrier within the eukaryotic cell that separates the genomic material from the cytoplasmic environment. Membrane-spanning nuclear pores are the only conduit of trafficking between the nucleus and the cytoplasm; nuclear pores are formed by specialized proteins termed nucleoporins (Nups). Early research initially demonstrated that there was a cytoplasmic redistribution of nuclear proteins in RV-B14 infected cells.[139,140] Further work has built on this to show that specific Nups are cleaved in RV infection, either by the 2Apro (Nup 98[141,142]) or the 3Cpro (Nup 153[141]). An RNA-binding protein, hnRNPA1, is normally shuttled between the nucleus and the cytoplasm and participates in the pretranslational regulation of cellular RNAs.[143] In RV infection, viral induced changes in the nuclear pore proteins correlate with the cytoplasmic retention of hnRNPA1, which can then participate in internal ribosomal entry site-mediated viral translation.[143] On the cytoplasmic and nuclear flanks of the NPC are Nups rich in Phenylalanine and Glycine (known as F-G Nups).[144] These Nups form both a physical barrier to free diffusion for large molecules through the nuclear pore and serve as the docking site for active nuclear transport carrier proteins.[145] Interestingly, the picornaviral proteases have been shown to selectively cleave the F-G Nups,[142] thereby disrupting active nuclear-cytoplasmic trafficking.[139,141] The cleavage of Nups by RV proteases confers virulence by the cytoplasmic redistribution of nuclear proteins, the sequestration of nuclear proteins for viral translation, and the shutdown of active nuclear transport.

2.6.1.2 Translation inhibition

The positive (messenger RNA) sense genomic RNA of RV is translated rapidly after uncoating of the capsid protein and release of the genome

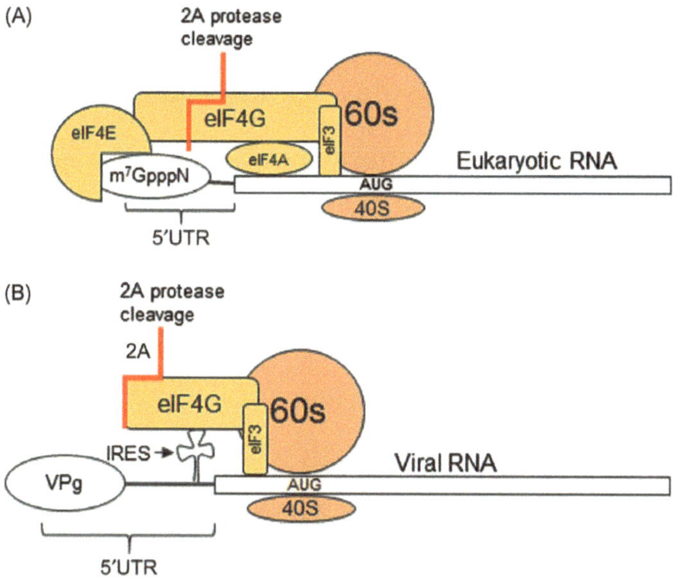

Figure 2.2 ***Cleavage of eIF4G by HRV 2A protease inhibits cellular but not viral translation.*** (A) Eukaryotic transcription. An eukaryotic transcription complex forms, including eukaryotic initiation factor (eIF)4E, which recognizes the 5′ cap (M^7GpppN); eIF4G, which links eIF4E to eIF4A; eIF4A, which binds the RNA; eIF3, which stabilizes the 60S and 40S ribosomal subunits. The RV 2A protease cleaves eIF4G, dissociating the 5′cap recognition protein (eIF4E) inhibiting cellular translation. (B) The internal ribosomal entry site (IRES), a secondary structure in the 5′untranslated region (UTR) of rhinovirus (RV) RNA stabilizes the cleaved eIF4G on the viral RNA and translation continues.

into the cytoplasm. After translation of the polyprotein, the viral proteases cleave the polyprotein and cleave eIFs to downregulate host translation and promote viral translation[123,124] (Fig. 2.2). Eukaryotic translation is initiated by the assembly of eIFs around the 5′cap (a methylated guanosine residue) on the end of eukaryotic mRNA[146,147]. The eIFs stabilize ribosomes on the RNA for protein elongation. The 5′cap is recognized by eIF4E, which is bridged to the 60S ribosome by eIF4G.[147] eIF4G is cleaved by the RV 2Apro, the kinetics of this cleavage being dependent on RV species.[20] The cleavage of eIF4G dissociates the 5′cap recognition protein (eIF4E), thus promoting cap-independent (viral) translation.[122–124] At the 5′untranslated end, the RV genome is capped by a genome-linked protein (VPg), negating the need for a cap-recognizing protein (eIF4E). Poly-A-binding protein (PABP) is a cellular protein that binds to the poly-A-tail of messenger RNAs, to circularize the RNA with the eIF/ribosome complex, allowing multiple rounds of translation.[148] PABP is cleaved by the RV 3Cpro,[125] destabilizing the recruitment of the RNA to the complex and promoting viral translation. Together, 2Apro and 3Cpro cleave key proteins in the host translation machinery, resulting in enhanced viral translation and little or no host translation.

2.6.1.3 *Modulation of apoptosis*

In response to detecting viral pathogen-associated molecular patterns (PAMPs), the infected cell can self-destruct in a noninflammatory manner, by apoptosis. In the context of apoptotic cell death, a pathogen-recognition receptor (PRR) detects a PAMP, subsequently a death-receptor complex forms, which is comprised of apoptotic adaptor proteins and procaspases. The procaspases (inactive) are cleaved into caspases (active), which in turn cleave other procaspases into caspases. Once cleaved into caspases, the executioner caspases (caspase 3, for example) cleave other cellular substrates contributing to the morphological, biochemical, or genomic changes associated with apoptosis. This apoptotic cell death is crucial to limiting the virus's ability to utilize the host cell machinery for replication, therefore limiting viral replication and dissemination. To overcome the restriction of apoptotic cell death on viral replication, as discussed below, the RV proteases cleave multiple apoptotic signaling molecules, acting as apoptosis antagonists. The action of these proteases towards apoptotic signaling molecules contributes to the virulence of the virus.

In RV infection, three important PRRs have been implicated in innate immune signaling in response to RV RNA within a cell. TLR-3, MDA-5 and RIG-I are PRRs that detect either dsRNA, an RV replication intermediate (TLR-3 and MDA-5), or viral 5′triphosphate RNA (RIG-I).[149] As expected, RV countermeasures to these signaling pathways have also been described (Fig. 2.3).

TLR-3 recognizes endosomal dsRNA following which adapter protein TIR-domain-containing adapter-inducing IFN-β (TRIF) is recruited to TLR signaling domain. TLR-3 apoptotic signaling involves the recruitment of proapoptotic proteins to TRIF, such as FADD, RIPK1, and Procaspase 8.[150–152] TRIF is cleaved in infection by multiple picornaviruses (enterovirus 71[153] and 68,[154] and hepatitis A virus[155]) or in vitro by the enterovirus 71 3Cpro,[153] stopping the downstream apoptotic responses. Essential to the apoptotic signaling of TLR-3 is RIPK1, which is cleaved by the RV-A16 and RV-B14 3Cpro,[156,157] and importantly,

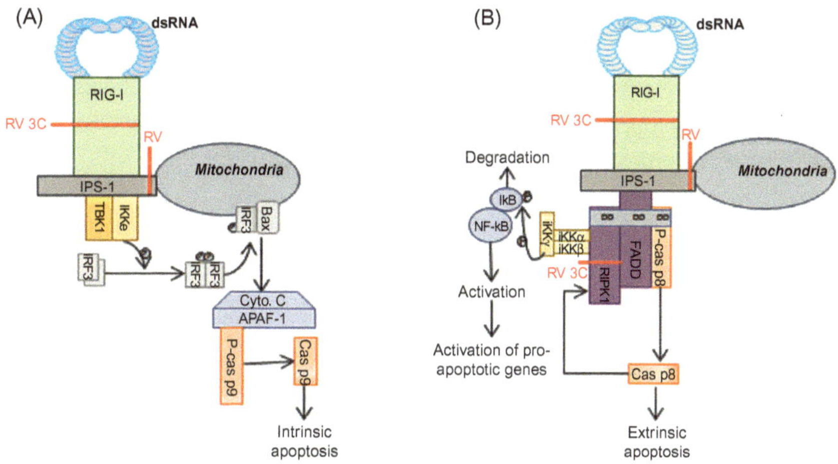

Figure 2.3 ***Effect of HRV infection on apoptosis pathways.*** (A) RIG-I and MDA-5, when activated, associate with IPS-1 and recruit TBK1 and IKKε into a death-receptor complex at the mitochondria. TBK1 and IKKε act by phosphorylating IRF-3 dimers, which then induce a pore in the mitochondrial membrane, allowing release of cytochrome c into the cytoplasm, inducing intrinsic apoptosis. RIG-I, MDA-5, and IPS-1 are cleaved during infection (indicated in *red*) by rhinovirus (RV) proteases, inhibiting intrinsic apoptosis. (B) Sensing of dsRNA by RIG-I/MDA-5 leads to the formation of a death-receptor complex at the mitochondria through the adaptor protein interferon beta promoter stimulator protein-1 (IPS-1). Subsequent ubiquitination of RIPK1 is recognized by the IKKγ (Nemo) subunit of the IKKα−IKKβ−IKKγ complex leading to phosphorylation and degradation of IκB and release of NF-κB that translocates into the nucleus to upregulate expression of proapoptotic genes leading to extrinsic apoptosis. Cleavage of RIG-I, IPS-1 and RIPK1 by RV 3C protease disrupts extrinsic apoptosis signaling and may impact NF-κB activation, attenuating transcriptional activation of proapoptotic factors.

this cleavage suppresses apoptotic cell death. RIG-I is also cleaved in major and minor group RV infection.[158] Cleavage of RIG-I was also observed on addition of the poliovirus 3Cpro to cells, indicating that 3Cpro is likely responsible for the cleavage seen in infection.[158] IFN-β promoter stimulator protein-1 (IPS-1) is a mitochondrial anchored adapter protein that bridges RIG-I and MDA-5[159,160] to apoptosis adapter proteins (FADD and RIPK1) and procaspase 8.[161] IPS-1 is cleaved in RV infection[162] and in vitro by the poliovirus 2Apro or 3Cpro,[162] and the hepatitis A virus 3CD precursor protease[163] indicating that this protein may be cleaved by multiple picornavirus proteases. The protease-mediated cleavage or viral infection, in all cases, resulted in dampened apoptotic responses.[162,163]

Together, the above discussion implicates the RV proteases as key factors in suppressing apoptotic responses to viral infection. This countermeasure to a restrictive response results in optimal viral replication. In addition, this suppression of caspase-dependent cell death may drive the cell toward a lytic cell death,[156,157,164,165] which in the context of bronchial epithelial cell infection could contribute to an inflammatory phenotype.

2.7 CONCLUSION

RVs are a major cause of all respiratory tract virus infections worldwide, causing a wide range of diseases, from common cold to asthma exacerbations. Given their year-round presence and global occurrence, clearly there is a need for development of therapeutic or preventative strategies. The main obstacle to any such development has been the diversity of RVs; at current count, there are 167 subtypes belonging to three species. A potentially viable strategy could be to identify relevant subtypes contributing to illness severity and then specifically target these. Clinical epidemiological studies clearly show a link between RV species and clinical illness, with RV-A and -C causing more severe disease, while RV-B is rarely isolated from patients. Importantly, the clinical data correlate well with laboratory cell culture data showing higher replication of RV-A and -C and higher inflammatory responses, relative to RV-B viruses. Inflammatory responses have an important role in RV disease. A meta-analysis of clinical studies where the isolated RVs were subtyped using molecular techniques demonstrates that there is a clear association of subtypes within each species with clinical disease. This confirms previous reports where occurrence of a small group of RV subtypes was investigated in clinical disease. That species/subtype association with clinical disease may be due to variation in the activities of the 2Apro and 3Cpro encoded by RVs is indicated by recent studies showing a clear subtype-specific cleavage of host factors. For example, cleavage of nucleoporins by 2Apro of RV-A subtypes leads to a different pattern of cleaved fragments from that by 2Apro of RV-B or -C subtypes. Similarly, 3Cpro cleaves a key intermediate in apoptotic and inflammatory signaling pathways in a subtype-specific manner. The RV protease-mediated disruption of nucleocytoplasmic trafficking inhibits antiviral pathways, including inflammatory responses, and contributes to RV pathogenesis. RVs are known to modulate apoptosis, an important innate response to infection, with effective modulation potentially related to increased pathogenesis. The accumulating evidence suggests an important role for species/subtype-specific protease activity that may be linked to clinical illness caused by RV infection. However, this is yet to be directly tested.

REFERENCES

1. Glanville N, Johnston SL. Challenges in developing a cross-serotype rhinovirus vaccine. *Curr Opin Virol*. 2015;11:83–88.
2. Palmenberg AC, Gern JE. Classification and evolution of human rhinoviruses. *Methods Mol Biol*. 2015;1221:1–10.
3. Simmonds P, McIntyre C, Savolainen-Kopra C, Tapparel C, Mackay IM, Hovi T. Proposals for the classification of human rhinovirus species C into genotypically assigned types. *J Gen Virol*. 2010;91(Pt 10):2409–2419.
4. International Committee on Taxonomy of Viruses. *Taxonomy*. <https://talk.ictvonline.org/taxonomy/>; 2017 Accessed 17.08.18.
5. Abraham G, Colonno RJ. Many rhinovirus serotypes share the same cellular receptor. *J Virol*. 1984;51(2):340–345.
6. Greve JM, Davis G, Meyer AM, et al. The major human rhinovirus receptor is ICAM-1. *Cell*. 1989;56(5):839–847.
7. Hofer F, Gruenberger M, Kowalski H, et al. Members of the low density lipoprotein receptor family mediate cell entry of a minor-group common cold virus. *Proc Natl Acad Sci USA*. 1994;91(5):1839–1842.
8. Casanova V, Sousa FH, Stevens C, Barlow PG. Antiviral therapeutic approaches for human rhinovirus infections. *Future Virol*. 2018;13(7):505–518.
9. Effects of Pleconaril Nasal Spray on Common Cold Symptoms and Asthma Exacerbations Following Rhinovirus Exposure (Study P04295). <https://clinicaltrials.gov/ct2/show/NCT00394914>; 2006 Accessed 2018.
10. Patick AK, Brothers MA, Maldonado F, et al. In vitro antiviral activity and single-dose pharmacokinetics in humans of a novel, orally bioavailable inhibitor of human rhinovirus 3C protease. *Antimicrob Agents Chemother*. 2005;49(6):2267–2275.
11. Jamieson KC, Warner SM, Leigh R, Proud D. Rhinovirus in the pathogenesis and clinical course of asthma. *Chest*. 2015;148(6):1508–1516.
12. Heinonen S, Jartti T, Garcia C, et al. Rhinovirus detection in symptomatic and asymptomatic children: value of host transcriptome analysis. *Am J Respir Crit Care Med*. 2016;193(7):772–782.
13. Ma X, Conrad T, Alchikh M, Reiche J, Schweiger B, Rath B. Can we distinguish respiratory viral infections based on clinical features? A prospective pediatric cohort compared to systematic literature review. *Rev Med Virol*. 2018;:e1997.
14. Iwane MK, Prill MM, Lu X, et al. Human rhinovirus species associated with hospitalizations for acute respiratory illness in young US children. *J Infect Dis*. 2011;204(11):1702–1710.
15. Aponte FE, Taboada B, Espinoza MA, et al. Rhinovirus is an important pathogen in upper and lower respiratory tract infections in Mexican children. *Virol J*. 2015;12:31.
16. Jin Y, Yuan XH, Xie ZP, et al. Prevalence and clinical characterization of a newly identified human rhinovirus C species in children with acute respiratory tract infections. *J Clin Microbiol*. 2009;47(9):2895–2900.
17. Lau SK, Yip CC, Lin AW, et al. Clinical and molecular epidemiology of human rhinovirus C in children and adults in Hong Kong reveals a possible distinct human rhinovirus C subgroup. *J Infect Dis*. 2009;200(7):1096–1103.

18. Lu QB, Wo Y, Wang LY, et al. Molecular epidemiology of human rhinovirus in children with acute respiratory diseases in Chongqing, China. *Sci Rep*. 2014;4:6686.
19. Wildenbeest JG, van der Schee MP, Hashimoto S, et al. Prevalence of rhinoviruses in young children of an unselected birth cohort from the Netherlands. *Clin Microbiol Infect*. 2016;22(8):736.e9–736.e15.
20. Watters K, Palmenberg AC. Differential processing of nuclear pore complex proteins by rhinovirus 2A proteases from different species and serotypes. *J Virol*. 2011;85 (20):10874–10883.
21. Jensen LM, Walker EJ, Jans DA, Ghildyal R. Proteases of human rhinovirus: role in infection. *Methods Mol Biol*. 2015;1221:129–141.
22. Jacobs SE, Lamson DM, St George K, Walsh TJ. Human rhinoviruses. *Clin Microbiol Rev*. 2013;26(1):135–162.
23. Herbert C, Do K, Chiu V, et al. Allergic environment enhances airway epithelial pro-inflammatory responses to rhinovirus infection. *Clin Sci (Lond)*. 2017;131 (6):499–509.
24. Miller EK, Williams JV, Gebretsadik T, et al. Host and viral factors associated with severity of human rhinovirus-associated infant respiratory tract illness. *J Allergy Clin Immunol*. 2011;127(4):883–891.
25. Lee WM, Lemanske Jr RF, Evans MD, et al. Human rhinovirus species and season of infection determine illness severity. *Am J Respir Crit Care Med*. 2012;186(9):886–891.
26. Palmenberg AC, Spiro D, Kuzmickas R, et al. Sequencing and analyses of all known human rhinovirus genomes reveal structure and evolution. *Science*. 2009;324 (5923):55–59.
27. Ledford RM, Patel NR, Demenczuk TM, et al. VP1 sequencing of all human rhinovirus serotypes: insights into genus phylogeny and susceptibility to antiviral capsid-binding compounds. *J Virol*. 2004;78(7):3663–3674.
28. Arden KE, McErlean P, Nissen MD, Sloots TP, Mackay IM. Frequent detection of human rhinoviruses, paramyxoviruses, coronaviruses, and bocavirus during acute respiratory tract infections. *J Med Virol*. 2006;78(9):1232–1240.
29. The Picornavirus Pages. <http://www.picornaviridae.com/enterovirus/rv-c/rv-c_seqs.htm>; 2018.
30. McErlean P, Shackelton LA, Lambert SB, Nissen MD, Sloots TP, Mackay IM. Characterisation of a newly identified human rhinovirus, HRV-QPM, discovered in infants with bronchiolitis. *J Clin Virol*. 2007;39(2):67–75.
31. Wisdom A, Leitch EC, Gaunt E, Harvala H, Simmonds P. Screening respiratory samples for detection of human rhinoviruses (HRVs) and enteroviruses: comprehensive VP4-VP2 typing reveals high incidence and genetic diversity of HRV species C. *J Clin Microbiol*. 2009;47(12):3958–3967.
32. Miller EK, Khuri-Bulos N, Williams JV, et al. Human rhinovirus C associated with wheezing in hospitalised children in the Middle East. *J Clin Virol*. 2009;46(1):85–89.
33. Annamalay AA, Khoo SK, Jacoby P, et al. Prevalence of and risk factors for human rhinovirus infection in healthy aboriginal and non-aboriginal Western Australian children. *Pediatr Infect Dis J*. 2012;31(7):673–679.
34. Linder JE, Kraft DC, Mohamed Y, et al. Human rhinovirus C: age, season, and lower respiratory illness over the past 3 decades. *J Allergy Clin Immunol*. 2013;131(1):69–77. e61–66.
35. Stenberg-Hammar K, Hedlin G, Soderhall C. Rhinovirus and preschool wheeze. *Pediatr Allergy Immunol*. 2017;28(6):513–520.
36. Ahn JG, Kim DS, Kim KH. Clinical characteristics and cytokine profiles of children with acute lower respiratory tract infections caused by human rhinovirus. *PLoS One*. 2018;13(7):e0198624.

37. Jackson DJ, Gangnon RE, Evans MD, et al. Wheezing rhinovirus illnesses in early life predict asthma development in high-risk children. *Am J Respir Crit Care Med.* 2008;178(7):667–672.
38. Khetsuriani N, Lu X, Teague WG, Kazerouni N, Anderson LJ, Erdman DD. Novel human rhinoviruses and exacerbation of asthma in children. *Emerg Infect Dis.* 2008;14 (11):1793–1796.
39. Mak RK, Tse LY, Lam WY, Wong GW, Chan PK, Leung TF. Clinical spectrum of human rhinovirus infections in hospitalized Hong Kong children. *Pediatr Infect Dis J.* 2011;30(9):749–753.
40. Bizzintino J, Lee WM, Laing IA, et al. Association between human rhinovirus C and severity of acute asthma in children. *Eur Respir J.* 2011;37(5):1037–1042.
41. Cox DW, Bizzintino J, Ferrari G, et al. Human rhinovirus species C infection in young children with acute wheeze is associated with increased acute respiratory hospital admissions. *Am J Respir Crit Care Med.* 2013;188(11):1358–1364.
42. Renwick N, Schweiger B, Kapoor V, et al. A recently identified rhinovirus genotype is associated with severe respiratory-tract infection in children in Germany. *J Infect Dis.* 2007;196(12):1754–1760.
43. Arden KE, Chang AB, Lambert SB, Nissen MD, Sloots TP, Mackay IM. Newly identified respiratory viruses in children with asthma exacerbation not requiring admission to hospital. *J Med Virol.* 2010;82(8):1458–1461.
44. Lee S, Nguyen MT, Currier MG, et al. A polyvalent inactivated rhinovirus vaccine is broadly immunogenic in rhesus macaques. *Nat Commun.* 2016;7:12838.
45. Louie JK, Yagi S, Nelson FA, et al. Rhinovirus outbreak in a long term care facility for elderly persons associated with unusually high mortality. *Clin Infect Dis.* 2005;41 (2):262–265.
46. Ren L, Yang D, Ren X, et al. Genotyping of human rhinovirus in adult patients with acute respiratory infections identified predominant infections of genotype A21. *Sci Rep.* 2017;7:41601.
47. Kiang D, Yagi S, Kantardjieff KA, Kim EJ, Louie JK, Schnurr DP. Molecular characterization of a variant rhinovirus from an outbreak associated with uncommonly high mortality. *J Clin Virol.* 2007;38(3):227–237.
48. Huang T, Wang W, Bessaud M, et al. Evidence of recombination and genetic diversity in human rhinoviruses in children with acute respiratory infection. *PLoS One.* 2009;4(7):e6355.
49. Ferrer-Orta C, Ferrero D, Verdaguer N. RNA-dependent RNA polymerases of picornaviruses: from the structure to regulatory mechanisms. *Viruses.* 2015;7 (8):4438–4460.
50. Rathe JA, Liu X, Tallon LJ, Gern JE, Liggett SB. Full-genome sequence and analysis of a novel human rhinovirus strain within a divergent HRV-A clade. *Arch Virol.* 2010;155(1):83–87.
51. Naughtin M, Sareth R, Sentilhes AC, et al. Genetic diversity of human rhinoviruses in Cambodia during a three-year period reveals novel genetic types. *Infect Genet Evol.* 2015;35:42–49.
52. Kim H, Kim K, Kim DW, et al. Identification of recombinant human rhinovirus A and C in circulating strains from upper and lower respiratory infections. *PLoS One.* 2013;8(6):e68081.
53. Midgley CM, Watson JT, Nix WA, et al. Severe respiratory illness associated with a nationwide outbreak of enterovirus D68 in the USA (2014): a descriptive epidemiological investigation. *Lancet Respir Med.* 2015;3(11):879–887.
54. Cassidy H, Poelman R, Knoester M, Van Leer-Buter CC, Niesters HGM. Enterovirus D68—the new polio? *Front Microbiol.* 2018;9:2677.

55. Meijer A, van der Sanden S, Snijders BE, et al. Emergence and epidemic occurrence of enterovirus 68 respiratory infections in The Netherlands in 2010. *Virology*. 2012;423(1):49–57.
56. Huang W, Wang G, Zhuge J, Nolan SM, Dimitrova N, Fallon JT. Whole-genome sequence analysis reveals the enterovirus D68 isolates during the United States 2014 outbreak mainly belong to a novel clade. *Sci Rep*. 2015;5:15223.
57. Tang JW, Lam TT, Zaraket H, et al. Global epidemiology of non-influenza RNA respiratory viruses: data gaps and a growing need for surveillance. *Lancet Infect Dis*. 2017;17(10):e320–e326.
58. Monto AS, Bryan ER, Ohmit S. Rhinovirus infections in Tecumseh, Michigan: frequency of illness and number of serotypes. *J Infect Dis*. 1987;156(1):43–49.
59. Faux CE, Arden KE, Lambert SB, et al. Usefulness of published PCR primers in detecting human rhinovirus infection. *Emerg Infect Dis*. 2011;17(2):296–298.
60. Bochkov YA, Grindle K, Vang F, Evans MD, Gern JE. Improved molecular typing assay for rhinovirus species A, B, and C. *J Clin Microbiol*. 2014;52(7):2461–2471.
61. Andeweg AC, Bestebroer TM, Huybreghs M, Kimman TG, de Jong JC. Improved detection of rhinoviruses in clinical samples by using a newly developed nested reverse transcription-PCR assay. *J Clin Microbiol*. 1999;37(3):524–530.
62. Bochkov YA, Palmenberg AC, Lee WM, et al. Molecular modeling, organ culture and reverse genetics for a newly identified human rhinovirus C. *Nat Med*. 2011;17 (5):627–632.
63. Tran DN, Trinh QD, Pham NT, et al. Human rhinovirus infections in hospitalized children: clinical, epidemiological and virological features. *Epidemiol Infect*. 2016;144 (2):346–354.
64. van der Linden L, Bruning AH, Thomas XV, et al. A molecular epidemiological perspective of rhinovirus types circulating in Amsterdam from 2007 to 2012. *Clin Microbiol Infect*. 2016;22(12). 1002 e1009–1002 e1014.
65. Piralla A, Baldanti F, Gerna G. Phylogenetic patterns of human respiratory picornavirus species, including the newly identified group C rhinoviruses, during a 1-year surveillance of a hospitalized patient population in Italy. *J Clin Microbiol*. 2011;49 (1):373–376.
66. Smuts HE, Workman LJ, Zar HJ. Human rhinovirus infection in young African children with acute wheezing. *BMC Infect Dis*. 2011;11:65.
67. Arakawa M, Okamoto-Nakagawa R, Toda S, et al. Molecular epidemiological study of human rhinovirus species A, B and C from patients with acute respiratory illnesses in Japan. *J Med Microbiol*. 2012;61(Pt 3):410–419.
68. Franco D, Delfraro A, Abrego L, et al. High genetic diversity and predominance of Rhinovirus A and C from Panamanian hospitalized children under five years with respiratory infections. *Virol J*. 2012;9:257.
69. Henquell C, Mirand A, Deusebis AL, et al. Prospective genotyping of human rhinoviruses in children and adults during the winter of 2009-2010. *J Clin Virol*. 2012;53 (4):280–284.
70. Cuevas MT, Molinero M, Pozo F, et al. Spread of different rhinovirus B genotypes in hospitalized children in Spain. *Influenza Other Respir Viruses*. 2013;7(5):623–628.
71. Espinola EE, Russomando G, Aquino C, Basualdo W. Phylogeny-based classification of human rhinoviruses detected in hospitalized children with acute lower respiratory infection in Paraguay, 2010-2011. *J Med Virol*. 2013;85(9):1645–1651.
72. Pierangeli A, Ciccozzi M, Chiavelli S, et al. Molecular epidemiology and genetic diversity of human rhinovirus affecting hospitalized children in Rome. *Med Microbiol Immunol*. 2013;202(4):303–311.

73. Sansone M, Andersson M, Brittain-Long R, et al. Rhinovirus infections in western Sweden: a four-year molecular epidemiology study comparing local and globally appearing types. *Eur J Clin Microbiol Infect Dis*. 2013;32(7):947–954.
74. Cutino-Moguel T, Lauinger IL, Srivastava S, Zuckerman M, Tong CY, Devereux S. Analysis of a potential cluster of rhinovirus infections in patients and staff on two haemato-oncology wards. *J Clin Virol*. 2014;60(1):57–59.
75. Flight WG, Bright-Thomas RJ, Tilston P, et al. Incidence and clinical impact of respiratory viruses in adults with cystic fibrosis. *Thorax*. 2014;69(3):247–253.
76. Marcone DN, Culasso A, Carballal G, Campos R, Echavarria M. Genetic diversity and clinical impact of human rhinoviruses in hospitalized and outpatient children with acute respiratory infection, Argentina. *J Clin Virol*. 2014;61(4):558–564.
77. Zlateva KT, de Vries JJ, Coenjaerts FE, et al. Prolonged shedding of rhinovirus and re-infection in adults with respiratory tract illness. *Eur Respir J*. 2014;44(1):169–177.
78. Bruning AHL, Thomas XV, van der Linden L, et al. Clinical, virological and epidemiological characteristics of rhinovirus infections in early childhood: a comparison between non-hospitalised and hospitalised children. *J Clin Virol*. 2015;73:120–126.
79. L'Huillier AG, Kaiser L, Petty TJ, et al. Molecular epidemiology of human rhinoviruses and enteroviruses highlights their diversity in sub-Saharan Africa. *Viruses*. 2015;7(12):6412–6423.
80. Prasetyo AA, Desyardi MN, Tanamas J, et al. Respiratory viruses and torque teno virus in adults with acute respiratory infections. *Intervirology*. 2015;58(1):57–68.
81. Richter J, Nikolaou E, Panayiotou C, Tryfonos C, Koliou M, Christodoulou C. Molecular epidemiology of rhinoviruses in Cyprus over three consecutive seasons. *Epidemiol Infect*. 2015;143(9):1876–1883.
82. Milanoi S, Ongus JR, Gachara G, Coldren R, Bulimo W. Serotype and genetic diversity of human rhinovirus strains that circulated in Kenya in 2008. *Influenza Other Respir Viruses*. 2016;10(3):185–191.
83. Ratnamohan VM, Zeng F, Donovan L, MacIntyre CR, Kok J, Dwyer DE. Phylogenetic analysis of human rhinoviruses collected over four successive years in Sydney, Australia. *Influenza Other Respir Viruses*. 2016;10(6):493–503.
84. Bochkov YA, Watters K, Ashraf S, et al. Cadherin-related family member 3, a childhood asthma susceptibility gene product, mediates rhinovirus C binding and replication. *Proc Natl Acad Sci USA*. 2015;112(17):5485–5490.
85. Hastings GZ, Francis MJ, Rowlands DJ, Chain BM. Antigen processing and presentation of human rhinovirus to CD4 T cells is facilitated by binding to cellular receptors for virus. *Eur J Immunol*. 1993;23(6):1340–1345.
86. Beffert U, Stolt PC, Herz J. Functions of lipoprotein receptors in neurons. *J Lipid Res*. 2004;45(3):403–409.
87. Lewis-Rogers N, Bendall ML, Crandall KA. Phylogenetic relationships and molecular adaptation dynamics of human rhinoviruses. *Mol Biol Evol*. 2009;26(5):969–981.
88. Ganjian H, Zietz C, Mechtcheriakova D, Blaas D, Fuchs R. ICAM-1 binding rhinoviruses enter HeLa cells via multiple pathways and travel to distinct intracellular compartments for uncoating. *Viruses*. 2017;9:4.
89. Conzemius R, Ganjian H, Blaas D, Fuchs R. ICAM-1 binding rhinoviruses A89 and B14 uncoat in different endosomal compartments. *J Virol*. 2016;90(17):7934–7942.
90. de Arruda 3rd E, Mifflin TE, Gwaltney Jr JM, Winther B, Hayden FG. Localization of rhinovirus replication in vitro with in situ hybridization. *J Med Virol*. 1991;34(1):38–44.
91. Griggs TF, Bochkov YA, Basnet S, et al. Rhinovirus C targets ciliated airway epithelial cells. *Respir Res*. 2017;18(1):84.

92. Jakiela B, Brockman-Schneider R, Amineva S, Lee WM, Gern JE. Basal cells of differentiated bronchial epithelium are more susceptible to rhinovirus infection. *Am J Respir Cell Mol Biol.* 2008;38(5):517–523.
93. Granados A, Luinstra K, Chong S, et al. Use of an improved quantitative polymerase chain reaction assay to determine differences in human rhinovirus viral loads in different populations. *Diagn Microbiol Infect Dis.* 2012;74(4):384–387.
94. Hendley JO, Edmondson Jr WP, Gwaltney Jr. JM. Relation between naturally acquired immunity and infectivity of two rhinoviruses in volunteers. *J Infect Dis.* 1972;125(3):243–248.
95. Papadopoulos NG, Sanderson G, Hunter J, Johnston SL. Rhinoviruses replicate effectively at lower airway temperatures. *J Med Virol.* 1999;58(1):100–104.
96. Ashraf S, Brockman-Schneider R, Bochkov YA, Pasic TR, Gern JE. Biological characteristics and propagation of human rhinovirus-C in differentiated sinus epithelial cells. *Virology.* 2013;436(1):143–149.
97. Gerna G, Piralla A, Rovida F, et al. Correlation of rhinovirus load in the respiratory tract and clinical symptoms in hospitalized immunocompetent and immunocompromised patients. *J Med Virol.* 2009;81(8):1498–1507.
98. Takeyama A, Hashimoto K, Sato M, et al. Rhinovirus load and disease severity in children with lower respiratory tract infections. *J Med Virol.* 2012;84(7):1135–1142.
99. Xiao Q, Zheng S, Zhou L, et al. Impact of human rhinovirus types and viral load on the severity of illness in hospitalized children with lower respiratory tract infections. *Pediatr Infect Dis J.* 2015;34(11):1187–1192.
100. Milano F, Campbell AP, Guthrie KA, et al. Human rhinovirus and coronavirus detection among allogeneic hematopoietic stem cell transplantation recipients. *Blood.* 2010;115(10):2088–2094.
101. Ogimi C, Xie H, Leisenring WM, et al. Initial high viral load is associated with prolonged shedding of human rhinovirus in allogeneic hematopoietic cell transplant Recipients. *Biol Blood Marrow Transplant.* 2018;24:2160–2163.
102. Lupo J, Schuffenecker I, Morel-Baccard C, et al. Disseminated rhinovirus C8 infection with infectious virus in blood and fatal outcome in a child with repeated episodes of bronchiolitis. *J Clin Microbiol.* 2015;53(5):1775–1777.
103. Urquhart GE, Stott EJ. Rhinoviraemia. *Br Med J.* 1970;4(5726):28–30.
104. Urquhart GE, Grist NR. Virological studies of sudden, unexplained infant deaths in Glasgow 1967-70. *J Clin Pathol.* 1972;25(5):443–446.
105. Xatzipsalti M, Kyrana S, Tsolia M, et al. Rhinovirus viremia in children with respiratory infections. *Am J Respir Crit Care Med.* 2005;172(8):1037–1040.
106. Esposito S, Daleno C, Scala A, et al. Impact of rhinovirus nasopharyngeal viral load and viremia on severity of respiratory infections in children. *Eur J Clin Microbiol Infect Dis.* 2014;33(1):41–48.
107. Fuji N, Suzuki A, Lupisan S, et al. Detection of human rhinovirus C viral genome in blood among children with severe respiratory infections in the Philippines. *PLoS One.* 2011;6(11):e27247.
108. Lu X, Schneider E, Jain S, et al. Rhinovirus viremia in patients hospitalized with community-acquired pneumonia. *J Infect Dis.* 2017;216(9):1104–1111.
109. Winther B, Gwaltney JM, Hendley JO. Respiratory virus infection of monolayer cultures of human nasal epithelial cells. *Am Rev Respir Dis.* 1990;141(4 Pt 1):839–845.
110. Lopez-Souza N, Dolganov G, Dubin R, et al. Resistance of differentiated human airway epithelium to infection by rhinovirus. *Am J Physiol Lung Cell Mol Physiol.* 2004;286(2):L373–381.
111. Wark PA, Grissell T, Davies B, See H, Gibson PG. Diversity in the bronchial epithelial cell response to infection with different rhinovirus strains. *Respirology.* 2009;14 (2):180–186.

112. Ashraf S, Brockman-Schneider R, Gern JE. Propagation of rhinovirus-C strains in human airway epithelial cells differentiated at air-liquid interface. *Methods Mol Biol.* 2015;1221:63–70.
113. Nakagome K, et al. Effects of rhinovirus species on viral replication and cytokine production. *J Allergy Clin Immunol.* 2014;134(2):332–341.
114. Watters K, Inankur B, Gardiner JC, et al. Differential disruption of nucleocytoplasmic trafficking pathways by rhinovirus 2A proteases. *J Virol.* 2017;91(8):e02472–02416.
115. Li C, Qiao Q, Hao SB, et al. Nonstructural protein 2A modulates replication and virulence of enterovirus 71. *Virus Res.* 2018;244:262–269.
116. Pang LL, Yuan XH, Shao CS, et al. The suppression of innate immune response by human rhinovirus C. *Biochem Biophys Res Commun.* 2017;490(1):22–28.
117. Palmenberg AC. Proteolytic processing of picornaviral polyprotein. *Annu Rev Microbiol.* 1990;44(1):603–623.
118. Cordingley MG, Register RB, Callahan PL, Garsky VM, Colonno RJ. Cleavage of small peptides in vitro by human rhinovirus 14 3C protease expressed in *Escherichia coli*. *J Virol.* 1989;63(12):5037–5045.
119. Matthews DA, Smith WW, Ferre RA, et al. Structure of human rhinovirus 3C protease reveals a trypsin-like polypeptide fold, RNA-binding site, and means for cleaving precursor polyprotein. *Cell.* 1994;77(5):761–771.
120. Gorbalenya AE, Donchenko AP, Blinov VM, Koonin EV. Cysteine proteases of positive strand RNA viruses and chymotrypsin-like serine proteases. *FEBS Lett.* 1989;243(2):103–114.
121. Cheah KC, Leong LE, Porter AG. Site-directed mutagenesis suggests close functional relationship between a human rhinovirus 3C cysteine protease and cellular trypsin-like serine proteases. *J Biol Chem.* 1990;265(13):7180–7187.
122. Gradi A, Svitkin YV, Imataka H, Sonenberg N. Proteolysis of human eukaryotic translation initiation factor eIF4GII, but not eIF4GI, coincides with the shutoff of host protein synthesis after poliovirus infection. *Proc Natl Acad Sci USA.* 1998;95(19):11089–11094.
123. Haghighat A, Svitkin Y, Novoa I, Kuechler E, Skern T, Sonenberg N. The eIF4G-eIF4E complex is the target for direct cleavage by the rhinovirus 2A proteinase. *J Virol.* 1996;70(12):8444–8450.
124. Svitkin YV, Gradi A, Imataka H, Morino S, Sonenberg N. Eukaryotic initiation factor 4GII (eIF4GII), but not eIF4GI, cleavage correlates with inhibition of host cell protein synthesis after human rhinovirus infection. *J Virol.* 1999;73(4):3467–3472.
125. Bushell M, Wood W, Carpenter G, Pain VM, Morley SJ, Clemens MJ. Disruption of the interaction of mammalian protein synthesis eukaryotic initiation factor 4B with the poly(A)-binding protein by caspase- and viral protease-mediated cleavages. *J Biol Chem.* 2001;276(26):23922–23928.
126. Walker E, Jensen L, Croft S, et al. Rhinovirus 16 2A protease affects nuclear localization of 3CD during infection. *J Virol.* 2016;90(24):11032–11042.
127. Petersen JF, Cherney MM, Liebig HD, Skern T, Kuechler E, James MN. The structure of the 2A proteinase from a common cold virus: a proteinase responsible for the shut-off of host-cell protein synthesis. *EMBO J.* 1999;18(20):5463–5475.
128. Voss T, Meyer R, Sommergruber W. Spectroscopic characterization of rhino viral protease 2a: Zn is essential for the structural integrity. *Protein Sci.* 1995;4(12):2526–2531.
129. Hellen CU, Lee CK, Wimmer E. Determinants of substrate recognition by poliovirus 2A proteinase. *J Virol.* 1992;66(6):3330–3338.
130. Lee C-K, Wimmeri E. Proteolytic processing of poliovirus polyprotein: elimination of 2Apro-mediated, alternative cleavage of polypeptide 3CD by in vitro mutagenesis. *Virology.* 1988;166(2):405–414.

131. Davis GJ, Wang QM, Cox GA, et al. Expression and purification of recombinant rhinovirus 14 3CD proteinase and its comparison to the 3C proteinase. *Arch Biochem Biophys*. 1997;346(1):125–130.
132. Long AC, Orr DC, Cameron JM, Dunn BM, Kay J. A consensus sequence for substrate hydrolysis by rhino virus 3C proteinase. *FEBS Lett*. 1989;258(1):75–78.
133. Binford SL, Maldonado F, Brothers MA, et al. Conservation of amino acids in human rhinovirus 3C protease correlates with broad-spectrum antiviral activity of rupintrivir, a novel human rhinovirus 3C protease inhibitor. *Antimicrob Agents Chemother*. 2005;49(2):619–626.
134. Binford SL, Weady PT, Maldonado F, Brothers MA, Matthews DA, Patick AK. In vitro resistance study of rupintrivir, a novel inhibitor of human rhinovirus 3C protease. *Antimicrob Agents Chemother*. 2007;51(12):4366–4373.
135. Matthews DA, Dragovich PS, Webber SE, et al. Structure-assisted design of mechanism-based irreversible inhibitors of human rhinovirus 3C protease with potent antiviral activity against multiple rhinovirus serotypes. *Proc Natl Acad Sci USA*. 1999;96(20):11000–11007.
136. Matthews DA, Dragovich PS, Webber SE, et al. Structure-assisted design of mechanism-based irreversible inhibitors of human rhinovirus 3C protease with potent antiviral activity against multiple rhinovirus serotypes. *Proc Natl Acad Sci USA*. 1999;96(20):11000–11007.
137. Hayden FG, Turner RB, Gwaltney JM, et al. Phase II, randomized, double-blind, placebo-controlled studies of ruprintrivir nasal spray 2-percent suspension for prevention and treatment of experimentally induced rhinovirus colds in healthy volunteers. *Antimicrob Agents Chemother*. 2003;47(12):3907–3916.
138. De Palma AM, Vliegen I, De Clercq E, Neyts J. Selective inhibitors of picornavirus replication. *Med Res Rev*. 2008 Nov;28(6):823–884. Available from: https://doi.org/10.1002/med.20125.
139. Gustin KE, Sarnow P. Inhibition of nuclear import and alteration of nuclear pore complex composition by rhinovirus. *J Virol*. 2002;76(17):8787–8796.
140. Gustin KE, Sarnow P. Effects of poliovirus infection on nucleo-cytoplasmic trafficking and nuclear pore complex composition. *EMBO J*. 2001;20(1–2):240–249.
141. Walker EJ, Younessi P, Fulcher AJ, et al. Rhinovirus 3C protease facilitates specific nucleoporin cleavage and mislocalisation of nuclear proteins in infected host cells. *PLoS One*. 2013;8(8):e71316.
142. Park N, Schweers NJ, Gustin KE. Selective removal of FG repeat domains from the nuclear pore complex by enterovirus 2Apro. *J Virol*. 2015;89(21):11069–11079.
143. Cammas A, Pileur F, Bonnal S, et al. Cytoplasmic relocalization of heterogeneous nuclear ribonucleoprotein A1 controls translation initiation of specific mRNAs. *Mol Biol Cell*. 2007;18(12):5048–5059.
144. Flather D, Semler BL. Picornaviruses and nuclear functions: targeting a cellular compartment distinct from the replication site of a positive-strand RNA virus. *Front Microbiol*. 2015;6:594.
145. Xu S, Powers MA. Nuclear pore proteins and cancer. *Semin Cell Dev Biol*. 2009;20(5):620–630.
146. López-Lastra M, Rivas A, Barría MI. Protein synthesis in eukaryotes: the growing biological relevance of cap-independent translation initiation. *Biol Res*. 2005;38(2-3):121–146.
147. Gingras A-C, Raught B, Sonenberg N. eIF4 initiation factors: effectors of mRNA recruitment to ribosomes and regulators of translation. *Annu Rev Biochem*. 1999;68(1):913–963.
148. Borman AM, Michel YM, Kean KM. Biochemical characterisation of cap–poly (A) synergy in rabbit reticulocyte lysates: the eIF4G–PABP interaction increases the

functional affinity of eIF4E for the capped mRNA 5′-end. *Nucleic Acids Res.* 2000;28(21):4068–4075.
149. Slater L, Bartlett NW, Haas JJ, et al. Co-ordinated role of TLR3, RIG-I and MDA5 in the innate response to rhinovirus in bronchial epithelium. *PLoS Pathog.* 2010;6(11):e1001178.
150. Estornes Y, Toscano F, Virard F, et al. dsRNA induces apoptosis through an atypical death complex associating TLR3 to caspase-8. *Cell Death Differ.* 2012;19(9):1482–1494.
151. Han K-J, Su X, Xu L-G, Bin L-H, Zhang J, Shu H-B. Mechanisms of the TRIF-induced interferon-stimulated response element and NF-κB activation and apoptosis pathways. *J Biol Chem.* 2004;279(15):15652–15661.
152. Kaiser WJ, Offermann MK. Apoptosis induced by the Toll-like receptor adaptor TRIF is dependent on its receptor interacting protein homotypic interaction motif. *J Immunol.* 2005;174(8):4942–4952.
153. Lei X, Sun Z, Liu X, Jin Q, He B, Wang J. Cleavage of the adaptor protein TRIF by enterovirus 71 3C inhibits antiviral responses mediated by Toll-like receptor 3. *J Virol.* 2011;85(17):8811–8818.
154. Xiang Z, Li L, Lei X, et al. Enterovirus 68 3C protease cleaves TRIF To attenuate antiviral responses mediated by Toll-like receptor 3. *J Virol.* 2014;88(12):6650–6659.
155. Qu L, Feng Z, Yamane D, et al. Disruption of TLR3 signaling due to cleavage of TRIF by the hepatitis A virus protease-polymerase processing intermediate, 3CD. *PLoS pathog.* 2011;7(9):e1002169.
156. Croft SN, Walker EJ, Ghildyal R. Human rhinovirus 3C protease cleaves RIPK1, concurrent with caspase 8 activation. *Sci Rep.* 2018;8(1):1569.
157. Lotzerich M, Roulin PS, Boucke K, Witte R, Georgiev O, Greber UF. Rhinovirus 3C protease suppresses apoptosis and triggers caspase-independent cell death. *Cell Death Dis.* 2018;9(3):272.
158. Barral PM, Sarkar D, Fisher PB, Racaniello VR. RIG-I is cleaved during picornavirus infection. *Virology.* 2009;391(2):171–176.
159. Kawai T, Takahashi K, Sato S, et al. IPS-1, an adaptor triggering RIG-I- and Mda5-mediated type I interferon induction. *Nat Immunol.* 2005;6(10):981–988.
160. Kumar H, Kawai T, Kato H, et al. Essential role of IPS-1 in innate immune responses against RNA viruses. *J Exp Med.* 2006;203(7):1795–1803.
161. Rajput A, Kovalenko A, Bogdanov K, et al. RIG-I RNA helicase activation of IRF3 transcription factor is negatively regulated by caspase-8-mediated cleavage of the RIP1 protein. *Immunity.* 2011;34(3):340–351.
162. Drahos J, Racaniello VR. Cleavage of IPS-1 in cells infected with human rhinovirus. *J Virol.* 2009;83(22):11581–11587.
163. Yang Y, Liang Y, Qu L, et al. Disruption of innate immunity due to mitochondrial targeting of a picornaviral protease precursor. *Proc Natl Acad Sci USA.* 2007;104(17):7253–7258.
164. Agol VI, Belov GA, Bienz K, et al. Competing death programs in poliovirus-infected cells: commitment switch in the middle of the infectious cycle. *J Virol.* 2000;74(12):5534–5541.
165. Belov GA, Romanova LI, Tolskaya EA, Kolesnikova MS, Lazebnik YA, Agol VI. The major apoptotic pathway activated and suppressed by poliovirus. *J Virol.* 2003;77(1):45–56.

CHAPTER 3

Ground zero—the airway epithelium

Andrew T. Reid[1,2], Erika N. Sutanto[3], Punnam Chander-Veerati[1,2], Kevin Looi[3], Ngan Fung Li[2,4], Thomas Iosifidis[3], Su-Ling Loo[2,4], Luke W. Garratt[3], Anthony Kicic[3,5,6,7,8] and on behalf AusREC[2,3,9]

[1]School of Medicine and Public Health, University of Newcastle, Callaghan, NSW, Australia
[2]Priority Research Centre for Healthy Lungs, Hunter Medical Research Institute, The University of Newcastle, Callaghan, NSW, Australia
[3]Telethon Kids Institute, The University of Western Australia, Nedlands, WA, Australia
[4]School of Biomedical Sciences and Pharmacy, University of Newcastle, Callaghan, NSW, Australia
[5]Occupation and Environment, School of Public Health, Curtin University, Bentley, WA, Australia
[6]Paediatrics, Medical School, Faculty of Healthy and Medical Science, The University of Western Australia, Nedlands, WA, Australia
[7]Department of Respiratory and Sleep Medicine, Perth Children's Hospital, Nedlands, WA, Australia
[8]Centre for Cell Therapy and Regenerative Medicine, School of Medicine and Pharmacology, The University of Western Australia, Nedlands, WA, Australia
[9]Robinson Research Institute, University of Adelaide, Adelaide, SA, Australia

3.1 INTRODUCTION

The airway epithelium remains the first line of defense against insult, since it lies at the interface between the external environment and the internal milieu. Although thought to play a simple barrier role in this capacity, it has now been recognized to respond to the external environment and infectious stimuli via mediator release that direct signals with immune and mesenchymal cells. Specifically, in response to viral infection, it has developed a multifaceted approach to trapping and eliminating virus. A compromised epithelial barrier in diseases such as asthma, chronic obstructive pulmonary disease (COPD), or cystic fibrosis (CF) dramatically increases the risk of pathogenic infection. During infection, airway epithelial cells become the primary site of replication for respiratory viruses such as rhinovirus (RV) and initiate the hosts' immune responses. This chapter will discuss the structure of the airway highlighting the now broad number of cell types that comprise it. It will discuss these in the context of health and inflammatory respiratory disease as well as their role in facilitating viral infection. In addition, this chapter will summarize the primary roles of the airway, including its mucociliary, barrier, and reparative functions in these settings. We will examine the interaction between the airway epithelium

Rhinovirus Infections
DOI: https://doi.org/10.1016/B978-0-12-816417-4.00003-2

and RV, which promotes viral attachment, subsequent entry, and replication, and elaborate on the innate response of the airway to viral infection. Finally, we review the literature on interventional studies conducted to combat RV infection.

3.2 STRUCTURE OF THE AIRWAY EPITHELIUM

The airway epithelium is an intricate and adaptive network of cells and cell products that functions as an effective barrier to environmental and pathogen-induced insult. The epithelium is slowly but constantly renewed to maintain a pseudostratified structure that consists of various cell types that play unique roles in defense of the host. Traditionally, the core cell types essential for a functional epithelium are basal cells, club cells, ciliated cells, and goblet cells (Fig. 3.1). However, it is now becoming clearer that the variety of cellular subtypes is greater than first thought.[1,2] Barrier

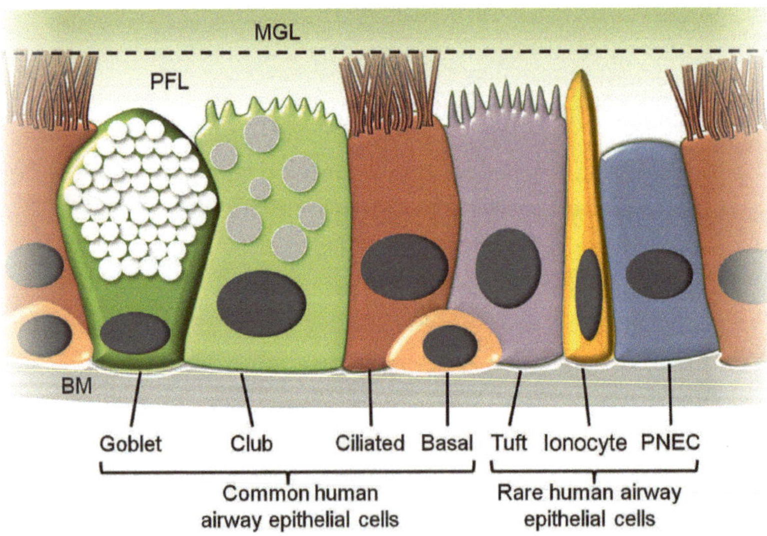

Figure 3.1 ***Schematic of common and rare cell types of the human airway epithelium.*** Airway epithelial cells are anchored to the BM and extend apically toward the airway lumen. Goblet cells and club cells, as well as submucosal glands (not shown), secrete mucins and other bioactive compounds into the PFL and the MGL to aid in airway barrier function. Basal, club, ciliated, and goblet cell types are regarded as common cell types making up human airway epithelium whilst PNECs, tuft cells, and the recently discovered pulmonary ionocytes are regarded as rare cells. Dotted line represents interface of PFL and MGL. *BM*, Basement membrane; *PFL*, periciliary fluid layer; *MGL*, mucus gel layer; *PNECs*, pulmonary neuroendocrine cells.

function of the epithelium is principally dependent on the formation of intercellular junctions that physically link neighboring cells as well as the apical mucous layer.[3] Epithelial features/cells such as basement membrane, submucosal glands (SMGs), pulmonary neuroendocrine cells (PNECs) and associated myeloid and plasmacytoid dendritic cells are also essential in establishing antiviral immunity.[4,5] To recognize and respond to pathogen-associated molecular patterns or damage-associated molecular patterns, epithelial cells express numerous pattern recognition receptors (PRRs).[6,7] Examples of PRRs include Toll-like receptors (TLRs), NOD-like receptors (NLRs), and retinoic acid inducible gene-I-like receptors (RLRs).[8] These responses are facilitated by secretion of various cytokines and chemokines that stimulate innate and adaptive immune cells. The epithelium is also highly plastic and capable of undergoing considerable remodeling in response to the detection of foreign molecules such as viral products. In chronic airway diseases, there is often constant remodeling that leads to an imbalance in epithelial cell type and number, which impairs the function of the epithelium as a whole.

3.2.1 Basal cells

Airway basal cells comprise anywhere from 6% to 30% of the human airway epithelium and are most commonly distinguished by high levels of tumor protein p63 (TP63) and cytokeratin 5 (KRT5).[9,10] They appear as cuboidal-shaped cells and their numbers are directly correlated to epithelial height, decreasing in abundance from trachea to bronchioles.[11] They are defined as either multipotent stem cells (basal stem cell), or committed precursor cells (basal luminal progenitor) that reside proximal to the basal lamina and proliferate or differentiate during development or in response to epithelial injury.[10–13] This differentiation occurs as a tightly controlled sequence of events. First, basal cells from the committed precursor population undergo differentiation to form intermediate club cells. Then, following stimulation of distinct transcriptional pathways, these cells undergo further differentiation into resultant goblet cells or ciliated cells.[1,10] In diseases such as asthma, increased basal cell proliferation and decreased basal cell differentiation occurs following epithelial damage and results in a compromised airway epithelial barrier.[14]

Basal cells are tightly bound to the surrounding cells and also offer structural support to the epithelium by anchoring to the underlying matrix (basement membrane) via hemidesmosomal attachment.[15]

Respiratory viruses have adapted to leverage these receptors for cellular uptake. For example, basal cells express high levels of intercellular adhesion molecule 1 (ICAM-1),[1] the key receptor for major group RVs.[16] Whereas basal cells of the nasal epithelium express low-density lipoprotein (LDL) receptor (LDLR), which is the key host protein that minor group RVs bind to.[17] A novel type of basal cell called the "hillock" basal cell, in reference to the clumps of "hillock" club cells that result following their differentiation, has recently been defined in areas of high epithelial cell turnover.[1] As such, these novel hillock basal cells may regulate the epithelium's rapid response to injury, as they express genes associated with immune regulation and asthma.[1]

3.2.2 Club cells

The club cell is often referred to as the second stem cell in the airway epithelium, having both regeneration and differentiation potential depending on their location on the tracheal–bronchial axis.[18] Club cells are recognized as secretory cells of major importance to the epithelium, whilst also being the precursor for ciliated and goblet cell types, a process driven by the junctional protein E-cadherin.[19] Club cells are dome-shaped cells studded with microvilli and contain dense cytoplasmic granules and their differentiation from basal cells is dependent on Notch signaling.[20] Club cells are the major secretory cell in the small airway epithelium occupying anywhere between 5% and 20% of total cells and secrete an array of bioactive compounds into the airway surface liquid (ASL) that protect the epithelium from damage, including the unique secretion of secretoglobin family 1A member 1 (SCGB1A1).[21] For this chapter we will refer to ASL as the combination of both periciliary fluid and overlying mucus gel layer.[22] SCGB1A1 is the most abundant protein found within human bronchiolar lavage fluid and is a key protective molecule in ASL, able to suppress inflammation and associated cell ageing, xenobiotic damage and bacterial infection.[23] Furthermore, augmentation of this protein in vitro suppresses smoking induced damage to airway epithelial cells in air–liquid interface (ALI) culture.[24] Due to the function and high production of SCGB1A1 by club cells, they are regarded as key protective cells of the airway epithelium.

In COPD, club cell numbers are reduced compared with matched normal smokers, reducing levels of SCGB1A1 in ASL and further compromising the airway epithelium.[25] Similarly, SCGB1A1 levels are significantly lower in serum of asthmatic patients compared with healthy donors, and this reduction was even more pronounced in chronic asthma.[26] Club cells have recently been shown to produce α1-antitrypsin, formerly thought of as being derived solely from the liver, and as such have antiprotease activity.[21] Club cells also express numerous cytokines such as CXCL17, CXCL1, and CXCL6 that attract dendritic cells as well as neutrophils. Of relevance to this chapter, club cells express high levels of CREB3L, a protein that plays a role in antiviral defense by preventing proliferation of infected cells.[27] Until recently, club cells were regarded as a single population of epithelial cells, however advances in RNA sequencing technology, namely, single-cell RNA sequencing combined with cell lineage tracing, have identified different functional classes of club cells based on their location.[1] The newly identified KRT13 + /KRT4 + hillock club cell appears to be involved in epithelial regeneration following injury, as high turnover as well as high expression of immunomodulatory (Galectin3) and barrier function (Claudin3 and Annexin A1) related genes was observed.

3.2.3 Ciliated cells

The major terminal points for differentiation of the precursor cells are ciliated cells. Conversion of club cells into ciliated cells involves downregulation of Notch signaling, a key determiner of cell fate.[28] They are the predominant cell type within the human airway epithelium, increasing from ~50% to ~75% of total cells when moving from the trachea to small airways respectively.[29,30] Characterized by a large number of cilia that extend into the airway lumen, ciliated cells are necessary for mucocilliary clearance, a function critical to a healthy lung. This facilitates the removal of inhaled foreign particles that are trapped by the epithelial mucus layer, whereby rhythmic ciliary beating forces mucus toward the pharynx, at which point the particles and mucus are swallowed or coughed out.[31] The presence of cilia greatly expands the surface area/volume ratio of the epithelium and transmembrane receptor proteins with functions in immune signaling are harbored on these cilia such as CX3C chemokine receptor 1 (CX3CR1), necessary for leukocyte attachment.[32] These also include the bitter taste receptors T2R4, T2R43, T2R38, and T2R46, which are believed to enable ciliated cells to sense and respond to noxious substances.[33] Numerous defects in ciliated cell function and morphology have been associated with COPD, asthma, and CF.[34–36] In severe asthma there is a significant reduction in the number of ciliated cells resulting in abnormal mucociliary clearance and increased risk of infection.[36] Ciliated cells of the asthmatic epithelium also feature reduced cilia beat frequency, cilia shortening, microtubular defects, mitochondrial damage, and cytoplasmic blebbing.[36,37] In COPD and CF, reduced or lost function of the cystic fibrosis transmembrane conductance regulator (CFTR) alters airway hydration, affecting the periciliary layer and obstructing cilia movement.[22] Finally, respiratory viruses are known to exploit cilia associated proteins, with RV-C using cadherin-related family member 3 (CDHR3) receptors upon cilia a prime example.[38]

3.2.4 Goblet cells

Named for their distinct goblet shape, the goblet cell is a secretory cell that is tightly packed with vesicle bound mucin granules and surfactant proteins and can be found in proximal airways within healthy individuals.[39,40] The primary role of goblet cells is to secrete mucins like mucin 5AC (MUC5AC) into the ASL to generate the mucus gel layer.[26] Excess goblet cell differentiation is associated with chronic inflammation in asthma, COPD, and CF.[41] A number of type-2 cytokines including Interleukin (IL)-4, -5, -9, and -13 are able to increase goblet cell differentiation as well as the overexpression of signal transducer and activator of transcription 6 (STAT6) and SAM-pointed domain-containing ETS transcription factor (SPDEF),[42,43] with SPDEF also able to upregulate goblet cells expression of MUC5AC.[44] This increase in goblet cell number in areas where goblet cells are commonly found is termed goblet cell hyperplasia. Alternatively, differentiation of goblet cells in areas where goblet cells are normally lacking is termed goblet cell metaplasia. Both of these phenomena occur in severe asthma and COPD while this is not as definitive for CF; in fact goblet cell hypertrophy (increase in goblet cell size) may occur instead.[45] In addition to mucins, goblet cells also express proteins with proposed antixenobiotic functions such as Trefoil factor 1 and 2 (*Tff1* and *Tff2*) and Demilune cell and parotid proteins (*Dcpp*) 1–3.[1] The recent single cell analysis of airway epithelium by Montoro et al. has also revealed three transcriptionally distinct, and potentially disease relevant, subtypes of goblet cell: immature goblet (*Gp2*$^{-}$), goblet-1 (Tff2^{+}), and goblet-2 (lipase F^{+}; Lipf^{+}).[1] Further investigations on the roles of these subtypes will reveal how these subtypes potentially function during airways disease and virus infection. Although not a primary location for viral replication, infection can affect goblet cells with recent work reporting RV infection increasing SPDEF and goblet cell differentiation through the transcription factor NOTCH3.[46,47]

3.2.5 Pulmonary neuroendocrine cells

Single PNECs are a rare epithelial cell type that are scattered in the respiratory epithelium, but can also appear clustered into neuroendocrine bodies. These cells were first recognized in the lungs almost 50 years ago for their endocrine properties and close association to with nerve fibers.[48] PNECs are innervated at the base and have microvilli on the apical cell surface that extend into the airway lumen. These cells are proposed to be the first specialized cell type to appear within the airway epithelium, implying that progressive cell-type specification in the lung may start with progenitors of neuroendocrine nature.[48,49] PNECs contain secretory granules and dense-core vesicles, which contents are released by physiological stimuli, such as hypoxia, and many of them have potent physiological effects. The main functions of PNECs are airway oxygen sensing, pulmonary blood flow regulation, bronchial tonus control and immune response modulation.[49,50] Despite being a rare population of cells in the airway, PNECs are able to amplify the airway allergen signal into mucosal type 2 responses. PNECs achieve that by secreting γ-aminobutyric acid (GABA) to stimulate airway epithelial mucus production.[51] In parallel, PNECs act through another product, calcitonin gene related peptide, to stimulate ILC2 production of cytokines, thus initiating downstream immune cell recruitment.[51] Interestingly, PNECs are thought to maintain a stem cell niche essential for club cell regeneration during lung injury.[52,53] However, deletion of PNECs did not affect the ability of airway progenitors to regenerate after naphthalene injury,[54] suggesting cross-talk between various cell types and the microenvironment may be at play during airway regeneration.

3.2.6 Tuft cells

Tuft cells, also termed brush or solitary chemosensory cells, were first described over 60 years ago as are part of the respiratory epithelial layer.[55] The cells are named after their characteristic tufts of blunt microvilli (about 130 per cell) on the apical cell surface. The microvilli contain internal filaments that stretch into the underlying cytoplasm, and have a distinctive shape with a wide base and a narrow microvillus apex.[55]

Similar to tuft cells present in gastrointestinal epithelia, airway tuft cells express transient receptor cation channel subfamily M member 5 and the taste-associated G protein, gustducin, and taste receptors.[56,57] Airway tuft epithelial cells were revealed to be the solitary source of IL-25, a member of the IL-17 cytokine family known to promote expression of mediators of type 2 immunity.[58,59] These cells have chemosensory properties that provide mucosal defense in response to aeroallergens and RV infections,[60,61] where IL-25 from tuft cells activate mucosal group 2 innate lymphoid cells to secrete IL-13, which feedbacks on epithelial crypt precursors to skew differentiation of small bowel epithelia toward mucus-producing goblet cells and additional tuft cells.[58,59] Following rhinoviral infection, the epithelial-derived levels of IL-25 are elevated, which can be induced further in subjects with asthma.[61]

3.2.7 Pulmonary ionocytes

The pulmonary ionocyte is a newly identified cell type of the airway epithelium and was discovered using single cell RNA sequencing (scRNAseq).[1,2] Plasschaert et al.[2] demonstrated that $Foxi1^+$ pulmonary ionocytes differentiate from basal cells under the control of Notch signaling. Whilst there is currently limited information on the functionality of the cell it is speculated from single cell transcriptomic data that pulmonary ionocytes are similar to ionocytes found in *Xenopus* and Zebrafish. Pulmonary ionocytes also specifically express vacuolar-type H^+-ATPase (V-ATPase) subunits as well as CFTR and likely regulate ion transport and fluid pH and potentially have a role in cell–cell communication.[1,2] Whether these rare cells play a role during innate or adaptive immunity is currently unknown and will no doubt be addressed in future studies. This breakthrough discovery of a new CFTR expressing airway epithelial cell has very large implications for epithelial cell biology and may open new avenues for treating CF as well as other chronic airways diseases such as asthma and COPD.

3.2.8 Airway basement membrane and submucosal glands

In addition to the primary cell types listed above there are a number of other important cells and cell products that play a role in establishing an effective barrier to xenobiotic insult. The basement membrane, which is formed at the basolateral surface of basal cells, is a tightly regulated structure that functions as both an anchor for epithelial cells and as a regulator of epithelial cell-specific functions.[62] The major components of the basement membrane are type IV collagen, nidogens, laminin, and select glycoproteins (e.g., fibronectin and tenascin). Basement membrane thickening occurs in asthma and is thought to be due to increased production of extracellular matrix from bronchial fibroblasts.

SMGs are found throughout the human airway (bronchial and nasal), decreasing in abundance from trachea to bronchioles and are responsible for the secretion of mucus into the airway lumen. SMGs produce a wide range of proteins with various functions, which include gel-forming mucins, surfactants, antimicrobials, bactericides, immune signaling proteins, chemotactic molecules, tissue remodeling proteins, and proinflammatory and antiinflammatory molecules.[63] SMGs are composed of two major cell types: mucous cells and serous cells. Mucous cells are responsible for the secretion of mucin glycoproteins, the major of which is the gel-forming MUC5B. In contrast, serous cells produce the majority of the remaining compounds listed above, the most abundant being lactoferrin and lysozyme in nasal SMGs. Total SMG volume is estimated to be 50 times the total volume of goblet cells in the airway and as such SMGs have the potential to produce much more mucus than other secretory regions of the epithelium. The secretion of xenoprotective proteins in combination with a variety of mucin glycoproteins is critical in maintaining an effective barrier to infection and xenobiotic exposure. Serous acinar cells of the SMG are rich in CFTR,[64] and until the discovery of the ionocyte, were thought to be the key target for CFTR therapeutics to ameliorate the symptoms of CF.

In summary, the airway epithelium is composed from a variety of cellular types and morphologies that together create a barrier to the external environment. How this barrier functions and is maintained through challenge and injury will be discussed.

3.3 FUNCTIONS OF THE AIRWAY EPITHELIUM

3.3.1 The airway mucus barrier

The first barrier encountered in the lower airways by inhaled particles is a dynamic layer of sticky mucus. By trapping and removing these particles, it provides an important role in reducing the number of direct challenges to the airway tissue itself. Airway mucus production is a tightly regulated process of major importance to maintaining effective epithelial barrier function. As briefly mentioned above, goblet cells, SMGs, and club cells all contribute to the production of ASL. This ASL is comprised of two structurally distinct phases: a periciliary fluid layer that extends approximately 7 μm (healthy epithelium) above the epithelial cell apical membrane in which cilia are able to beat freely, and a more viscoelastic mucus gel layer responsible for trapping foreign particles.[22,40] The periciliary layer is a near frictionless mixture of water, ions and tethered mucins (MUC1, MUC4, and MUC20).[63] Periciliary fluid height changes depending on its location in the lung and is most tightly controlled by the changes in osmotic pressure that result from chloride ion movement. In contrast, the mucus gel layer is formed above the periciliary layer and contains water, salts, gel-forming mucins (MUC5B and MUC5AC), enzymes, lipids, and cell debris. The mucus gel layer is a highly viscoelastic substance that carries a large negative charge coupled with high levels of defensive proteins. The global negative charge is able to prevent the transport of negatively charged proteins through the gel, such as those present on many biological surfaces (bacterial, fungal, and viral). Furthermore, this net negative charge allows for the prolonged association of key positively charged defensive proteins such as lysozyme.[65]

Healthy mucus contains a delicate balance of the mucin glycoproteins MUC5AC and MUC5B mainly secreted by goblet cells and SMGs respectively. A normal mucus gel layer from a healthy individual has a pore size of roughly 100–500 nm.[66] To put this into context, pollen vary from 6 to 100 μm, a *Pseudomonas* spp. bacterium is approximately 1 μm wide and 1–5 μm long, whereas the RV capsid size is approximately 30 nm,[67] much smaller than even the minimum mucus pore size measured. As such, mucus is able to capture inhaled bacteria, pollen, and other foreign insults but is poor at trapping viral particles. There are no current measurements of the pore size of mucus from an individual with airways disease and increased MUC5AC. Mucin molecules are present in tightly packed dehydrated granules prior to release from secretory cells and upon release greatly expand in volume by approximately 1000-fold. MUC5B expansion and assembly into nets occurs following an increase in pH and the exchange of Ca^{2+} ions for Na^{+} ions, which is facilitated by a bicarbonate rich environment. The increase in Na^{+} ions then stimulates mucin hydration.[68] MUC5AC is also believed to expand via an equivalent mechanism.[69]

MUC5AC and MUC5B have been scrupulously investigated and found to adopt unique tertiary structures and exhibit different roles. In the pig trachea MUC5B appears to exist as linear arrangements of mucin molecules following secretion from SMGs whilst MUC5AC appears as thin wisps or sheets that are routinely found to remain tethered to goblet cells following secretion.[70] These MUC5AC conglomerates are believed to coat and anchor the long bundles of MUC5B that pass by and as such regulate mucus adherence; a function necessary for the trapping and eliminating of foreign material by resident macrophages for antigen presentation. An increase in MUC5AC:MUC5B ratio is therefore likely to result in more adherence of mucus to the airway. Interestingly, mice that do not express *Muc5ac* are viable and exhibit normal mucociliary clearance, whereas mice lacking *Muc5b* exhibit dysregulated mucociliary clearance resulting in material accumulation in upper and lower airways leading to chronic bacterial infection.[71] In asthma, COPD, and CF, dysregulated mucin production can arise from goblet cell hyperplasia or metaplasia, SMG hypertrophy, or chloride channel dysregulation that also results in an imbalanced MUC5AC:MUC5B ratio in the lumen.[72–74] There is some correlation between severity and degree of ratio imbalance, for example in severe asthma it is well established that there is an much larger increase in MUC5AC than MUC5B.[75]

3.3.2 Barrier function of the airway epithelium

To form the tightest barrier possible across the surface of the conducting airway, whilst also permitting dynamic movement required for respiration, the epithelial cells are attached together by the coordinated action of an assortment of intercellular junction proteins (Fig. 3.2). Visualization of epithelial intercellular junctions was first performed by Farquhar and Palade in 1963.[76] Adherens and tight junctions constitute a major role in this intercellular junction bordering the apico-basolateral membranes between adjacent cells. Under ultrathin sections by electron microscopy, a series of discrete fusion sites are located in a region termed a "tight junction" and show a near obliteration of the intercellular space and convergence of lateral membranes between adjacent cells. Hence, this region is also termed the zona occludens. Tight junctions extend to a belt-like network within each adjacent cell to form a continuous seal across the epithelial layer and are recognized as the major structural determinant for paracellular permeability. Electron dense tracers and molecule such as hemoglobin, colloidal lanthanum, and ruthenium red have been shown to stop at the tight junction region whilst freely diffuse along the intercellular space.[76–78] Tight junctions provide a semipermeable, size- and ion- selective

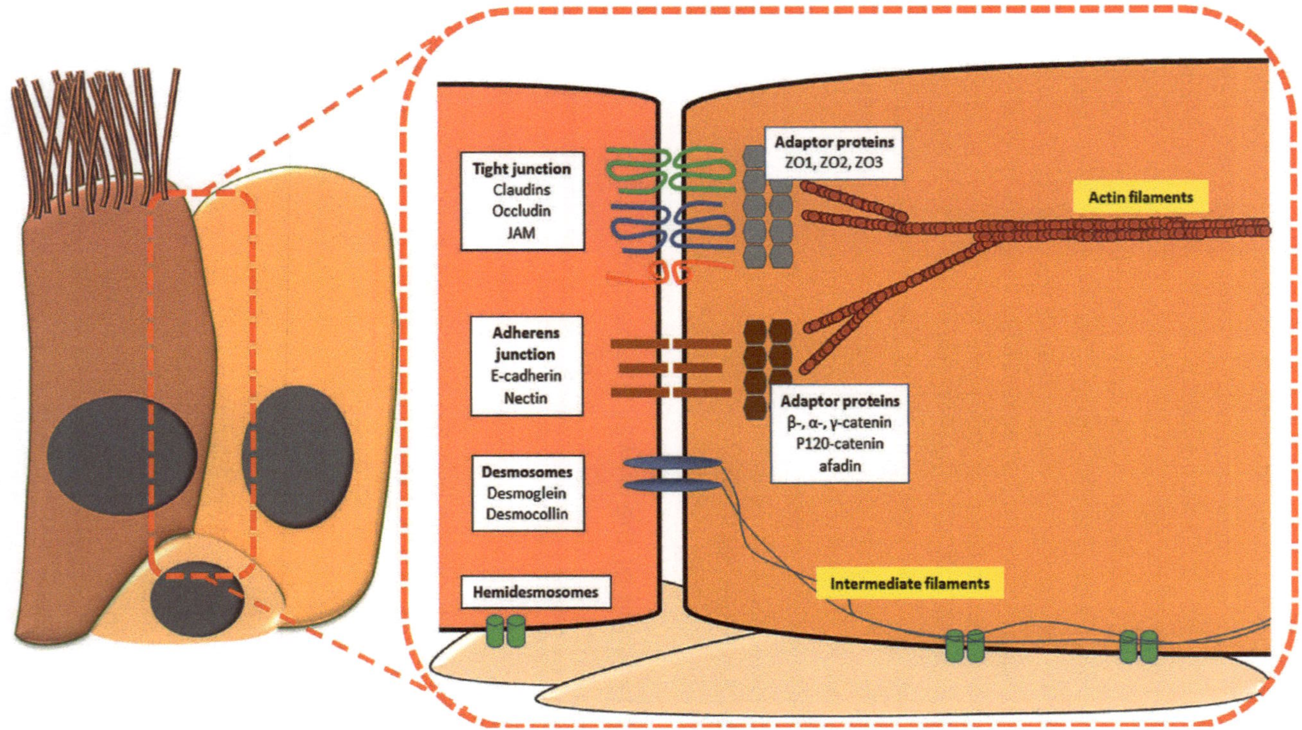

Figure 3.2 ***Composition of intercellular junctions in differentiated airway epithelium.*** Schematic of intercellular junctions in differentiated airway epithelium. Intercellular junctions constitute TJs, AJs, and desmosomes, which together mediate interepithelial adhesion with a paracellular barrier/fence function. TJs are made up of at least three types of transmembrane proteins: occludin, claudins, and the IgG-like family of JAMs with cytoplasmic adapter proteins such as zonula occludens -1,-2, and -3. AJs are positioned directly below TJs and are comprised of two adhesive units: the nectin–afadin complex and the classical cadherin–catenin complex. Both TJs and AJs are closely associated with the actin cytoskeleton. While cell adhesion attachment is also facilitated by desmosomes, hemidesmosomes expressed by basal cells ensure attachment to the basement membrane. *TJs*, Tight junctions; *AJs*, adherens junctions; *JAMs*, junctional adhesion molecules.

barrier in the intercellular space, depending on the molecular composition specifically expressed in individual types of epithelia (reviewed extensively by Anderson et al.).[79] In addition to this external barrier function, tight junctions also act as a physical boundary to limit free passage of macromolecules between the compositionally distinct apical and basolateral domains. Three types of structural transmembrane proteins enriched at tight junction are known to mediate cell−cell adhesion: the IgG-like family of junctional adhesion molecules, and the claudin and occludin families of four transmembrane spanning molecules.

Adherens junctions are positioned immediately below tight junctions. This region is characterized by two opposing membranes separated 20 nm apart, running over the distance of 0.2−0.5 μm in parallel below tight junction region. Adherens junctions are crucial in initiation and maintenance of intercellular adhesion. In fact, earlier studies showed that the assembly of the tight junction is dependent on the formation of the adherens junction. This is demonstrated by interaction between components of adherens and tight junctions, such as zona occludens 1 (ZO1) and α-catenin or afadin. For instance, recruitment of ZO1 to cadherin-containing intercellular contacts provides an early scaffold to signal for formation of tight junction.[80] The adherens junction is made up of two adhesive units: the nectin−afadin complex and the classical cadherin−catenin complex. Cell-specific expression of cadherins and nectins determines the strength and adhesive specificity of the adherens junctions. Collectively, the adherens and tight junctions are named the apical−junctional complex and despite their formation by different molecules, both are closely associated with the circumferential belt of actin and both regulate cellular permeability and polarity. In addition to providing structural integrity, apical−junctional complex has been identified to function as landmarks to spatially confine signaling molecules[81] and establish polarity cues for specialized cell function.[82]

Occludin was the first junction protein identified that is most ubiquitously expressed at the apical region and has been associated with the regulation of permeability properties of the tight junction seal, particularly size selection of molecules undergoing paracellular diffusion.[83] Although TJs without occludin are quite rare, the physiological function of occludin in the apical junctional complex remains elusive due to contrasting observations in studies performed on immortalized cell lines, mice or in epithelial tissues lacking occludin. Claudins are another integral membrane proteins located in the apical regions of all epithelia and since the discovery of the first claudin as a strand-forming tight junction protein in 1998,[84] at least 27 members of the claudin family have been reported. Despite numerous studies demonstrating their importance in the maintenance of epithelial integrity,[85–88] their roles in regulating epithelial permeability currently remain unclear because the functional characteristics of the majority of claudins remain relatively unknown at present.

Supporting the tetrapass membrane proteins, namely claudins and occludin is the cytoplasmic plaque consisting of a network of the densely packed peripheral adaptor proteins ZO1, ZO2, and ZO3. The most studied plaque component is the peripheral adaptor protein ZO1, which binds to a number of cytoskeletal, signaling, and membrane proteins.[89,90] Again, supporting the tight junctions are the adherens junctions, consisting of the cadherin family of calcium-dependent adhesion proteins, which perform various roles including the initiation and stabilization of cell to cell adhesion.[91] Located around the midsection of the cell are intercellular junctions such as desmosomes, which provide adhesive bonds between epithelial cells to give mechanical strength to the epithelium. Further down at the basolateral side are gap junctions, which are specialized cell to cell channels that permit the diffusion of molecules and solutes between adjacent cells.[92] Overall, this collective group of junctional proteins serve to provide a barrier, regulate solute and water flow through the paracellular space, and separate the apical from the basolateral domain to establish cell polarity. What is becoming more apparent and requires more investigation is that junctional proteins can also participate in the signal transduction process to regulate gene expression, epithelial cell proliferation, differentiation, and morphogenesis.[93]

Abnormalities in function or expression of tight junction proteins may underlie or reflect pathology of airway diseases, as numerous targeted studies of the airway epithelium have demonstrated altered barrier function and innate immunity.[94–96] For example, the physical barrier of the airway epithelium in asthma is disrupted with evidence of loss of tight junction proteins and increase in epithelial permeability.[95,96] Adherens junction protein expression may also be decreased along with a reduction in desmosome length.[97,98] Together the evidence supports the concept that the epithelial barrier in asthma is structurally and functionally different. Since asthma is a disease that involves both gene and environment interactions, there have been postulations of the key role of genetic susceptibility in a dysregulated epithelial barrier in asthma. Evidence from past genome-wide association studies that many of the asthma susceptibility genes are expressed in the airway epithelium have placed the epithelium at the center stage of asthma pathogenesis. Furthermore, recent approaches using unbiased transcriptomic analysis have allowed for greater in-depth assessments of epithelial gene expression profiles. This continues to provide evidence of molecular mechanisms that will allow for the categorization of the clinical syndrome of asthma into defined endotypes that are different in pathogenesis, disease progression, and response to therapy.[99–101] Despite the use of large-scale transcriptomic approaches to analyze gene expression profiles in asthma, further comprehension of epithelial barrier dysfunction can only be achieved from the integration of these large-scale data set findings with functional assessments.

3.3.3 Reparative function of the airway epithelium

In response to damage, the airway epithelium has to rapidly repair and regenerate to restore its integrity and functions, particularly providing a protective barrier of the body to the external environment as well as the efficient gas exchange[102–104] (Fig. 3.3). The main sources of airway epithelial injury are infectious pathogens and noninfectious pollutants/allergens.[105–110] The type of wounding stimulus has been known to affect the outcome of airway wound repair and resolution,[111–113] where long-term and/or systemic complications may also arise.[114] Following the initial wounding event, there is a transient mucous release, rapid shedding of ciliated epithelial cells, as well as secretion by the local epithelium of a number of danger signals including alarmins like heat shock and S100 proteins.[115] The secretory epithelial cells at the wound edge have the capacity to dedifferentiate, and ciliated cells can transdifferentiate to squamous cells to transiently cover the denuded area. Neighboring basal cells are able to spread and then migrate very rapidly to reepithelialize the wound, followed by proliferation and differentiation.[116,117] These repair processes conclude with the complete restitution of a pseudostratified airway epithelium with intact barrier integrity and mucociliary clearance properties.

Impairment of the bronchial epithelium to repair and regenerate following injury has been observed in chronic airway diseases like asthma and CF, which results in structural and functional consequences. People with asthma or CF have an altered epithelial phenotype characterized by inadequate barrier function, poor wound repair, and deficient innate immune responses at baseline and postinfection.[118–127] In fact, airway

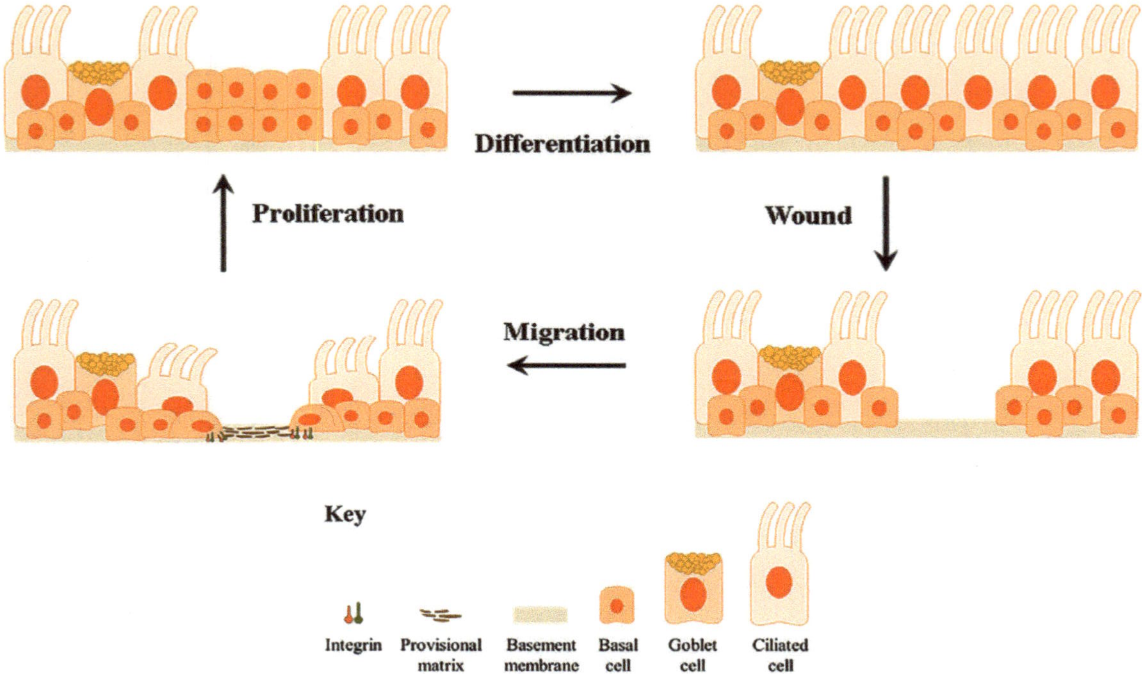

Figure 3.3 ***Repair and the airway epithelium.*** The airway epithelium forms a tight barrier against the external environment comprising of a variety of epithelial cells including basal progenitor cells, mucus-producing goblet cells, and terminally differentiated ciliated columnar cells. Following damage, airway epithelial repair involves a rapid and coordinated series of events following damage including cell spreading, migration, proliferation, and differentiation. Initially, leading edge cells flatten, spread, and migrate to cover the wound site. In addition, ciliated columnar epithelial cells can transdifferentiate to squamous cells to transiently cover the denuded area. Leading edge epithelial cell migration is facilitated by the interaction of extracellular matrix transmembrane receptors, integrin, with the wound provisional matrix, such as fibronectin, collagens, and laminin. Subsequent proliferation and differentiation restore the epithelial structure and barrier function.

epithelial repair in asthmatic subjects is dysregulated despite a higher proliferative capacity of airway epithelial cells[121,128–132] compared with healthy controls. It was also reported that cells from asthmatics expressed and secreted more plasminogen activator inhibitor -1, IL-1β, -10, -13, which have been shown to play pivotal roles in physiological proliferation, differentiation, and repair of the airway epithelium.[128,132] It has been postulated before that airway remodeling, including basement membrane thickening, subepithelial fibrosis, increased angiogenesis, and smooth muscle cell hyperplasia, could be secondary phenomena as a consequence of dysfunctional epithelial repair to some extent.[106,133–139]

In short, the function of the airway epithelium can be summarized as a tight barrier with flexible adaptive ability, whose primary response upon challenge is to ensure not only that the barrier integrity is maintained, but also that the challenge is effectively dealt with. Of these challenges to the airway epithelium, RV is perhaps the most common and well studied.

3.4 RHINOVIRUS TARGETING OF AIRWAY EPITHELIAL CELLS

Human RVs are icosahedral (30 nm in diameter), nonenveloped viruses with a (+) ssRNA genome of approximately 7100 bases. Within the genus *Enterovirus* of the family *Picornaviridae*, RV is classified into three species: A (80 serotypes), B (32 serotypes), and C (55 serotypes), determined primarily through nucleotide sequence identity.[140] Amongst RV-A and RV-B, they are further classified into major and minor groups based on specificity to host cell receptors exploited for attachment. Major group RVs enter cells through binding to cell-surface molecule ICAM-1, while minor group RVs use members of the LDLR, very-LDLR (VLDLR), and LDLR-related protein 1 (LRP-1).[141–144] Some major group RVs also use heparin sulfate proteoglycans as an additional receptor.[143,145,146] In addition to ICAM-1 and LDLR, CDHR3, a member of the cadherin family of transmembrane proteins, mediates recently identified RV-C entry into host cells.[147]

3.4.1 Viral attachment

3.4.1.1 Intercellular adhesion molecule 1 (ICAM-1)

The physiological function and cell type–specific expression of these receptors are different. Intercellular cell adhesion molecule-1 (ICAM-1/CD54) is a glycoprotein of the immunoglobulin supergene family expressed on the surface of several cell types including leukocytes,[148] fibroblasts, endothelial,[149] and epithelial cells.[150] Generally, expression of ICAM-1 is low and readily upregulates in the presence of inflammatory cytokines such as IL-1, TNF-α and interferon (IFN)-γ,[151–153] and pathogens like RV.[154] Depending on cell type, ICAM-1 elicits different actions; its role in trafficking and adhesion of inflammatory cells during an inflammatory response has been well characterized. Endothelial ICAM-1 interacts with leukocyte-associated integrins such as macrophage 1 antigen (MAC-1) and lymphocyte function-associated antigen 1 (LFA-1) to initiate adhesion and emigration of leukocytes into inflamed tissues.[155,156] Similar to endothelium, airway epithelial cells also elicit high ICAM-1 expression and emigrate inflammatory cells into the lung lumen, especially in inflammatory allergic airway diseases.[157] However, the expression of ICAM-1 is low in the airway in normal conditions like noninflamed nose,[158] nasopharynx,[159] and airway epithelium.[150,160]

In 1989, ICAM-1 was identified as the major RV receptor collectively by several groups using monoclonal antibodies against ICAM-1[161,162] or cloning strategies.[163] RV undergoes replication in polarized epithelial cells in both the upper and lower airways.[164–166] Attempted infection with RV-A and -B on differentiated airway epithelial cells results in minimal numbers of ciliated cells becoming infected suggesting a higher resistance to RV infection compared with undifferentiated cells.[167] In differentiated human primary bronchial epithelial cells, ICAM-1 is exclusively expressed in the basal cells when compared with other epithelial cell types.[168,169] Jakiela et al.[168] demonstrated increased viral RNA and positive cells staining for VP2 in basal layer compared with suprabasal layers from a fully differentiated primary epithelial cells following partial trypsinization. They found 1 in 3 basal cells expressed ICAM-1, compared with 1 in 12 nonbasal cells. However, since minor group RVs also replicate in basal cells, it is likely that ICAM-1 expression is not the only factor in promoting viral replication.[168]

Epithelial ICAM-1 expression profiles differ depending on disease states. RV14 viral titers from human nasal epithelial cells from allergic subjects in submerged culture were significantly higher than the titers of the cells obtained from nonallergic subjects. The epithelial cells obtained from allergic subjects expressed higher levels of ICAM-1, suggesting that receptor expression may be a reason for susceptibility to infection with RV in asthmatic airway epithelial cells.[170] In contrast, Lopez-Souza et al. found that RV16 replicated more easily in bronchial cells grown at the ALI compared with nasal cells but found no difference between asthmatic and healthy subjects.[171] Furthermore, cells from healthy donors express higher levels of ICAM-1 mRNA compared with cells from asthmatic donors at baseline and in the absence of infection.[172] While monocytes are not generally permissive to RV,[173] the airway epithelial cell-line A549 was demonstrated to upregulate ICAM-1 in a monocytic cell line THP-1 during RV16 infection.[174] As such it appears that infection of the airway epithelium itself may influence the expression of ICAM-1 on nonepithelial cell types.

3.4.1.2 Low-density lipoprotein receptor (LDLR)

Minor group RVs use members of the LDL, VLDLR, and LRP-1 for binding and subsequent infection.[141–144] A number of structural modules are common to this family of membrane proteins. These include direct repeats (~40aa) located at the extracellular N terminus, which are similar to the complement component C9. In addition, are several epidermal growth factor precursor repeats, as well as a plasma membrane–associated highly *O*-glycosylated region. Finally, the transmembrane region is followed by a relatively short cytoplasmic tail.[175,176] The function of the LDLR is to uptake cholesterol-carrying lipoprotein particles into cells by clathrin-mediated endocytosis and the release of these particles upon delivery to the low pH milieu of the endosome; maintaining cholesterol homeostasis.[177] LDLR family members are endocytic recycling receptors found in cytoplasmic/recycling endosomes and to a lesser extent at the plasma membrane.[178,179] The VLDL receptor functions as a peripheral lipoprotein receptor in association with lipoprotein lipase in heart, muscle, adipose tissue, and macrophages.[180] LRP is a multiligand receptor binds and mediates the transport of ligands and some proteinase–proteinase inhibitor complexes to lysosomes for their degradation.[144] Immunofluorescence staining for LDLR and LRP1 of normal nasal tissue revealed that these receptors are present at the apical surface in both ciliated and basal cells.[17]

3.4.1.3 Cadherin-related family member 3 (CDHR3)

CDHR3 is a member of cadherin superfamily of transmembrane glycoproteins with yet unknown biological function. Members of this family such as classical cadherins are mainly responsible for calcium-dependent cell–cell communication.[181] As mentioned above, cadherins are also the principal components of adherens junctions and desmosomes.[182]

CDHR3 was identified as the first and currently only known RV-C entry factor.[147,183] Little is known about the biochemical function of CDHR3, although it is highly expressed in differentiated airway epithelium in vitro as well as in human lung tissues and bronchial epithelium in vivo.[184,185] This receptor was shown by Griggs et al. to be regulated during development and mainly expressed in ciliated cells in ALI cultures.[38] CDHR3 has been found to be genetically associated with early childhood severe and recurrent asthma exacerbation in early genome-wide association susceptibility (GWAS) studies.[184] Although CDHR3 is the only cellular factor currently known to enable RV-C entry and propagation in normally unsusceptible host cells, loss-of-function and direct receptor–virus interaction have not yet been confirmed.

3.4.1.4 Serotype variance

Productive entry pathways of RV-A and RV-B have been well characterized in established cell lines such as HeLa,[17,186] whereas the molecular mechanism of RV-C entry and receptor-mediated signaling remain unclear. In general, RV entry and uptake is dependent on the receptor type (major or minor group), and this can occur through clathrin-dependent[187] or -independent endocytosis or via macropinocytosis.[188] Upon binding to ICAM-1, uncoating of viral particles is mediated by ICAM-1 itself[189] and this process is facilitated by the low endosomal pH[190] and is temperature-dependent.[191,192]

By inhibiting clathrin-coated pit budding through overexpression of dynamin-2 mutant, Kirchhausen et al. proposed that the major RVs,[193] exemplified by RV-14 requires an active clathrin-mediated endocytic pathways for successful infection in HeLa-H1 cells. However, ICAM-1 lacks a clathrin localization signal and remains functional as a viral receptor when its cytoplasmic tail is reanchored with a glycosylphosphatidylinositol.[194] This suggested that other clathrin-independent entry pathways could also be involved for productive viral uptake. In rhabdomyosarcoma cells transfected human ICAM-1, entry of RV-14 was mediated in a clathrin-, caveolin-, and flottillin-independent manner.[188]

The minor group RVs, exemplified by RV-2, are internalized by clathrin-mediated endocytosis into early endosomes.[187] RV2 was shown to utilize clathrin- and dynamin-independent pathways when this pathway was blocked via cytosol acidification or overexpression of nonfunctional dynamin.[187,195] Low endosomal pH also plays an important role in major group,[190] and particularly minor group conformational modifications of virions during receptor-mediated endocytosis.[186,196] After uptake, the virus subsequently undergoes conformational changes that yield hydrophobic subviral particles, which do not have the capsid protein VP4 but still contain the RNA genome. The RNA genome enters the cytosol through a pore formed by viral proteins or following membrane rupture. Viral RNA is translated into a polyprotein that is cleaved by the viral proteinases $2A^{pro}$, $3C^{pro}$, and its precursor $3CD^{pro}$ into structural proteins VP1, VP2 and VP3, and the nonstructural proteins required for virus replication.[197] After assembly and packaging of the virions, they are released through cell lysis.

3.5 AIRWAY EPITHELIAL CELL RESPONSES TO RHINOVIRUS INFECTION

3.5.1 Innate immune response induction

To identify and limit viral replication, the epithelial cell employs PRRs such as TLRs, RLRs, C-type lectin receptors (CLRs), and NLRs. This interaction leads to the activation of antiviral host defense and inflammatory signaling pathways, resulting in the production of antiviral molecules [IFNs and β-defensins] as well as cytokines and chemokines. These initial host innate immune responses include increased level of proinflammatory cytokines [IL-6, IL-8, IFN-γ-induced protein 10 (IP-10) and chemokine (c-c motif) ligand 5 (RANTES)], release of antiviral IFN-β, IFN-λ, and apoptosis. However, in cases of asthma, CF, and COPD, the ability of airway epithelial cells to mount host responses is deficient (Table 3.1). One early study showing a RV-triggered increase in asthma exacerbations, demonstrated that airway epithelial cells from donors with asthma had greater virus replication, deficient caspase 3/7 activity, lower IFN-β production, and a pathological decrease in apoptosis.[198] Further studies on asthma cohort have also demonstrated deficient IFN-λ production, which was associated with severity of RV-induced asthma exacerbation and virus load.[199,200] Recently, a study using airway epithelial cells from children with mild asthma reported similar finding with increased proinflammatory cytokines including IL-6, IL-8, IP-10, RANTES, and IL-1β but lower IFN-β and IFN-λ production.[122]

Similarly, RV infection of airway epithelial cells from COPD donors stimulates increased expression of genes involved in oxidative stress[201] as well as greater proinflammatory and antiviral responses compared with healthy airway epithelial cells.[202] However, both cohorts have similar

Table 3.1 Airway epithelial cell responses to in vitro rhinovirus (RV) infection in different phenotypes.

Study	Phenotype	Age (range or mean ± SD; year)	RV serotype (family)	Apoptosis	Cytokines/mRNA expression	
					Proinflammatory	Antiviral
Wark et al.[198]	Asthma	21–58	RV-A1, RV-A16	↓	-IL-6, -RANTES	↓ IFN-β,
Edwards et al.[200]	Asthma	2–15	RV-A1 RV-A16	NM	-IL-8, ↓ TLR3, ↓ RIG-1, ↓ MDA-5	↓ IFN-β, IFN-λ1–3
Kicic et al.[122]	Asthma	2.6–14.8	RV-A1, RV-B14	↓	↑ IL-6, ↑IL-8, ↑IP-10, ↑RANTES, ↑IL-1b	↓ IFN-β, IFN-λ1–3
Schneider et al.[201]	COPD	45–65	RV-A39	NM	↑ IL-6, ↑IL-8, ↑CXCL1, ↑IP-10, ↑IFN-λ1, ↑IFN-λ2	NM
Baines et al.[202]	COPD	66 ± 26	RV-A1	NM	↑ IL-6, ↑TNF-α,	-CCL5 -CXCL10, -IFN-λ1,
Chattoraj et al.[203]	CF	16–33	RV-A39	NM	↑ IL-8	-IFN-β, -IFN-λ1–3
Sutanto et al.[127]	CF	1–7	RV-A1	↓	↑ IL-6, ↑IL-8,	↓ IFN-β,
Kieninger et al.[205]	CF	0.2–45.7	RV-A1, RV-A16	–	-IL-6, -IL-8, -IP-10, -MCP, -RANTES, -CXCL1, -TLR3, -RIG-1, -MDA-5	-IFN-β, -IFN-λ1–3
Schogler et al.[204]	CF	1–11	RV-A1, RV-A16	NM	-IL-6, -IL-8, -IP-10	↓ IFN-β, ↓IFN-λ1–3
Dauletbaev et al.[206]	CF	19–41	RV-A16	NM	-IL-8,	-IFN-β

Note: NM not measured in study, ↓ lower measured in target phenotype, ↑ elevated measured in target phenotype, - no change measured in the target phenotype compared with controls. *IL*, Interleukin; *COPD*, chronic obstructive pulmonary disease; *CF*, cystic fibrosis; *IP*, IFN-γ-induced protein; *TLR3*, Toll-like receptor 3; *IFN*, interferon; *MCP*, methyl-accepting chemotaxis protein.

levels of viral replication and airway epithelial cells from COPD donors were susceptible to apoptosis. Validated microarray analysis showed greater expression of calgranulins at baseline as well as increased pellino-1 (*PELI1*) and IL-1 receptor associated kinase 2 post−RV infection. These increases potentially contribute to the dysregulated immune responses in COPD. In a different disease setting, several studies interrogating the effects of RV infection on individuals with CF reported elevated release of inflammatory cytokines[204,205] especially at the high multiplicity of infection.[127] Other features such as dampened apoptosis-[204] and increased necrosis[205] were also found compared with healthy cells as well as decreased production of IFN-β.[204] Dauletbaev et al. have reported similar levels of IFN-β in airway epithelial cells from individuals with CF following RV16 infection.[206] These conflicting observations might be in part due to age of the subjects involved, the different RV strain used, dose, and length of infection.

Indeed, airway epithelial cells respond differently depending on major and minor group virus infection, and also depending on the strains of RV. For example, infection of healthy airway epithelial cells with minor group RV strains resulted in greater release of IFN β, IL-6, and IP-10 compared with major group RV.[207] A study by Nakagome et al. also reported that infection with RV group B was less severe than RV group A and C.[208] Clinical data corroborated this finding about pathogenicity of the different strains of RV. A study involving children with acute asthma and wheezing episode on presentation to the hospital found that RVC is associated with more severe asthma,[208] acute wheezing exacerbations and increased risk of subsequent wheezing, and hospital admission due to respiratory-related illnesses.[209] The same group has also reported higher association of RVC with acute respiratory illnesses in tertiary pediatric intensive care unit.[209]

To further understand the potential molecular mechanisms of RV infection, pathway analysis of airway epithelial cells following RV infection demonstrated that genes involved in antiviral signaling, RIG-I/MDA5 signaling, antigen processing and presentation, and apoptosis were significantly altered.[210] This was corroborated in a study by which demonstrated differential expression of unique RV-induced genes involved in immune responses (IL-1β, IL-1F9, IL-24, and IFI44) and airway remodeling (LOXL2, MMP10, FN1).[211] Further upstream regulator analysis identified IFN regulatory factor (IRF7) as a major molecular driver of antiviral response.[212] Gene silencing experiments targeting IRF7 in airway epithelial cells confirmed that knockdown of IRF7 reduced antiviral response genes whilst increased the expression of genes involved in inflammation and oxidative stress.[213] To determine whether this deficient innate immune response in epithelial cells from donors with asthma was due to the abnormal signaling, gene silencing experiments using short interfering RNAs against melanoma differentiation-associated protein 5 (siMDA5) and TLR3 (siTLR3) were conducted. This data revealed that healthy airway epithelial cells had upregulated expression of TLR3 and MDA5 upon RV infection and that silencing these factors inhibited type I and III IFN production.[214] On the other hand, despite upregulation of TLR3 and MDA5 airway epithelial cells from asthma failed to mount an appropriate IFN response. This suggests that the defective antiviral responses in asthmatic cells are due to lack of activation of types I and III IFN signaling, which could increase susceptibility to RV infection. Overall this finding validates the role of IFN in airway epithelial cell antiviral response and reveals a potentially viable therapeutic approach.

As suggested by Bochkov et al., asthmatic airway epithelial cells display unique patterns of gene expression involved in immune responses, wound repair, and remodeling that contribute to host innate immune responses.[211] A study of asthmatic airway epithelial cell responses following RV infection showed that infection caused delayed wound repair in healthy cells and further diminished capacity of repair in asthmatic cells, which could be due to reduced cell proliferation.[122] Indeed, studies using the BEAS-2B airway epithelial cell-line demonstrated that RV infection decreased cell proliferation and this contributed to the ability of wound repair.[215] Another study using BEAS-2B cells found that viral mimic and TLR agonist poly(I:C) induced a marked inhibition of wound repair.[216] Furthermore, injured cells treated with supernatant collected from cells infected with RV (RV1B) displayed decreased regeneration, and increased TGF-β mRNA expression.[216] The effect on RV on cellular integrity and function was also investigated in polarized 16HBE cells, which demonstrated that RV infection caused a substantial reduction in transepithelial resistance and increased cellular permeability due to the loss of ZO1 from the tight junction complexes.[217] A follow-up study identified RV disrupted epithelial barrier by inducing the production of reactive oxygen species,[218] which required the presence of NADPH-oxidase activity.[219] The role of tight junction and impaired barrier was further examined in primary airway epithelial cells and it was found that RV significantly decreased protein expression of tight junctions ZO1, E-cadherin, claudin

1, and occludin,[220] which also correlates with increased epithelial permeability to different sizes of dextran molecules.[221] These studies identified mechanisms by which RV disrupts epithelial barrier integrity and function. This finding is relevant especially on the effect of RV on cell repair in chronic airway disease like asthma and CF. The respiratory epithelium is important in maintaining tissue homeostasis and as such when it is compromised and damaged, a rapid repair response is essential to restore this homeostasis. As part of repair, cells must be able to proliferate and/or migrate to cover the wound and repolarize again to establish the barrier integrity. In asthma, however, the epithelium is abnormal compared with the healthy epithelium and it is characterized by higher proportion of basal cells,[119] which are more susceptible to RV infection.[168] Therefore, infection with RV may interfere with cell repolarization following cell injury. A study using an injury/regeneration model of airway epithelial cells confirmed that RV lengthens the barrier dysfunction and affect cell differentiation by inducing decreased expression of tight junction proteins (E-cadherin, occluding, ZO1, claudins 1 and 4, and Crumb3) whilst increasing a mesenchymal cell marker vimentin.[222] Overall, this combination of dysregulated innate immune response and the inability to repair damaged epithelium may contribute to continuous cycle of injury in the asthmatic epithelium following RV infection. This also confirms that it is essential for the airway epithelial cell to have an effective innate immune system in response to RV infection.

3.5.2 Treatment strategies to combat rhinovirus infection

To date, there are limited potential therapies for RV infection in respiratory diseases. As it stands currently there are no vaccines against RV due to the complexity of antigenic diversity across the three strains and the challenge to find a suitable model to produce the vaccines. Therefore, efforts have been made to restore antiviral activity and limit inflammation using other methods based on what has been known about RV infection. As type I and type III IFNs have been identified as a possible contributor for defective apoptosis and greater virus replication, supplement of exogenous IFN-β or IFN-λ to the cells had been assessed and shown to reverse these effects.[198,223] A study by Cakebread et al. demonstrated that in addition to reduced viral load, exogenous IFN-β also significantly reduced RV-induced IP-10 and RANTES in asthmatic airway epithelial cells.[224] However, a study testing both type I and III IFNs reported that addition of IFNs enhanced production of IP-10 and RANTES. The authors argue that this might be due to differences in experimental setup such as viral strains, phenotypes, and source of cells from explant versus brushing. In vivo administration of IFN-β has also been assessed recently and despite increased antiviral activity, it did not reduce viral induced exacerbations in asthma.[225] The use of exogenous IFNs, especially IFN-λ, warrants further research as type III IFN receptors are restricted to epithelial and dendritic cells,[226] potentially having less systemic side effects.

As the majority of RV binds to ICAM-1 receptor, therapies aimed at blocking virus entry through this receptor have the potential to prevent RV-induced exacerbations. An early study has looked at the use of inhaled recombinant soluble ICAM-1.[227] Despite promising results of reduced symptom severity, the high-cost dosing regimen is cost prohibitive to use this as a therapy. An investigation has been conducted on the administration of anti-ICAM-1 antibody in vivo into mice and the data demonstrated that this compound was able to prevent viral entry into the cells, and reduced proinflammatory cytokines and viral load.[228] In addition, the use of macrolides including azithromycin has also been assessed as potential therapy in both CF and asthma settings. Pretreatment of airway epithelial cells with azithromycin prior to RV infection was able to reduce viral load and replication as well as increased expression of IFN stimulated genes (ISGs) and PRRs.[201,229] Recently, a screening of novel macrolides was conducted to test their antiviral activity and the results were similar to previous data of diminished viral load and increased ISGs.[230] Interestingly administration of azithromycin did not have any effects on RV-induced proinflammatory response. Overall, further studies are required in each of these potential therapies to elucidate their clinical impacts.

3.6 CONCLUSION

In conclusion, we have identified that the airway epithelium is positioned and structured accordingly to directly interact with RV to contain its replication and minimalize resultant infection. This interaction far exceeds the traditional and passive barrier role that the airway epithelium has been labeled with and we have illustrated its dynamic nature. This may be in part due to the greater number of morphologically distinct cells types now evident in the airway architecture; each with their own distinct structural, biochemical, and functional properties. We have illustrated that in inflammatory airway diseases, there are inherent differences in these cells that prevent traditional innate responses including apoptosis that typically contain infection. The inability to mount an appropriate innate immune response then facilitates a cycle of viral replication, cell death, and inflammation. Concomitant to this a defective repair response resultantly triggers a vicious cycle of infection and inflammation that can cause airway exacerbation. Research has been begun to target these defective responses to mitigate the effects of RV infection. Although results have not been completely restorative, it is hoped that future investigations utilizing emerging cutting edge technology will significantly advance this field and the pipeline of new interventional therapies that specifically target the airway epithelium.

REFERENCES

1. Montoro DT, Haber AL, Biton M, et al. A revised airway epithelial hierarchy includes CFTR-expressing ionocytes. *Nature*. 2018;560:319–324.
2. Plasschaert LW, Zilionis R, Choo-Wing R, et al. A single-cell atlas of the airway epithelium reveals the CFTR-rich pulmonary ionocyte. *Nature*. 2018;560: 377–381.
3. Tam A, Wadsworth S, Dorscheid D, Man SFP, Sin DD. The airway epithelium: more than just a structural barrier. *Ther Adv Respir Dis*. 2011;5(4):255–273.
4. Jahnsen FL, Strickland DH, Thomas JA, et al. Accelerated antigen sampling and transport by airway mucosal dendritic cells following inhalation of a bacterial stimulus. *J Immunol*. 2006;177(9):5861–5867.
5. Condon TV, Sawyer RT, Fenton MJ, Riches DW. Lung dendritic cells at the innate-adaptive immune interface. *J Leukoc Biol*. 2011;90(5):883–895.
6. Pichavant M, Taront S, Jeannin P, et al. Impact of bronchial epithelium on dendritic cell migration and function: modulation by the bacterial motif KpOmpA. *J Immunol*. 2006;177(9):5912–5919.
7. Hansel TT, Johnston SL, Openshaw PJ. Microbes and mucosal immune responses in asthma. *Lancet*. 2013;381(9869):861–873.
8. Kawai T, Akira S. The role of pattern-recognition receptors in innate immunity: update on Toll-like receptors. *Nat Immunol*. 2010;11:373.
9. Rock JR, Randell SH, Hogan BLM. Airway basal stem cells: a perspective on their roles in epithelial homeostasis and remodeling. *Dis Model Mech*. 2010;3(9-10):545–556.
10. Watson Julie K, Rulands S, Wilkinson Adam C, et al. Clonal dynamics reveal two distinct populations of basal cells in slow-turnover airway epithelium. *Cell Rep*. 2015;12(1):90–101.
11. Evans MJ, Van Winkle LS, Fanucchi MV, Plopper CG. Cellular and molecular characteristics of basal cells in airway epithelium. *Exp Lung Res*. 2001;27(5):401–415.
12. Rock JR, Onaitis MW, Rawlins EL, et al. Basal cells as stem cells of the mouse trachea and human airway epithelium. *Proc Natl Acad Sci USA*. 2009;106(31):12771–12775.
13. Hackett TL, Shaheen F, Johnson A, et al. Characterization of side population cells from human airway epithelium. *Stem Cells*. 2008;26(10):2576–2585.

14. Hallstrand TS, Hackett TL, Altemeier WA, Matute-Bello G, Hansbro PM, Knight DA. Airway epithelial regulation of pulmonary immune homeostasis and inflammation. *Clin Immunol.* 2014;151(1):1−15.
15. Evans MJ, Cox RA, Shami SG, Wilson B, Plopper CG. The role of basal cells in attachment of columnar cells to the basal lamina of the trachea. *Am J Respir Cell Mol Biol.* 1989;1(6):463−469.
16. Bochkov YA, Gern JE. Rhinoviruses and their receptors: implications for allergic disease. *Curr Allergy Asthma Rep.* 2016;16(4):30.
17. Fuchs R, Blaas D. Productive entry pathways of human rhinoviruses. *Adv Virol.* 2012;:13. 2012.
18. Rawlins EL, Okubo T, Xue Y, et al. The role of Scgb1a1 + Clara cells in the long-term maintenance and repair of lung airway, but not alveolar, epithelium. *Cell Stem Cell.* 2009;4(6):525−534.
19. Ceteci F, Ceteci S, Zanucco E, et al. E-cadherin controls bronchiolar progenitor cells and onset of preneoplastic lesions in mice. *Neoplasia.* 2012;14(12):1164−1177.
20. Rock Jason R, Gao X, Xue Y, Randell Scott H, Kong Y-Y, Hogan Brigid LM. Notch-dependent differentiation of adult airway basal stem cells. *Cell Stem Cell.* 2011;8(6):639−648.
21. Zuo WL, Shenoy SA, Li S, et al. Ontogeny and biology of human small airway epithelial club cells. *Am J Respir Crit Care Med.* 2018;198:1375−1388.
22. Button B, Cai LH, Ehre C, et al. A periciliary brush promotes the lung health by separating the mucus layer from airway epithelia. *Science.* 2012;337(6097):937−941.
23. Laucho-Contreras ME, Polverino F, Rojas-Quintero J, Wang X, Owen CA. Club cell protein 16 (Cc16) deficiency increases inflamm-aging in the lungs of mice. *Physiol Rep.* 2018;6(15):e13797.
24. Laucho-Contreras ME, Polverino F, Tesfaigzi Y, Pilon A, Celli BR, Owen CA. Club cell protein 16 (CC16) augmentation: a potential disease-modifying approach for chronic obstructive pulmonary disease (COPD). *Expert Opin Ther Targets.* 2016;20 (7):869−883.
25. Gamez AS, Gras D, Petit A, et al. Supplementing defect in club cell secretory protein attenuates airway inflammation in COPD. *Chest.* 2015;147(6):1467−1476.
26. Shijubo N, Itoh Y, Yamaguchi T, et al. Clara cell protein-positive epithelial cells are reduced in small airways of asthmatics. *Am J Respir Crit Care Med.* 1999;160 (3):930−933.
27. Denard B, Seemann J, Chen Q, et al. The membrane-bound transcription factor CREB3L1 is activated in response to virus infection to inhibit proliferation of virus-infected cells. *Cell Host Microbe.* 2011;10(1):65−74.
28. Lafkas D, Shelton A, Chiu C, et al. Therapeutic antibodies reveal Notch control of transdifferentiation in the adult lung. *Nature.* 2015;528(7580):127−131.
29. Knight DA, Holgate ST. The airway epithelium: structural and functional properties in health and disease. *Respirology.* 2003;8(4):432−446.
30. Raman T, O'Connor TP, Hackett NR, et al. Quality control in microarray assessment of gene expression in human airway epithelium. *BMC Genomics.* 2009;10:493.
31. Ganesan S, Comstock AT, Sajjan US. Barrier function of airway tract epithelium. *Tissue Barriers.* 2013;1(4):e24997.
32. Jeong K-I, Piepenhagen PA, Kishko M, et al. CX3CR1 is expressed in differentiated human ciliated airway cells and co-localizes with respiratory syncytial virus on cilia in a G protein-dependent manner. *PLoS One.* 2015;10(6):e0130517.
33. Shah AS, Ben-Shahar Y, Moninger TO, Kline JN, Welsh MJ. Motile cilia of human airway epithelia are chemosensory. *Science (New York, NY).* 2009;325(5944):1131−1134.
34. Yaghi A, Dolovich M. Airway epithelial cell cilia and obstructive lung disease. *Cells.* 2016;5(4):40.

35. Knowles MR, Boucher RC. Mucus clearance as a primary innate defense mechanism for mammalian airways. *J Clin Invest*. 2002;109(5):571–577.
36. Thomas B, Rutman A, Hirst RA, et al. Ciliary dysfunction and ultrastructural abnormalities are features of severe asthma. *J Allergy Clin Immunol*. 2010;126(4):722–729. e722.
37. Cloonan SM, Lam HC, Ryter SW, Choi AM. "Ciliophagy": the consumption of cilia components by autophagy. *Autophagy*. 2014;10(3):532–534.
38. Griggs TF, Bochkov YA, Basnet S, et al. Rhinovirus C targets ciliated airway epithelial cells. *Respir Res*. 2017;18(1):84.
39. Ross M, Pawlina W. *Respiratory system: micrographs. Histology*. 5th ed. Baltimore, MD: Lippincott Williams & Wilkins; 2006:620.
40. Fahy JV, Dickey BF. Airway mucus function and dysfunction. *N Engl J Med*. 2010;363(23):2233–2247.
41. Reid AT, Veerati PC, Gosens R, et al. Persistent induction of goblet cell differentiation in the airways: therapeutic approaches. *Pharmacol Ther*. 2018;185:155–169.
42. Park KS, Korfhagen TR, Bruno MD, et al. SPDEF regulates goblet cell hyperplasia in the airway epithelium. *J Clin Invest*. 2007;117(4):978–988.
43. Kuperman D, Schofield B, Wills-Karp M, Grusby MJ. Signal transducer and activator of transcription factor 6 (Stat6)-deficient mice are protected from antigen-induced airway hyperresponsiveness and mucus production. *J Exp Med*. 1998;187(6):939–948.
44. Song J, Cano Rodriguez D, Winkle M, et al. Targeted epigenetic editing of SPDEF reduces mucus production in lung epithelial cells. *Am J Physiol Lung Cell Mol Physiol*. 2017;312(3):L334–L347.
45. Hays SR, Fahy JV. Characterizing mucous cell remodeling in cystic fibrosis: relationship to neutrophils. *Am J Respir Crit Care Med*. 2006;174(9):1018–1024.
46. Korfhagen TR, Kitzmiller J, Chen G, et al. SAM-pointed domain ETS factor mediates epithelial cell-intrinsic innate immune signaling during airway mucous metaplasia. *Proc Natl Acad Sci USA*. 2012;109(41):16630–16635.
47. Jing Y, Gimenes JA, Mishra R, et al. NOTCH3 contributes to rhinovirus-induced goblet cell hyperplasia in COPD airway epithelial cells. *Thorax*. 2018;.
48. Van Lommel A. Pulmonary neuroendocrine cells (PNEC) and neuroepithelial bodies (NEB): chemoreceptors and regulators of lung development. *Paediatr Respir Rev*. 2001;2(2):171–176.
49. Linnoila RI. Functional facets of the pulmonary neuroendocrine system. *Lab Invest*. 2006;86(5):425–444.
50. Domnik NJ, Cutz E. Pulmonary neuroepithelial bodies as airway sensors: putative role in the generation of dyspnea. *Curr Opin Pharmacol*. 2011;11(3):211–217.
51. Sui P, Wiesner DL, Xu J, et al. Pulmonary neuroendocrine cells amplify allergic asthma responses. *Science*. 2018;360:6393.
52. Reynolds SD, Giangreco A, Power JH, Stripp BR. Neuroepithelial bodies of pulmonary airways serve as a reservoir of progenitor cells capable of epithelial regeneration. *Am J Pathol*. 2000;156(1):269–278.
53. Reynolds SD, Hong KU, Giangreco A, et al. Conditional clara cell ablation reveals a self-renewing progenitor function of pulmonary neuroendocrine cells. *Am J Physiol Lung Cell Mol Physiol*. 2000;278(6):L1256–L1263.
54. Song H, Yao E, Lin C, Gacayan R, Chen MH, Chuang PT. Functional characterization of pulmonary neuroendocrine cells in lung development, injury, and tumorigenesis. *Proc Natl Acad Sci USA*. 2012;109(43):17531–17536.
55. Gerbe F, Legraverend C, Jay P. The intestinal epithelium tuft cells: specification and function. *Cell Mol Life Sci*. 2012;69(17):2907–2917.
56. Lee RJ, Kofonow JM, Rosen PL, et al. Bitter and sweet taste receptors regulate human upper respiratory innate immunity. *J Clin Invest*. 2014;124(3):1393–1405.

57. Tizzano M, Cristofoletti M, Sbarbati A, Finger TE. Expression of taste receptors in solitary chemosensory cells of rodent airways. *BMC Pulm Med.* 2011;11:3.
58. Gerbe F, Sidot E, Smyth DJ, et al. Intestinal epithelial tuft cells initiate type 2 mucosal immunity to helminth parasites. *Nature.* 2016;529(7585):226–230.
59. von Moltke J, Ji M, Liang HE, Locksley RM. Tuft-cell-derived IL-25 regulates an intestinal ILC2-epithelial response circuit. *Nature.* 2016;529(7585):221–225.
60. Tamachi T, Maezawa Y, Ikeda K, et al. IL-25 enhances allergic airway inflammation by amplifying a TH2 cell-dependent pathway in mice. *J Allergy Clin Immunol.* 2006;118(3):606–614.
61. Beale J, Jayaraman A, Jackson DJ, et al. Rhinovirus-induced IL-25 in asthma exacerbation drives type 2 immunity and allergic pulmonary inflammation. *Sci Transl Med.* 2014;6(256). 256ra134.
62. Merker H-J. Morphology of the basement membrane. *Microsc Res Tech.* 1994;28(2):95–124.
63. Widdicombe JH, Wine JJ. Airway gland structure and function. *Physiol Rev.* 2015;95(4):1241–1319.
64. Lee RJ, Foskett JK. cAMP-activated Ca2 + signaling is required for CFTR-mediated serous cell fluid secretion in porcine and human airways. *J Clin Invest.* 2010;120(9):3137–3148.
65. Bansil R, Turner BS. The biology of mucus: composition, synthesis and organization. *Adv Drug Deliv Rev.* 2018;124:3–15.
66. Schuster BS, Suk JS, Woodworth GF, Hanes J. Nanoparticle diffusion in respiratory mucus from humans without lung disease. *Biomaterials.* 2013;34(13):3439–3446.
67. Panjwani A, Strauss M, Gold S, et al. Capsid protein VP4 of human rhinovirus induces membrane permeability by the formation of a size-selective multimeric pore. *PLoS Pathog.* 2014;10(8):e1004294.
68. Trillo-Muyo S, Nilsson HE, Recktenwald CV, et al. Granule-stored MUC5B mucins are packed by the noncovalent formation of N-terminal head-to-head tetramers. *J Biol Chem.* 2018;293(15):5746–5754.
69. Ridley C, Thornton DJ. Mucins: the frontline defence of the lung. *Biochem Soc Trans.* 2018;.
70. Ermund A, Meiss LN, Rodriguez-Pineiro AM, et al. The normal trachea is cleaned by MUC5B mucin bundles from the submucosal glands coated with the MUC5AC mucin. *Biochem Biophys Res Commun.* 2017;492(3):331–337.
71. Roy MG, Livraghi-Butrico A, Fletcher AA, et al. Muc5b is required for airway defence. *Nature.* 2014;505(7483):412–416.
72. Welsh KG, Rousseau K, Fisher G, et al. MUC5AC and a glycosylated variant of MUC5B alter mucin composition in children with acute asthma. *Chest.* 2017;152(4):771–779.
73. Lachowicz-Scroggins ME, Yuan S, Kerr SC, et al. Abnormalities in MUC5AC and MUC5B protein in airway mucus in asthma. *Am J Respir Crit Care Med.* 2016;194(10):1296–1299.
74. Fernández-Blanco JA, Fakih D, Arike L, et al. Attached stratified mucus separates bacteria from the epithelial cells in COPD lungs. *JCI Insight.* 2018;3(17):120994.
75. Bonser LR, Erle DJ. Airway mucus and asthma: the role of MUC5AC and MUC5B. *J Clin Med.* 2017;6:12.
76. Farquhar MG, Palade GE. Junctional complexes in various epithelia. *J Cell Biol.* 1963;17:375–412.
77. Martinez-Palomo A, Erlij D, Bracho H. Localization of permeability barriers in the frog skin epithelium. *J Cell Biol.* 1971;50(2):277–287.
78. Whittembury G, Rawlins FA. Evidence of a paracellular pathway for ion flow in the kidney proximal tubule: electromicroscopic demonstration of lanthanum precipitate in the tight junction. *Pflügers Archiv.* 1971;330(4):302–309.

79. Anderson JM, Van Itallie CM. Physiology and function of the tight junction. *Cold Spring Harb Perspect Biol.* 2009;1(2):a002584.
80. Itoh M, Nagafuchi A, Moroi S, Tsukita S. Involvement of ZO-1 in cadherin-based cell adhesion through its direct binding to α catenin and actin filaments. *J Cell Biol.* 1997;138(1):181–192.
81. Matter K, Balda MS. Signalling to and from tight junctions. *Nat Rev Mol Cell Biol.* 2003;:4.
82. Nelson WJ. Adaptation of core mechanisms to generate cell polarity. *Nature.* 2003;422(6933):766–774.
83. Balda MS, Flores-Maldonado C, Cereijido M, Matter K. Multiple domains of occludin are involved in the regulation of paracellular permeability. *J Cell Biochem.* 2000;78 (1):85–96.
84. Furuse M, Fujita K, Hiiragi T, Fujimoto K, Tsukita S. Claudin-1 and -2: novel integral membrane proteins localizing at tight junctions with no sequence similarity to occludin. *J Cell Biol.* 1998;141:1539–1550.
85. Tsukita S, Furuse M. Claudin-based barrier in simple and stratified cellular sheets. *Curr Opin Cell Biol.* 2002;14(5):531–536.
86. Coyne CB, Gambling TM, Boucher RC, Carson JL, Johnson LG. Role of claudin interactions in airway tight junctional permeability. *Am J Physiol Lung Cell Mol Physiol.* 2003;285(5):L1166–L1178.
87. Findley MK. Regulation and roles for claudin-family tight junction proteins. *Int Union Biochem Mol Biol Life.* 2009;61(4):431.
88. Liu S, Yang W, Shen L, Turner JR, Coyne CB, Wang T. Tight junction proteins claudin-1 and occludin control hepatitis C virus entry and are downregulated during Infection to prevent superinfection. *J Virol.* 2009;83(4):2011–2014.
89. Guillemot L. The cytoplasmic plaque of tight junctions: a scaffolding and signalling center. *Biochim Biophys Acta.* 2008;1778(3):601–613.
90. Fanning AS, Anderson JM. Zonula occludens-1 and -2 are cytosolic scaffolds that regulate the assembly of cellular junctions. *Ann N Y Acad Sci.* 2009;1165(1):113–120.
91. Liu Y, Nusrat A, Schnell FJ, et al. Human junction adhesion molecule regulates tight junction resealing in epithelia. *J Cell Sci.* 2000;113(13):2363–2374.
92. Mese G, Richard G, White TW. Gap junctions: basic structure and function. *J Invest Dermatol.* 2007;127(11):2516–2524.
93. Balda MS, Matter K. Tight junctions and the regulation of gene expression. *Biochim Biophys Acta, Biomembr.* 2009;1788(4):761–767.
94. Lackie PM, Baker JE, Günthert U, Holgate ST. Expression of CD44 isoforms is increased in the airway epithelium of asthmatic subjects. *Am J Respir Cell Mol Biol.* 1997;16(1):14–22.
95. Xiao C, Puddicombe SM, Field S, et al. Defective epithelial barrier function in asthma. *J Allergy Clin Immunol.* 2011;128(3):549–556. e512.
96. Looi K, Buckley A, Rigby P, et al. Effects of human rhinovirus on epithelial barrier integrity and function in children with asthma. *Clin Exp Allergy.* 2018;48 (5):513–524.
97. Shahana S, Björnsson E, Lúdvíksdóttir D, et al. Ultrastructure of bronchial biopsies from patients with allergic and non-allergic asthma. *Respir Med.* 2005;99 (4):429–443.
98. de Boer WI, Sharma HS, Baelemans SM, Hoogsteden HC, Lambrecht BN, Braunstahl GJ. Altered expression of epithelial junctional proteins in atopic asthma: possible role in inflammation. *Can J Physiol Pharmacol.* 2008;86(3):105–112.
99. Lötvall J, Akdis CA, Bacharier LB, et al. Asthma endotypes: a new approach to classification of disease entities within the asthma syndrome. *J Allergy Clin Immunol.* 2011;127(2):355–360.

100. Modena BD, Bleecker ER, Busse WW, et al. Gene expression correlated with severe asthma characteristics reveals heterogeneous mechanisms of severe disease. *Am J Respir Crit Care Med.* 2017;195(11):1449–1463.
101. Modena BD, Tedrow JR, Milosevic J, et al. Gene expression in relation to exhaled nitric oxide identifies novel asthma phenotypes with unique biomolecular pathways. *Am J Respir Crit Care Med.* 2014;190(12):1363–1372.
102. Meban C. Thickness of the air-blood barriers in vertebrate lungs. *J Anat.* 1980;131 (Pt 2):299–307.
103. Musah S, Chen J, Hoyle GW. Repair of tracheal epithelium by basal cells after chlorine-induced injury. *Respir Res.* 2012;13:107.
104. Erjefält JS, Erjefält I, Sundler F, Persson CG. In vivo restitution of airway epithelium. *Cell Tissue Res.* 1995;281(2):305–316.
105. D'Amato G, Liccardi G, D'Amato M, Holgate S. Environmental risk factors and allergic bronchial asthma. *Clin Exp Allergy.* 2005;35(9):1113–1124.
106. Holgate ST. Pathogenesis of asthma. *Clin Exp Allergy.* 2008;38(6):872–897.
107. Holgate ST, Arshad HS, Roberts GC, Howarth PH, Thurner P, Davies DE. A new look at the pathogenesis of asthma. *Clin Sci (Lond).* 2009;118(7):439–450.
108. Holgate ST. A brief history of asthma and its mechanisms to modern concepts of disease pathogenesis. *Allergy Asthma Immunol Res.* 2010;2(3):165–171.
109. Holgate ST. A look at the pathogenesis of asthma: the need for a change in direction. *Discov Med.* 2010;9(48):439–447.
110. Holgate ST. Has the time come to rethink the pathogenesis of asthma? *Curr Opin Allergy Clin Immunol.* 2010;10(1):48–53.
111. Gaillard D, Jouet JB, Egreteau L, et al. Airway epithelial damage and inflammation in children with recurrent bronchitis. *Am J Respir Crit Care Med.* 1994;150 (3):810–817.
112. Amatngalim GD, Broekman W, Daniel NM, De Mooij-Eijk Y, Hiemstra PS. Cigarette smoke amplifies airway epithelial cell innate immune responses during wound repair. *Am J Respir Crit Care Med.* 2014;189:A4897.
113. Coraux C, Hajj R, Lesimple P, Puchelle E. [Repair and regeneration of the airway epithelium]. *Med Sci (Paris).* 2005;21(12):1063–1069.
114. Vareille M, Kieninger E, Edwards MR, Regamey N. The airway epithelium: soldier in the fight against respiratory viruses. *Clin Microbiol Rev.* 2011;24(1):210–229.
115. Bianchi ME. DAMPs, PAMPs and alarmins: all we need to know about danger. *J Leukoc Biol.* 2007;81(1):1–5.
116. Warburton D, Tefft D, Mailleux A, et al. Do lung remodeling, repair, and regeneration recapitulate respiratory ontogeny? *Am J Respir Crit Care Med.* 2001;164(10 Pt 2):S59–S62.
117. Kim SH, Matthay MA, Mostov K, Hunt CA. Simulation of lung alveolar epithelial wound healing in vitro. *J R Soc Interface.* 2010;7(49):1157–1170.
118. Trinh NTN, Privé A, Maillé E, Noël J, Brochiero E. EGF and K + channel activity control normal and cystic fibrosis bronchial epithelia repair. *Am J Physiol Lung Cell Mol Physiol.* 2008;295(5):L866–L880.
119. Kicic A, Sutanto EN, Stevens PT, Knight DA, Stick SM. Intrinsic biochemical and functional differences in bronchial epithelial cells of children with asthma. *Am J Respir Crit Care Med.* 2006;174(10):1110–1118.
120. Stevens PT, Kicic A, Sutanto EN, Knight DA, Stick SM. Dysregulated repair in asthmatic paediatric airway epithelial cells: the role of plasminogen activator inhibitor-1. *Clin Exp Allergy.* 2008;38(12):1901–1910.
121. Kicic A, Hallstrand TS, Sutanto EN, et al. Decreased fibronectin production significantly contributes to dysregulated repair of asthmatic epithelium. *Am J Respir Crit Care Med.* 2010;181(9):889–898.

122. Kicic A, Stevens PT, Sutanto EN, et al. Impaired airway epithelial cell responses from children with asthma to rhinoviral infection. *Clin Exp Allergy*. 2016;46(11):1441–1455.
123. Schiller KR, Maniak PJ, O'Grady SM. Cystic fibrosis transmembrane conductance regulator is involved in airway epithelial wound repair. *Am J Physiol Cell Physiol*. 2010;299(5):C912–C921.
124. Trinh NTN, Bardou O, Privé A, et al. Improvement of defective cystic fibrosis airway epithelial wound repair after CFTR rescue. *Eur Respir J*. 2012;40(6):1390–1400.
125. Garratt LW, Sutanto EN, Ling KM, et al. Alpha-1 antitrypsin mitigates the inhibition of airway epithelial cell repair by neutrophil elastase. *Am J Respir Cell Mol Biol*. 2016;54(3):341–349.
126. Garratt LW, Sutanto EN, Ling K, et al. Defining protease inhibition of airway epithelial cell function for assessment of anti-protease therapy. *Pediatr Pulmonol*. 2015;50:239.
127. Sutanto EN, Kicic A, Foo CJ, et al. Innate inflammatory responses of pediatric cystic fibrosis airway epithelial cells: effects of nonviral and viral stimulation. *Am J Respir Cell Mol Biol*. 2011;44(6):761–767.
128. Stevens PT, Kicic A, Sutanto EN, Knight DA, Stick SM. Dysregulated repair in asthmatic paediatric airway epithelial cells: the role of plasminogen activator inhibitor-1. *Clin Exp Allergy*. 2008;38(12):1901–1910.
129. Haj-Salem I, Fakhfakh R, Berube JC, et al. MicroRNA-19a enhances proliferation of bronchial epithelial cells by targeting TGFbetaR2 gene in severe asthma. *Allergy*. 2015;70(2):212–219.
130. Cohen L, Xueping E, Tarsi J, et al. Epithelial cell proliferation contributes to airway remodeling in severe asthma. *Am J Respir Crit Care Med*. 2007;176(2):138–145.
131. Hastie AT, Kraft WK, Nyce KB, et al. Asthmatic epithelial cell proliferation and stimulation of collagen production: human asthmatic epithelial cells stimulate collagen type III production by human lung myofibroblasts after segmental allergen challenge. *Am J Respir Crit Care Med*. 2002;165(2):266–272.
132. Freishtat RJ, Watson AM, Benton AS, et al. Asthmatic bronchial epithelium is intrinsically inflammogenic, mitotically dyssynchronous, and is rescued by glucocorticoids. *J Invest Med*. 2010;58(3):590.
133. Saglani S, Molyneux C, Gong H, et al. Ultrastructure of the reticular basement membrane in asthmatic adults, children and infants. *Eur Respir J*. 2006;28(3):505–512.
134. Roche WR, Beasley R, Williams JH, Holgate ST. Subepithelial fibrosis in the bronchi of asthmatics. *Lancet*. 1989;1(8637):520–524.
135. Karjalainen EM, Lindqvist A, Laitinen LA, et al. Airway inflammation and basement membrane tenascin in newly diagnosed atopic and nonatopic asthma. *Respir Med*. 2003;97(9):1045–1051.
136. Pohunek P, Warner JO, Turzikova J, Kudrmann J, Roche WR. Markers of eosinophilic inflammation and tissue re-modelling in children before clinically diagnosed bronchial asthma. *Pediatr Allergy Immunol*. 2005;16(1):43–51.
137. Campbell AM. Bronchial epithelial cells in asthma. *Allergy*. 1997;52(5):483–489.
138. Holgate ST. Epithelium dysfunction in asthma. *J Allergy Clin Immunol*. 2007;120(6):1233–1246.
139. Kumar RK, Siegle JS, Kaiko GE, Herbert C, Mattes JE, Foster PS. Responses of airway epithelium to environmental injury: role in the induction phase of childhood asthma. *J Allergy (Cairo)*. 2011;:257017. 2011.
140. Ledford RM, Patel NR, Demenczuk TM, et al. VP1 sequencing of all human rhinovirus serotypes: insights into genus phylogeny and susceptibility to antiviral capsid-binding compounds. *J Virol*. 2004;78(7):3663–3674.

141. Olson NH, Kolatkar PR, Oliveira MA, et al. Structure of a human rhinovirus complexed with its receptor molecule. *Proc Natl Acad Sci USA*. 1993;90(2):507–511.
142. Hofer F, Gruenberger M, Kowalski H, et al. Members of the low density lipoprotein receptor family mediate cell entry of a minor-group common cold virus. *Proc Natl Acad Sci USA*. 1994;91(5):1839–1842.
143. Vlasak M, Roivainen M, Reithmayer M, et al. The minor receptor group of human rhinovirus (HRV) Includes HRV23 and HRV25, but the presence of a lysine in the VP1 HI loop is not sufficient for receptor binding. *J Virol*. 2005;79(12):7389–7395.
144. Marlovits TC, Abrahamsberg C, Blaas D. Very-low-density lipoprotein receptor fragment shed from HeLa Cells inhibits human rhinovirus infection. *J Virol*. 1998;72 (12):10246–10250.
145. Khan AG, Pickl-Herk A, Gajdzik L, Marlovits TC, Fuchs R, Blaas D. Entry of a heparan sulphate-binding HRV8 variant strictly depends on dynamin but not on clathrin, caveolin, and flotillin. *Virology*. 2011;412(1):55–67.
146. Khan AG, Pichler J, Rosemann A, Blaas D. Human rhinovirus type 54 infection via heparan sulfate is less efficient and strictly dependent on low endosomal pH. *J Virol*. 2007;81(9):4625–4632.
147. Bochkov YA, Watters K, Ashraf S, et al. Cadherin-related family member 3, a childhood asthma susceptibility gene product, mediates rhinovirus C binding and replication. *Proc Natl Acad Sci USA*. 2015;112(17):5485–5490.
148. Dougherty GJ, Murdoch S, Hogg N. The function of human intercellular adhesion molecule-1 (ICAM-1) in the generation of an immune response. *Eur J Immunol*. 1988;18(1):35–39.
149. Bentley AM, Durham SR, Robinson DS, et al. Expression of endothelial and leukocyte adhesion molecules intercellular adhesion molecule-1, E-selectin, and vascular cell adhesion molecule-1 in the bronchial mucosa in steady-state and allergen-induced asthma. *J Allergy Clin Immunol*. 1993;92(6):857–868.
150. Vignola AM, Campbell AM, Chanez P, et al. HLA-DR and ICAM-1 expression on bronchial epithelial cells in asthma and chronic bronchitis. *Am Rev Respir Dis*. 1993;148(3):689–694.
151. Dustin ML, Rothlein R, Bhan AK, Dinarello CA, Springer TA. Induction by IL 1 and interferon-gamma: tissue distribution, biochemistry, and function of a natural adherence molecule (ICAM-1). *J Immunol*. 1986;137(1):245–254.
152. Colletti LM, Cortis A, Lukacs N, Kunkel SL, Green M, Strieter RM. Tumor necrosis factor up-regulates intercellular adhesion molecule 1, which is important in the neutrophil-dependent lung and liver injury associated with hepatic ischemia and reperfusion in the rat. *Shock*. 1998;10(3):182–191.
153. Tosi MF, Stark JM, Smith CW, Hamedani A, Gruenert DC, Infeld MD. Induction of ICAM-1 expression on human airway epithelial cells by inflammatory cytokines: effects on neutrophil-epithelial cell adhesion. *Am J Respir Cell Mol Biol*. 1992;7(2):214–221.
154. Papi A, Johnston SL. Rhinovirus infection induces expression of its own receptor intercellular adhesion molecule 1 (ICAM-1) via increased NF-kappaB-mediated transcription. *J Biol Chem*. 1999;274(14):9707–9720.
155. Diamond MS, Staunton DE, de Fougerolles AR, et al. ICAM-1 (CD54): a counter-receptor for Mac-1 (CD11b/CD18). *J Cell Biol*. 1990;111(6):3129.
156. Smith CW, Marlin SD, Rothlein R, Toman C, Anderson DC. Cooperative interactions of LFA-1 and Mac-1 with intercellular adhesion molecule-1 in facilitating adherence and transendothelial migration of human neutrophils in vitro. *J Clin Invest*. 1989;83(6):2008–2017.
157. Canonica GW, Ciprandi G, Pesce GP, Buscaglia S, Paolieri F, Bagnasco M. ICAM-1 on epithelial cells in allergic subjects: a hallmark of allergic inflammation. *Int Arch Allergy Immunol*. 1995;107(1–3):99–102.

158. Ciprandi G, Pronzato C, Ricca V, Passalacqua G, Bagnasco M, Canonica GW. Allergen-specific challenge induces intercellular adhesion molecule 1 (ICAM-1 or CD54) on nasal epithelial cells in allergic subjects. Relationships with early and late inflammatory phenomena. *Am J Respir Crit Care Med.* 1994;150(6 Pt 1):1653–1659.
159. Winther B, Greve JM, Gwaltney Jr JM, et al. Surface expression of intercellular adhesion molecule 1 on epithelial cells in the human adenoid. *J Infect Dis.* 1997;176 (2):523–525.
160. Gosset P, Tillie-Leblond I, Janin A, et al. Expression of E-selectin, ICAM-1 and VCAM-1 on bronchial biopsies from allergic and non-allergic asthmatic patients. *Int Arch Allergy Immunol.* 1995;106(1):69–77.
161. Staunton DE, Merluzzi VJ, Rothlein R, Barton R, Marlin SD, Springer TA. A cell adhesion molecule, ICAM-1, is the major surface receptor for rhinoviruses. *Cell.* 1989;56(5):849–853.
162. Greve JM, Davis G, Meyer AM, et al. The major human rhinovirus receptor is ICAM-1. *Cell.* 1989;56(5):839–847.
163. Tomassini JE, Graham D, DeWitt CM, Lineberger DW, Rodkey JA, Colonno RJ. cDNA cloning reveals that the major group rhinovirus receptor on HeLa cells is intercellular adhesion molecule 1. *Proc Natl Acad Sci USA.* 1989;86 (13):4907–4911.
164. Mosser AG, Brockman-Schneider R, Amineva S, et al. Similar frequency of rhinovirus-infectible cells in upper and lower airway epithelium. *J Infect Dis.* 2002;185(6):734–743.
165. Winther B. Rhinovirus infections in the upper airway. *Proc Am Thorac Soc.* 2011;8:79–89.
166. Papadopoulos NG, Bates PJ, Bardin PG, et al. Rhinoviruses infect the lower airways. *J Infect Dis.* 2000;181(6):1875–1884.
167. Lopez-Souza N, Dolganov G, Dubin R, et al. Resistance of differentiated human airway epithelium to infection by rhinovirus. *Am J Physiol—Lung Cell Mol Physiol.* 2004;286(2):L373–L381.
168. Jakiela B, Brockman-Schneider R, Amineva S, Lee WM, Gern JE. Basal cells of differentiated bronchial epithelium are more susceptible to rhinovirus infection. *Am J Respir Cell Mol Biol.* 2008;38(5):517–523.
169. Zhu J, Rogers AV, Burke-Gaffney A, Hellewell PG, Jeffery PK. Cytokine-induced airway epithelial ICAM-1 upregulation: quantification by high-resolution scanning and transmission electron microscopy. *Eur Respir J.* 1999;13(6):1318–1328.
170. Yamaya M, Nomura K, Arakawa K, et al. Increased rhinovirus replication in nasal mucosa cells in allergic subjects is associated with increased ICAM-1 levels and endosomal acidification and is inhibited by L-carbocisteine. *Immun Inflamm Dis.* 2016;4(2):166–181.
171. Lopez-Souza N, Favoreto S, Wong H, et al. Greater in vitro susceptibilty to rhinovirus infection of bronchial than nasal airway epithelial cells from human subjects. *J Allergy Clin Immunol.* 2009;123(6):1384–1390.e2.
172. Lopez-Souza N, Favoreto S, Wong H, et al. In vitro susceptibility to rhinovirus infection is greater for bronchial than for nasal airway epithelial cells in human subjects. *J Allergy Clin Immunol.* 2009;123(6):1384–1390.e2.
173. Gern JE, Dick EC, Lee WM, et al. Rhinovirus enters but does not replicate inside monocytes and airway macrophages. *J Immunol.* 1996;156(2):621–627.
174. Zhou X, Zhu L, Lizarraga R, Chen Y. Human airway epithelial cells direct significant rhinovirus replication in monocytic cells by enhancing ICAM1 expression. *Am J Respir Cell Mol Biol.* 2017;57(2):216–225.
175. Li Y, Cam J, Bu G. Low-density lipoprotein receptor family: endocytosis and signal transduction. *Mol Neurobiol.* 2001;23(1):53–67.

176. Jeon H, Blacklow SC. Structure and physiologic function of the low-density lipoprotein receptor. *Annu Rev Biochem.* 2005;74:535–562.
177. Beglova N, Blacklow SC. The LDL receptor: how acid pulls the trigger. *Trends Biochem Sci.* 2005;30(6):309–317.
178. Kang RS, Fölsch H. ARH cooperates with AP-1B in the exocytosis of LDLR in polarized epithelial cells. *J Cell Biol.* 2011;193(1):51–60.
179. Pietiäinen V, Vassilev B, Blom T, et al. NDRG1 functions in LDL receptor trafficking by regulating endosomal recycling and degradation. *J Cell Sci.* 2013;126(17):3961–3971.
180. Takahashi S. Triglyceride rich lipoprotein-LPL-VLDL receptor and Lp(*a*)-VLDL receptor pathways for macrophage foam cell formation. *J Atheroscler Thromb.* 2017;24(6):552–559.
181. Nelson WJ, Dickinson DJ, Weis WI. Roles of cadherins and catenins in cell-cell adhesion and epithelial cell polarity. *Prog Mol Biol Transl Sci.* 2013;116:3–23.
182. Sotomayor M, Gaudet R, Corey DP. Sorting out a promiscuous superfamily: towards cadherin connectomics. *Trends Cell Biol.* 2014;24(9):524–536.
183. Bochkov YA, Palmenberg AC, Lee WM, et al. Molecular modeling, organ culture and reverse genetics for a newly identified human rhinovirus C. *Nat Med.* 2011;17(5):627–632.
184. Bonnelykke K, Sleiman P, Nielsen K, et al. A genome-wide association study identifies CDHR3 as a susceptibility locus for early childhood asthma with severe exacerbations. *Nat Genet.* 2014;46(1):51–55.
185. Yanai I, Benjamin H, Shmoish M, et al. Genome-wide midrange transcription profiles reveal expression level relationships in human tissue specification. *Bioinformatics.* 2005;21(5):650–659.
186. Fuchs R, Blaas D. Uncoating of human rhinoviruses. *Rev Med Virol.* 2010;20(5):281–297.
187. Snyers L, Zwickl H, Blaas D. Human rhinovirus type 2 is internalized by clathrin-mediated endocytosis. *J Virol.* 2003;77(9):5360–5369.
188. Khan AG, Pickl-Herk A, Gajdzik L, Marlovits TC, Fuchs R, Blaas D. Human rhinovirus 14 enters rhabdomyosarcoma cells expressing ICAM-1 by a clathrin-, caveolin-, and flotillin-independent pathway. *J Virol.* 2010;84(8):3984–3992.
189. Bayer N, Prchla E, Schwab M, Blaas D, Fuchs R. Human rhinovirus HRV14 uncoats from early endosomes in the presence of bafilomycin. *FEBS Lett.* 1999;463(1):175–178.
190. Nurani G, Lindqvist B, Casasnovas JM. Receptor priming of major group human rhinoviruses for uncoating and entry at mild low-pH environments. *J Virol.* 2003;77(22):11985–11991.
191. Greve JM, Forte CP, Marlor CW, et al. Mechanisms of receptor-mediated rhinovirus neutralization defined by two soluble forms of ICAM-1. *J Virol.* 1991;65(11):6015–6023.
192. Hoover-Litty H, Greve JM. Formation of rhinovirus-soluble ICAM-1 complexes and conformational changes in the virion. *J Virol.* 1993;67(1):390–397.
193. DeTulleo L, Kirchhausen T. The clathrin endocytic pathway in viral infection. *EMBO J.* 1998;17(16):4585–4593.
194. Staunton DE, Gaur A, Chan PY, Springer TA. Internalization of a major group human rhinovirus does not require cytoplasmic or transmembrane domains of ICAM-1. *J Immunol.* 1992;148(10):3271.
195. Bayer N, Schober D, Huttinger M, Blaas D, Fuchs R. Inhibition of clathrin-dependent endocytosis has multiple effects on human rhinovirus serotype 2 cell entry. *J Biol Chem.* 2001;276(6):3952–3962.
196. Blaas D, Fuchs R. Mechanism of human rhinovirus infections. *Mol Cell Pediatr.* 2016;3(1):21.

197. Macnaughton MR. The structure and replication of rhinoviruses. *Curr Top Microbiol Immunol.* 1982;97:1–26.
198. Wark PA, Johnston SL, Bucchieri F, et al. Asthmatic bronchial epithelial cells have a deficient innate immune response to infection with rhinovirus. *J Exp Med.* 2005;201 (6):937–947.
199. Contoli M, Message SD, Laza-Stanca V, et al. Role of deficient type III interferon-l production in asthma exacerbations. *Nat Med.* 2006;12(9):1023–1026.
200. Edwards MR, Regamey N, Vareille M, et al. Impaired innate interferon induction in severe therapy resistant atopic asthmatic children. *Mucosal Immunol.* 2013;6 (4):797–806.
201. Schneider D, Ganesan S, Comstock AT, et al. Increased cytokine response of rhinovirus-infected airway epithelial cells in chronic obstructive pulmonary disease. *Am J Respir Crit Care Med.* 2010;182(3):332–340.
202. Baines KJ, Hsu AC, Tooze M, Gunawardhana LP, Gibson PG, Wark PA. Novel immune genes associated with excessive inflammatory and antiviral responses to rhinovirus in COPD. *Respir Res.* 2013;14:15.
203. Chattoraj SS, Ganesan S, Jones AM, Helm JM, Comstock AT, Bright-Thomas R, et al. Rhinovirus infection liberates planktonic bacteria from biofilm and increases chemokine responses in cystic fibrosis airway epithelial cells. *Thorax.* 2011;66: 333–339.
204. Schogler A, Kopf BS, Edwards MR, et al. Novel antiviral properties of azithromycin in cystic fibrosis airway epithelial cells. *Eur Respir J.* 2015;45(2):428–439.
205. Kieninger E, Vareille M, Kopf BS, et al. Lack of an exaggerated inflammatory response on virus infection in cystic fibrosis. *Eur Respir J.* 2012;39(2):297–304.
206. Dauletbaev N, Das M, Cammisano M, et al. Rhinovirus load is high despite preserved interferon-beta response in cystic fibrosis bronchial epithelial cells. *PLoS One.* 2015;10(11):e0143129.
207. Wark PA, Grissell T, Davies B, See H, Gibson PG. Diversity in the bronchial epithelial cell response to infection with different rhinovirus strains. *Respirology.* 2009;14 (2):180–186.
208. Nakagome K, Bochkov YA, Ashraf S, et al. Effects of rhinovirus species on viral replication and cytokine production. *J Allergy Clin Immunol.* 2014;134 (2):332–341.
209. Cox DW, Bizzintino J, Ferrari G, et al. Human rhinovirus species C infection in young children with acute wheeze is associated with increased acute respiratory hospital admissions. *Am J Respir Crit Care Med.* 2013;188(11):1358–1364.
210. Proud D, Turner RB, Winther B, et al. Gene expression profiles during in vivo human rhinovirus infection: insights into the host response. *Am J Respir Crit Care Med.* 2008;178(9):962–968.
211. Bochkov YA, Hanson KM, Keles S, Brockman-Schneider RA, Jarjour NN, Gern JE. Rhinovirus-induced modulation of gene expression in bronchial epithelial cells from subjects with asthma. *Mucosal Immunol.* 2010;3(1):69–80.
212. Bosco A, Ehteshami S, Panyala S, Martinez FD. Interferon regulatory factor 7 is a major hub connecting interferon-mediated responses in virus-induced asthma exacerbations in vivo. *J Allergy Clin Immunol.* 2012;129(1):88–94.
213. Bosco A, Wiehler S, Proud D. Interferon regulatory factor 7 regulates airway epithelial cell responses to human rhinovirus infection. *BMC Genomics.* 2016;17:76.
214. Parsons KS, Hsu AC, Wark PA. TLR3 and MDA5 signalling, although not expression, is impaired in asthmatic epithelial cells in response to rhinovirus infection. *Clin Exp Allergy.* 2014;44(1):91–101.
215. Bossios A, Psarras S, Gourgiotis D, et al. Rhinovirus infection induces cytotoxicity and delays wound healing in bronchial epithelial cells. *Respir Res.* 2005;6:114.

216. Lewandowska-Polak A, Brauncajs M, Jarzebska M, et al. Toll-like receptor agonists modulate wound regeneration in airway epithelial cells. *Int J Mol Sci.* 2018;19(8): E2456.
217. Sajjan U, Wang Q, Zhao Y, Gruenert DC, Hershenson MB. Rhinovirus disrupts the barrier function of polarized airway epithelial cells. *Am J Respir Crit Care Med.* 2008;178(12):1271–1281.
218. Unger BL, Ganesan S, Comstock AT, Faris AN, Hershenson MB, Sajjan US. Nod-like receptor X-1 is required for rhinovirus-induced barrier dysfunction in airway epithelial cells. *J Virol.* 2014;88(7):3705–3718.
219. Comstock AT, Ganesan S, Chattoraj A, et al. Rhinovirus-induced barrier dysfunction in polarized airway epithelial cells is mediated by NADPH oxidase 1. *J Virol.* 2011;85(13):6795–6808.
220. Yeo NK, Jang YJ. Rhinovirus infection-induced alteration of tight junction and adherens junction components in human nasal epithelial cells. *Laryngoscope.* 2010;120 (2):346–352.
221. Looi K, Troy NM, Garratt LW, et al. Effect of human rhinovirus infection on airway epithelium tight junction protein disassembly and transepithelial permeability. *Exp Lung Res.* 2016;:1–16.
222. Faris AN, Ganesan S, Chattoraj A, et al. Rhinovirus delays cell repolarization in a model of injured/regenerating human airway epithelium. *Am J Respir Cell Mol Biol.* 2016;55(4):487–499.
223. Gulraiz F, Bellinghausen C, Dentener MA, et al. Efficacy of IFN-lambda1 to protect human airway epithelial cells against human rhinovirus 1B infection. *PLoS One.* 2014;9(4):e95134.
224. Cakebread JA, Xu Y, Grainge C, et al. Exogenous IFN-beta has antiviral and anti-inflammatory properties in primary bronchial epithelial cells from asthmatic subjects exposed to rhinovirus. *J Allergy Clin Immunol.* 2011;127(5):1148–1154.e9.
225. Djukanovic R, Harrison T, Johnston SL, et al. The effect of inhaled IFN-beta on worsening of asthma symptoms caused by viral infections. A randomized trial. *Am J Respir Crit Care Med.* 2014;190(2):145–154.
226. de Weerd NA, Nguyen T. The interferons and their receptors—distribution and regulation. *Immunol Cell Biol.* 2012;90(5):483–491.
227. Turner RB, Wecker MT, Pohl G, et al. Efficacy of tremacamra, a soluble intercellular adhesion molecule 1, for experimental rhinovirus infection: a randomized clinical trial. *JAMA.* 1999;281(19):1797–1804.
228. Traub S, Nikonova A, Carruthers A, et al. An anti-human ICAM-1 antibody inhibits rhinovirus-induced exacerbations of lung inflammation. *PLoS Pathog.* 2013;9(8): e1003520.
229. Gielen V, Johnston SL, Edwards MR. Azithromycin induces antiviral responses in bronchial epithelial cells. *Eur Respir J.* 2010;36(3):646–654.
230. Porter JD, Watson J, Roberts LR, et al. Identification of novel macrolides with antibacterial, anti-inflammatory and type I and III IFN-augmenting activity in airway epithelium. *J Antimicrob Chemother.* 2016;71(10):2767–2781.

CHAPTER 4

Immunity to rhinoviruses

Sai P. Narla[1,2] **and John W. Upham**[1,2]
[1]Diamantina Institute, The University of Queensland, Brisbane, QLD, Australia
[2]Department of Respiratory Medicine, Princess Alexandra Hospital, Brisbane, QLD, Australia

4.1 INTRODUCTION

The human body is capable of eliciting a very rapid immune response to rhinovirus (RV) infection. Following viral entry and replication, epithelial cells (ECs) release "call to arms" danger signals consisting of interferons, chemokines and cytokines, initiating immediate innate immune responses and delayed adaptive immune responses simultaneously. Like most viruses, RV infection triggers the recruitment of a variety of immune cells including, dendritic cells (DCs), macrophages, monocytes, granulocytes, and lymphocytes making up the cell mediated immune response. Eventually, RV also elicits a humoral immune response with the synthesis of serotype-specific neutralizing antibodies that are highly effective in promoting viral clearance and protecting from re-infection with that seroptype. Although RV is usually as an innocuous virus, it can cause considerable lower respiratory tract morbidity, and even be life-threatening, especially in immunocompromised hosts.[1] Effective defense against RV involves close cooperation between airway mucosal ECs (as reviewed elsewhere in this book), and migratory immune cells, as described below.

4.2 INNATE IMMUNE RESPONSE

Our traditional perception of immune response is built upon specialized and separate roles for innate and adaptive immunity. The innate arm of the immune response is orchestrated by lung structural cells, DCs, macrophages, natural killer (NK) cells, neutrophils, and innate lymphoid cells (ILCs), which lay a foundation for the subsequent adaptive responses executed by differentiated T and B lymphocytes. However, recent conceptual advances have shown that this "division of labor" is too simplistic, with both innate and adaptive interacting with the microenvironment in a

Rhinovirus Infections
DOI: https://doi.org/10.1016/B978-0-12-816417-4.00004-4

coordinated manner. This has led to a substantial refinement of the way in which we view the immune system. However, in this chapter, innate and adaptive immune cell population will be discussed separately for the sake of clarity.

4.2.1 Dendritic cells

DCs initiate both innate and adaptive immune responses. In the lung, conventional or myeloid DCs are strategically positioned close to the basal lamina beneath the epithelial layer and use specialized "dendrites" to sample various antigens including bacteria, toxins, allergens, and viruses. They are also recruited by potent chemoattractants such as chemokine C−X−C motif ligand (CXCL)-10 [also known as interferon (IFN)-γ-inducible protein (IP-10)] and C−C ligand (CCL)-2 (also known as monocyte chemoattractant protein 1) released by RV infected airway ECs. RVs enter DCs but do not replicate.[2] Instead, they upregulate antigen presenting markers major histocompatibility complex II, cluster of differentiation (CD)-80 and CD86 advancing the DC to a mature phenotype.[3] The same study also found that in response to RV these DCs were able to upregulate the cytokines interleukin (IL)-15 (NK cell activator), IL-12 (Th1 cell primer), IL-10, and transforming growth factor (TGF)-β (immune suppressing) laying the foundation for an efficient antiviral immune response. More importantly, in response to RV infection, DCs release IFNs, a family of cytokines known to "interfere" with viral replication. They are classified into three major types: type I, IFN-α and IFN-β; type II, IFN-γ; and type III, IFN-λ1 and IFN-λ2/3. In particular, types I and III play a major role in protecting against most RV invasions.[4,5] In vitro stimulation of peripheral blood mononuclear cells (PBMCs) from healthy volunteers with RV produce both type I and type III IFNs demonstrating their role in maintaining immune homeostasis.[6] Likewise, deficient type I and type III IFN production has been associated with increased susceptibility to diseases such as asthma.[7]

Whereas early studies demonstrated that RV triggered innate immune responses through production of type I IFN in monocyte derived DCs,[2] it is now recognized that plasmacytoid DCs (pDCs) are the cell population most adept at secreting type I IFN in response to RV. This rare cell type comprises over 90% of the type I IFN producing cells whereas monocytes and myeloid DCs contribute only 10% and 1%, respectively.[8] Moreover, it has recently become apparent that pDCs and type I IFNs have an important regulatory role by curbing type II immune responses to RV in healthy volunteers.[9] This finding has particular significance in asthma, wherein deficient pDC function and deficient type I IFN response can lead to type II antiviral immune responses. Recently, we further demonstrated that pDCs are critical in regulating antiviral immune responses to RV.[10] By comparing intact PBMCs and pDC depleted PBMCs, we observed a major reduction in IFN-α production along with deficient activation of pathways downstream from IFN.

Although the antiviral role of pDCs is well established, their status in relation to RV and particularly asthma is yet to be fully understood. Chairakaki et al. showed that high pDC numbers in asthmatic sputum correlate with inflammation intensity and Lie et al. showed that pDCs upregulate the gene ORMDL3, associated with the 17q21 asthma risk allele genotype, in leukocytes via cell−cell interaction upon exposure to RV.[11,12] pDCs also express FcεRI, the high-affinity receptor for IgE antibodies that drive type I hypersensitivity reactions in allergic individuals.[13] Durrani et al. have shown that high numbers of pDCs expressing FcεRI is inversely proportional to immune responses to RV. Cross-linking of FcεRI prior to RV infection suppresses IFN-α and IFN-λ1 release in allergic asthmatic children,[14] highlighting one mechanism by which allergic inflammatory pathways can inhibit antiviral immunity. Finally, RV has also been shown to inhibit DC adaptive immune functions by upregulating the coinhibitory checkpoint molecule programmed death-ligand 1.[15] Whether RV demonstrates a similar effect on other immune checkpoint ligands such as inducible T Cell costimulator ligand (ICOS-L), glucocorticoid-induced TNFR-related ligand (GITR-L), and poliovirus receptor (PVR) is yet to be discovered. More studies are needed to help us elucidate the role of pDC responses to RV, especially in the context of various airway diseases.

4.2.2 Macrophages

Macrophages are the predominant immune cell in the airway lumen. In addition to their phagocytic function to clear bacteria, debris, and apoptotic cells, they also have defined antiviral functions. There is no clear evidence that RV can infect primary human macrophages: whereas in vitro studies have shown a minor degree of viral replication in blood monocyte-derived macrophages, viral replication is not detectable in bronchoalveolar lavage (BAL)-derived macrophages.[16,17] Macrophages derived from THP-1 cells, a human monocytic cell line, showed RV replication and deficient type I IFN production but monocyte derived macrophages showed limited RV replication and robust type I IFN production.[17] However, in both macrophage types, RV induced activation of the master transcription factor NF-κB, inducing release of tumor necrosis factor (TNF)-α, a proinflammatory cytokine. Although macrophages can host RV, their potent antiviral functions restrict viral replication.

Activation by RV can stimulate various chemokines, as well as proinflammatory or antiviral cytokines, depending on the macrophage activation state and serotype of RV infection.[18] While cytokines such as IL-1, IL-8 (CXCL-8), TNF-α, IFN-γ, and macrophage inflammatory protein−1α promote inflammation, chemokines such as CXCL-10 and CXCL-11 (eotaxin-1) recruit adaptive immune cells that guide viral clearance.[19] In addition, RV stimulated macrophages boost antiviral responses of airway ECs via CXCL-10.[20] However, Karta et al. observed that allergen exposure dampened the macrophage anti-RV response by inhibiting CXCL-10 and eotaxin-1 and enhanced the secretion of inflammatory cell recruiting chemokine CCL-2.[18] Using murine models, it has been shown that RV infection potentiates lung inflammation through macrophages that express Toll-like receptor (TLR)-2 and CD68, by upregulating CCL-2 expression and thereby inducing eotaxin-1 release causing eosinophilia and airway hyperresponsiveness.[21−23]

In addition, $CD68^+$ and $CD11b^+$ airway wall macrophages from asthmatic participants experimentally infected with RV clearly have the ability to ingest RV.[24] RV also induces extensive changes in genetic expression pattern in IFN-γ-polarized M1 macrophages, greater than that seen in IL-4-polarized M2 macrophages.[25] This shows that, depending on the cytokine milieu into which macrophages are recruited, RV can elicit varying inflammatory responses.

In addition to modifying antiviral responses, RV also interferes with antibacterial innate responses of macrophages. Oliver et al. demonstrated that RV reduces the capacity of alveolar macrophages to secrete TNF-α and IL-8 following lipopolysaccharide stimulation.[26] This study is significant as it shows that by impeding macrophage function, RV may facilitate an environment that promotes bacterial infections.

4.2.3 Neutrophils

Neutrophils are the most abundant immune cells in circulation and are amongst the first to respond to RV invasion by localizing initially in the nasal mucosa, especially in patients with respiratory tract infections.[27–29] IL-8, a chemokine released by ECs and fibroblasts, is a major recruiter of neutrophils and has been shown to be significantly elevated following RV stimulation.[30] In addition, macrophage derived CXCL-10 and growth factors such as granulocyte (G)−monocyte (GM) colony stimulating factor (GM-CSF) and G-CSF also enhance neutrophil recruitment, survival and activation. Experimental inoculation of allergic asthmatic subjects with RV showed increased bronchial and circulating neutrophils along with IL-8 and G-CSF.[31] RV play a dualistic role ranging from inflammatory to protective.[32]

Neutrophils execute vital antimicrobial functions through phagocytosis and degranulation. Under normal conditions, degranulated products, such as reactive oxygen species and matrix metalloproteinase-9, aid in resolution of inflammation and enhanced viral clearance.[33] Conversely, viral infection in subjects with compromised immune systems leads to excessive neutrophilic infiltration and inflammation. Chronic obstructive pulmonary disease (COPD) and asthma patients infected with RV have greater airway neutrophil counts, which correlated with viral induced exacerbations.[34,35] Similarly, Rhode et al. reported an increased BAL IL-8 and neutrophil count in RV infected asthma patients. They further identified that, in response to RV stimulated IL-8, neutrophils increased secretion of antimicrobial peptides human neutrophil peptide (HNP) 1-3, causing enhanced airway neutrophilia.[36]

Neutrophils can release neutrophilic extracellular traps (NETs), containing dsDNA and elastase, into the extracellular space through a process called NETosis, which is an effective immune response against viruses.[37] Recently, a groundbreaking study by Toussaint et al. demonstrated that RV infection induces amplified dsDNA release in asthmatics and that this is related to type-II cytokine induction and exacerbation severity.[38] This finding is significant as inhibiting NETosis could be a potential treatment strategy for virus induced asthma exacerbations. Challenging the paradigm that neutrophils are predominantly proinflammatory, Tang et al. discovered that neutrophils abrogated IL-6 and IL-8 secretion from monocytes in response to RV infection.[39]

4.2.4 Eosinophils

Although neutrophils constitute the majority of granulocytes in the lungs, experimental RV infection can induce eosinophil infiltration, and this is seen more dramatically during acute asthma exacerbations.[31] During RV infection, eosinophilic infiltrates are more prominent and persist for longer in allergic asthmatics compared with healthy controls.[40,41] RV infected ECs secrete potent eosinophil recruiting chemokines, CCL-5 (RANTES) and eotaxin-1.[42,43]

Similar to neutrophils, eosinophils have both antiviral and proinflammatory roles in response to pathogens. Eosinophilic degranulation results in elastases and ribonucleases such as eosinophil-derived neurotoxin and eosinophilic cationic protein, which possess antiviral properties and are increased during RV infection.[42,44] Early investigations demonstrated contradictory roles of virus-induced eosinophils. Whereas greater sputum eosinophil counts predicted increased vulnerability to RV infections, others have reported that elevated eosinophil numbers prior to RV inoculation correlates with better protection against severe colds.[45,46] Interestingly, Grissell et al. noted that during acute asthma exacerbations, eosinophilic influx was suppressed and antiinflammatory IL-10 expression levels were increased.[47] Eosinophilic inflammation is a major pathological condition in asthma and this study reveals an important anti-eosinophilic function of IL-10, which can be vital in controlling RV-induced asthma exacerbations. These varying reports highlight the complex nature of inflammation in asthma, both in steady state and during virus-induced exacerbations.

RV is also known to augment type II immune responses (discussed below) including the cytokine IL-5, which is an important promoter of eosinophil recruitment and activation, whose overexpression causes eosinophilia.[48,49] Recently, Hatchwell et al. have observed that IL-5 induced eosinophilia in asthmatics correlated inversely with antiviral TLR7 and type-III IFN expression.[50] Further studies are warranted to fully understand the role of RV in IL-5 mediated eosinophilia. Finally, eosinophils have also been shown to promote antiviral adaptive responses by imitating APC cell function by binding to RV and presenting antigen to RV specific T cells stimulating type-II IFN production.[51]

4.2.5 Basophils and mast cells

Circulating basophils and tissue-based mast cells play a central role in the development of allergic inflammation. Although they have unique inflammatory functions in type-II immunity, they both express FcεRI, secrete type II cytokines and degranulate on activation, releasing histamine promoting hypersensitivity reactions.[52] Basophils are recruited by chemokines eotaxin-1, RANTES, CCL-2, and are activated by IL-3.[53] GM-CSF, IL-3, IL-5, and eotaxin-1 promote mast cell chemotaxis and survival.[19] The EC cytokines, IL-33, IL-25, and thymic stromal lymphopoietin (TSLP) have also been found to activate basophils and mast cells to extend type II immune response.[54,55]

There are no convincing reports of RV replication in basophils or mast cells to date. Recently derived mast cell lines were shown to allow RV replication and release.[56] Earlier reports observed that RV had no effect on basophils and mast cells numbers or functions in healthy volunteers.[28,29] However, later investigations showed that RV inoculation activated basophils and mast cells causing increased histamine secretion in healthy and allergic volunteers.[40,57,58] In vitro infection of IgE stimulated human mast cell line and human basophilic leukocyte cell line with RV increases the production of inflammatory cytokines and chemokines including IL-4, IL-6, IL-8, GM-CSF, and histamine.[59]

IgE binding to FcεRI mediates basophilic inflammation and increased IgE levels are very common in subjects with atopic asthma and allergic rhinitis.[14,52] Agrawal et al. using an in vivo RV challenge model showed that RV infection enhances basophil receptivity to IgE in atopic asthmatics.[60] IgE upregulated TSLP receptor (TSLPR) and in response to RV, further increased TSLPR expression on basophils after 3 weeks. This study proposes that in response to RV infection, basophils may have "late" response role. Mast cells also contribute to bronchial inflammation by producing eicosanoid mediators, including leukotrienes and prostaglandins, in response to pathogens.[61] Seymoor et al. have observed that healthy volunteers infected with RV have increased mast cell numbers correlating with cycloxygenase and lipoxygenase positive cells.[62] However, additional work is needed to fully understand the role of RV induced eicosanoid mediators, as they could be potential targets for therapeutic intervention studies, especially in asthma.

4.2.6 Natural killer cells

Cytopathology is usually not as severe in RV infections, relative to other respiratory viruses such as influenza and respiratory syncytial virus (RSV). Although, RV infection fails to affect NK cell activation and cytotoxicity, it induces antiviral IFN-γ release.[63,64] RV infection recruits NK cells and T cells via mediated release of RANTES and CXCL-10 from infected ECs via the T cell receptor CXCR3.[65,66] In addition, Wark et al. also observed high serum levels of CXCL-10 during acute asthma exacerbations, especially virus induced exacerbations.[66] IL-15 is another novel NK cell attracting cytokine.[67] Jayaraman et al. have demonstrated that RV infection increases IL-15 expression in the nasal mucosa in humans, and shown in a mouse model that IL-15 has a critical role in recruiting NK cells and $CD8^+$ T cells and inducing IFN-γ production during RV infection.[68] Taken together, NK cells execute important antiviral functions against RV. Whether RV can also initiate NK cell-mediated antibody-dependent cellular cytotoxicity is currently unknown.

4.2.7 Innate lymphoid cells

ILC2 are recently described tissue resident cells that can secrete the type II cytokines IL-5 and IL-13 in response to a variety of stimuli. In the context of the current discussion, it is noteworthy that influenza virus can induce robust ILC2 airway inflammation, phenotypically identical to asthma.[69–71] ILC2s are recruited by the EC-derived cytokines such as IL-33, IL-25, and TSLP along with TNF-α and TGF-β.[72] Given that ILC2 drive type II immunity and disease, it seems plausible that they will also play a role on RV-induced immune responses, especially in asthma.[73] Although still in its infancy, exciting studies of RV infection on ILC2 function have emerged. Independent works on murine models have shown that RV infection promotes airway hyperresponsiveness and mucous metaplasia through EC derived IL-25 recruitment of ILC2.[55,74] In addition, Jackson et al. have observed that coculturing ILC2 cells with RV infected ECs increases IL-33 dependent Th2 cytokine profile.[75] Type I IFNs are known to regulate type II cytokine synthesis.[9] Concurrently, Duerr et al. have reported that IFN-β abrogated ILC2 mediated Th2 immune response and could be an attractive antiviral treatment.[76] ILC2 function in response to RV infection is predominantly inflammatory and in the context of RV induced asthma is likely to be an important mediator of tissue pathology. Hence it is imperative to elucidate the mechanisms behind inappropriate ILC2 responses during RV-induced asthma exacerbations.

4.3 ADAPTIVE IMMUNE RESPONSE

Under normal conditions, the innate immune system is capable of clearing RV within 5 days postinfection, well in advance of the 14-day period required for the activation and expansion of adaptive T cell and B cells.

If the innate immune system fails to clear the RV invasion, the adaptive immune response becomes more important, consisting mainly of preexisting memory T cells (Tmems) and antibodies specific to the serotype of RV. In this part of the chapter we will discuss the two kinds of adaptive immunity to RV: cell-mediated and humoral immune response.

4.3.1 Type I and type II immunity

T lymphocytes respond to RV infection and infiltrate the airway epithelium and submucosa. The decline in blood T cell numbers postinfection strongly correlates with increase in severity of RV induced symptoms suggesting that T cells are likely to play a major role in response to RV invasion.[40,77] Generally, DCs take up pathogen-derived antigens, mature, migrate, and prime naive T cells in regional lymph nodes, initiating a T cell response. However, whether DCs can prime T cells in response to RV is still being examined and is not fully understood. Steinke et al. provide the latest evidence that RV challenge propelled DC maturation and proliferation of $CD4^+$ and $CD8^+$ T cells implying that RV are capable of eliciting a classic adaptive response.[3] Interestingly, RV has also been shown to directly activate $CD4^+$ and $CD8^+$ T cells to produce proinflammatory cytokines and eosinophil degranulation, in the absence of APCs.[78] Although both studies were performed in vitro, the findings imply that RV infections, particularly in asthma or COPD, may be capable of triggering inappropriate T cell responses that escalate airway inflammation.

The balance between type I (classical response to virus) and type II (response to parasitic worms and allergens) immunity to RV infection is important in the case of diseases like asthma. Traditionally, type I immunity, mainly comprised of $CD4^+$ type I helper T cells (Th1 cells) secreting IFN-γ, is the most efficient response against intracellular microbes such as viruses, whereas CD4 + type II helper T cells (Th2 cells) driven type II immunity (defined by production of type II cytokines IL-4, IL-5, IL-9 and IL-13, eosiniophilic inflammation, IgE production and mast cell/basophil activation) is important in helminthic worm infections as well as allergic inflammation.

Early investigations revealed that symptom severity in RV infected individuals was inversely and directly correlated with IFN-γ and IL-5 levels, respectively.[79,80] In addition, in response to RV infection, healthy controls responded with intact IFN-γ release whilst asthmatics had considerably lower levels of IFN-γ, IL-12 and higher concentrations of IL-4 and IL-10.[81] As discussed earlier, a subset of asthmatics appear to have deficient type I IFN production to RV accompanied by a prominent type II response.[9] Deficiency of IFN-driven type I immunity may partly explain why many asthmatics have delayed viral clearance and type II inflammation including eosinophil infiltration. The consensus of most investigators is that under healthy conditions, a type I immune response to RV predominates. However, in some situations, especially in asthma and perhaps in immune deficiency, suboptimal type I and type III IFN responses and elevated type II response become more apparent, leading to airway hyper-reactivity and virus-induced exacerbations although Type III (IFN-λ) appears be increased in some studies of acute asthma.[7] This apparent discrepancy for deficient versus increased type III IFN during RV infections in asthma is likely due to timing of sampling with increased viral replication potentially driving increased type III IFN expression at later time times. Our lab found that in vitro challenge of PBMC with RV and IFN-λ increased IL-5 levels, and this was blunted in the presence of IFN-β.[82] Further investigations revealed that asthmatics had reduced interferon regulatory factor (IRF), signal transducer and activator of transcription (STAT), and NF-κB expression in conjunction impaired IFN responses to RV.[83] Taken together, the evidence points toward the antiviral protective role of type I immune response (initiated by production of type I- and type III-IFN) and impairment of such responses in diseases like asthma leads to an exaggerated disease phenotype.

Preexisting type II response conditions may differentially affect immune responses to RV. It is well established that type II cells and their cytokines inhibit Th1 cell expansion.[84] In addition, IL-5 recruited eosinophils also inhibit type I immune responses by inducing apoptosis of IFN-γ producing T-cells through indoleamine 2,3-dioxygenase.[85] It would be interesting to see if such a mechanism is indeed mediated by RV responsive eosinophils. The expression of the major RV receptor intercellular adhesion molecule (ICAM)-1 on ECs is known to be increased by type II cytokines and RV infection further enhances ICAM-1 expression on ECs.[86,87] In addition, IL-4 and IL-5 have also been shown to prime RV-induced Th2 immune responses by inhibiting TLR3 expression on ECs and consequentially downregulating IFN production.[88] The hallmark EC derived polarizing cytokines IL-33, IL-25, and TSLP are major promoters of Th2 cell recruitment and expansion.[89] As mentioned earlier, Beale et al. have shown that RV induces type II responses through IL-25,[55] and

TSLP has been reported to be upregulated in both murine models and in asthmatic children with RV infection.[90] TSLP has also been shown to promote type II responses by modulating DC function to selectively promote Th2 cell differentiation.[91] Very recently we have reported that IL-33 aggravates type II responses via ILC2s in asthmatic subjects.[92]

All the above reports strengthen the notion that RV infection, under circumstances of reduced type I immunity along with inappropriate type II responses, may elevate severity of accompanying airway diseases such as asthma.

4.3.2 Type III immune response

Type III immunity is mediated by Th17 cells and its cytokines IL-17 and IL-22, promoting both lung inflammation and viral clearance during influenza infections.[93,94] Increased IL-17 levels have been associated with neutrophilic inflammation, and RV infection promotes neutrophilic infiltration to the airways as noted previously.[44,95] Consistent with this, Wiehler et al. have found that IL-17A presence modifies RV infected ECs to selectively enhance secretion of the neutrophil chemoattractant IL-8 while suppressing the production of eosinophil recruiting factor RANTES.[96] In addition, Choy et al. have shown that Th17 and type II cytokine profiles were inversely correlated in asthmatics.[97] Although unconfirmed in vivo, we can speculate that presence of a Th17/IL-17 environment during RV infections may suppress type II cytokine production and eosinophilic inflammation and favor a neutrophilic response; the significance of these observations for eosinophilic asthma during a RV infection remain to be determined. However, Perez et al. have noted that severe RV infection in premature children is directly associated with high Th17 and type II cytokine levels and increased early life respiratory morbidity, an important risk factor for later stage asthma development.[98] These data indicate that Th17 immune responses to RV infections are dichotomous and further studies elucidating their role in relation to type I/II immune responses are needed.

4.3.3 Regulatory T cells

Regulatory T cells (Tregs) play an important role in maintaining homeostasis and suppressing inflammatory responses by secreting and maintaining the cytokines IL-10, TGF-β, and IL-35.[99] Respiratory syncytial virus (RSV) has been shown to attenuate Treg function causing an increased

Th2 profile and asthmatic response in mice.[100] However, data on direct RV and Treg interactions is limited. Recently, Gaido et al. reported that asthmatic children have lower blood Treg numbers and better Treg response to RV in vitro.[101] Further confirmation studies are needed to dissect the Treg response to RV. In contrast, Treg cytokines have been long associated with RV infections in both healthy and asthmatic individuals.[102–105] Like other T cells, Tregs can also be modulated by DC function. Through an immune checkpoint dependent signal, RV infected DCs can induce robust IL-35 production from Tregs along with suppression of T cell activity.[106] Recently Bielor et al. observed that RV infection of PBMCs ex vivo increased the expression of TGF-β receptor *TGFBRII* thus blocking TGF-β release and promoting Th1 response and viral clearance.[107] This mechanism was impaired in asthmatics, wherein exogenous increase in exogenous TGF-β diminished Th1 cell numbers facilitating viral persistence.

4.3.4 Memory T cells

In the event of an incomplete innate immune response, individuals with immune memory to the specific RV strain are able to clear virus within a few days, facilitated by the presence of preexisting Tmems. In contrast, in the naïve host, mobilization of an effective cell mediated immune response takes up to 14 days.[108] RV proteins contain several T cell epitopes that are evolutionarily conserved over multiple serotypes.[109] With over 150 serotypes of RV discovered, it is now clear that different RV serotypes trigger recurrent infections in the same individual.[110] Stienke et al. have shown that circulating $CD4^+$ and $CD8^+$ Tmems from healthy subjects were able to recognize and proliferate rapidly in response to RV infection in vitro along with increased levels of IFN-γ and IL-4.[3] In an important study, several HLA-DR binding $CD4^+$ T cell epitopes were identified that were conserved across all RV species.[111] Upon intranasal RV infection, increased numbers of circulatory epitope-specific Tmems were identified in healthy subjects, comprising both Th1 and T follicular helper cells. Recently, Gaido et al. reported several immunodominant peptides of VP1 conserved across RV-A and RV-C species.[112] RV-A and RV-C specific synthetic peptides were shown to successfully stimulate T cell responses to both RV species with similar potency, irrespective of preexisting antibody titers. These cross-reactive peptides could be vital to developing successful strategic anti-RV vaccines that improve protective immunity against multiple RV serotypes.

4.3.5 Neutralizing antibodies

In addition to the cell mediated responses discussed above, humoral immune responses are also activated to prevent RV infection, with the existence of RV specific neutralizing antibodies confirmed since the 1950s.[113] It is well known that older children and adults have greater titers of neutralizing antibodies from previous RV exposure, and that preexisting antibodies prevent further experimental RV infections.[114,115] B cell derived neutralizing antibodies have been observed in the nasal secretions of individuals infected with RV. The antibody responses are strictly RV serotype-specific with increased mucosal-derived IgA present after 3 days and serum-derived IgG detectable after 7–8 days postinfection and persisting for over a year.[116,117] High titers of these serotype-specific antibodies have been positively correlated with greater resistance to later experimental infections by that RV serotype, suggesting that humoral responses play an important role in protective immunity against recurrent RV infections.[118] Interestingly, Iwasaki et al. observed that asthmatic children have greater RV-A and RV-B specific antibody titers in comparison to nonasthmatic controls,[119] perhaps as a consequence of more frequent or more severe previous infections. RV-specific neutralizing antibodies contribute to direct viral clearance by restricting viral motility, by blocking cellular attachment ligands and through facilitating phagocytic antigen presentation by opsonizing RV. Furthermore, they also activate the classical complement pathway and initiate NK cell mediated cytotoxic response promoting effective cell mediated immune responses.[110,120]

Unlike Tmems, neutralizing antibodies have demonstrated very limited cross-reactivity across different RV strains, representing a major obstacle for vaccine development.[121] In addition, an antibody response to RV is not detectable in all infected individuals and interestingly, studies in asthmatics show that higher antibody response was associated with greater burden of respiratory symptoms.[122] Taken together with the extensive genetic diversity of RV, it is not a surprise that progress toward a clinically effective vaccine has been very slow. However, a few promising studies on animal models have shown potential. Purified recombinant VP1, derived from *Escherichia coli*, has been successfully used to immunize mice and rabbits by inducing neutralizing antibodies that cross-reacted to multiple RV strains.[123] The elicited antibodies were misdirected toward a non-neutralizing epitope, presenting a mechanism through which RV may evade efficient neutralization.[124] McLean et al. provided the first in vivo evidence of effective cross-serotype neutralizing antibody responses to

VP1 in a mice model subjected to repeated intranasal RV infections.[125] Similarly, mice immunized with VP0 also generated similar responses along with increased Tmem levels and greater viral clearance compared with nonimmunized mice.[126] These results indicate some progress is now being made toward the ultimate goal of a successful anti-RV vaccine. In an exciting new study, a polyvalent inactivated vaccine covering 50 RV serotypes demonstrated potent antibody responses in rhesus macaques against multiple RV subtypes.[127] Although this vaccine lacks RV-C antigens, it is an important demonstration that inactivated virus titers can also be a potentially therapeutic strategy to achieve a clinically effective vaccine.

4.4 CONCLUSION

While there has been enormous progress in understanding the immune response to RV in healthy individuals, and in those with a variety of lung diseases such as asthma, much still needs to be uncovered and progress toward a vaccine has been very slow. In asthma, it is still unclear whether the mechanisms by which RV induces exacerbations can be attributed to altered type I or type III IFN leading to high viral loads, a dysregulated and excessive immune response, or both. Indeed, as researchers and clinicians have come to understand that asthma is more than one disease, and that asthma exacerbations involve different triggers and different inflammatory pathways, it is likely that the immune response to RV is heterogeneous and will require a variety of approaches if progress is to be made in preventing and treating asthma exacerbations.

FUNDING SOURCE

John W. Upham receives grant support from the National Health and Medical Research Council of Australia.

REFERENCES

1. Pitkaranta A, Hayden FG. Rhinoviruses: important respiratory pathogens. *Ann Med.* 1998;30(6):529–537.
2. Schrauf C, Kirchberger S, Majdic O, et al. The ssRNA genome of human rhinovirus induces a type I IFN response but fails to induce maturation in human monocyte-derived dendritic cells. *J Immunol.* 2009;183(7):4440–4448.

3. Steinke JW, Liu LX, Turner RB, Braciale TJ, Borish L. Immune surveillance by rhinovirus-specific circulating CD4(+) and CD8(+) T lymphocytes. *PLoS One*. 2015;10:1.
4. Sen GC. Viruses and interferons. *Annu Rev Microbiol*. 2001;55:255–281.
5. Forbes RL, Gibson PG, Murphy VE, Wark PA. Impaired type I and III interferon response to rhinovirus infection during pregnancy and asthma. *Thorax*. 2012;67 (3):209–214.
6. Khaitov MR, Laza-Stanca V, Edwards MR, et al. Respiratory virus induction of alpha-, beta- and lambda-interferons in bronchial epithelial cells and peripheral blood mononuclear cells. *Allergy*. 2009;64(3):375–386.
7. Contoli M, Message SD, Laza-Stanca V, et al. Role of deficient type III interferon-lambda production in asthma exacerbations. *Nat Med*. 2006;12(9):1023–1026.
8. Xi Y, Finlayson A, White OJ, Carroll ML, Upham JW. Rhinovirus stimulated IFN-alpha production: how important are plasmacytoid DCs, monocytes and endosomal pH? *Clin Transl Immunol*. 2015;:4.
9. Pritchard AL, Carroll ML, Burel JG, White OJ, Phipps S, Upham JW. Innate IFNs and Plasmacytoid dendritic cells constrain Th2 cytokine responses to rhinovirus: a regulatory mechanism with relevance to asthma. *J Immunol*. 2012;188(12):5898–5905.
10. Xi Y, Troy NM, Anderson D, et al. Critical role of plasmacytoid dendritic cells in regulating gene expression and innate immune responses to human rhinovirus-16. *Front Immunol*. 2017:8.
11. Chairakaki AD, Saridaki MI, Pyrillou K, et al. Plasmacytoid dendritic cells drive acute asthma exacerbations. *J Allergy Clin Immunol*. 2018;142(2):542–556.e12.
12. Liu YP, Rajamanikham V, Baron M, et al. Association of ORMDL3 with rhinovirus-induced endoplasmic reticulum stress and type I interferon responses in human leucocytes. *Clin Exp Allergy*. 2017;47(3):371–382.
13. Kinet JP. The high-affinity IgE receptor (Fc epsilon RI): from physiology to pathology. *Annu Rev Immunol*. 1999;17:931–972.
14. Durrani SR, Montville DJ, Pratt AS, et al. Innate immune responses to rhinovirus are reduced by the high-affinity IgE receptor in allergic asthmatic children. *J Allergy Clin Immun*. 2012,130(2):489–495.
15. Kirchberger S, Majdic O, Steinberger P, et al. Human rhinoviruses inhibit the accessory function of dendritic cells by inducing sialoadhesin and B7-H1 expression. *J Immunol*. 2005;175(2):1145–1152.
16. Gern JE, Galagan DM, Dick EC. Rhinovirus enters but does not replicate inside airway macrophages. *J Allergy Clin Immun*. 1994;93(1):203.
17. Laza-Stanca V, Stanciu LA, Message SD, Edwards MR, Gern JE, Johnston SL. Rhinovirus replication in human macrophages induces NF-kappa B-dependent tumor necrosis factor alpha production. *J Virol*. 2006;80(16):8248–8258.
18. Karta MR, Wickert LE, Curran CS, et al. Allergen challenge in vivo alters rhinovirus-induced chemokine secretion from human airway macrophages. *J Allergy Clin Immun*. 2014;133(4). 1227–1230.
19. Kelly JT, Busse WW. Host immune responses to rhinovirus: mechanisms in asthma. *J Allergy Clin Immun*. 2008;122(4):671–682.
20. Korpi-Steiner NL, Valkenaar SM, Bates ME, Evans MD, Gern JE, Bertics PJ. Human monocytic cells direct the robust release of CXCL10 by bronchial epithelial cells during rhinovirus infection. *Clin Exp Allergy*. 2010;40(8):1203–1213.
21. Nagarkar DR, Bowman ER, Schneider D, et al. Rhinovirus infection of allergen-sensitized and -challenged mice induces eotaxin release from functionally polarized macrophages. *J Immunol*. 2010;185(4):2525–2535.
22. Schneider D, Hong JY, Bowman ER, et al. Macrophage/epithelial cell CCL2 contributes to rhinovirus-induced hyperresponsiveness and inflammation in a mouse model

of allergic airways disease. *Am J Physiol Lung Cell Mol Physiol.* 2013;304(3): L162–L169.
23. Han M, Chung Y, Young Hong J, et al. Toll-like receptor 2-expressing macrophages are required and sufficient for rhinovirus-induced airway inflammation. *J Allergy Clin Immunol.* 2016;138(6):1619–1630.
24. Bentley JK, Sajjan US, Dzaman MB, et al. Rhinovirus colocalizes with CD68-and CD11b-positive macrophages following experimental infection in humans. *J Allergy Clin Immun.* 2013;132(3):758–761.
25. Rajput C, Walsh MP, Eder BN, Metitiri EE, Popova AP, Hershenson MB. Rhinovirus infection induces distinct transcriptome profiles in polarized human macrophages. *Physiol Genomics.* 2018;50(5):299–312.
26. Oliver BGG, Lim S, Wark P, et al. Rhinovirus exposure impairs immune responses to bacterial products in human alveolar macrophages. *Thorax.* 2008;63(6):519–525.
27. Turner RB. The role of neutrophils in the pathogenesis of rhinovirus infections. *Pediatr Infect Dis J.* 1990;9(11):832–835.
28. Winther B, Farr B, Turner RB, Hendley JO, Gwaltney JM, Mygind N. Histopathologic examination and enumeration of polymorphonuclear leukocytes in the nasal-mucosa during experimental rhinovirus colds. *Acta Oto-Laryngol.* 1984:19–24.
29. Naclerio RM, Proud D, Lichtenstein LM, et al. Kinins are generated during experimental rhinovirus colds. *J Infect Dis.* 1988;157(1):133–142.
30. Zhu Z, Tang WL, Gwaltney JM, Wu Y, Elias JA. Rhinovirus stimulation of interleukin-8 in vivo and in vitro: role of NF-kappa B. *Am J Physiol-Lung C.* 1997;273(4):L814–L824.
31. Jarjour NN, Gern JE, Kelly EAB, Swenson CA, Dick CR, Busse WW. The effect of an experimental rhinovirus 16 infection on bronchial lavage neutrophils. *J Allergy Clin Immun.* 2000;105(6):1169–1177.
32. Drescher B, Bai F. Neutrophil in viral infections, friend or foe? *Virus Res.* 2013;171 (1):1–7.
33. Mantovani A, Cassatella MA, Costantini C, Jaillon S. Neutrophils in the activation and regulation of innate and adaptive immunity. *Nat Rev Immunol.* 2011;11 (8):519–531.
34. Mallia P, Message SD, Gielen V, et al. Experimental rhinovirus infection as a human model of chronic obstructive pulmonary disease exacerbation. *Am J Resp Crit Care.* 2011;183(6):734–742.
35. Denlinger LC, Sorkness RL, Lee WM, et al. Lower airway rhinovirus burden and the seasonal risk of asthma exacerbation. *Am J Resp Crit Care.* 2011;184 (9):1007–1014.
36. Rohde G, Message SD, Haas JJ, et al. CXC chemokines and antimicrobial peptides in rhinovirus-induced experimental asthma exacerbations. *Clin Exp Allergy.* 2014;44 (7):930–939.
37. Schonrich G, Raftery MJ. Neutrophil extracellular traps go viral. *Front Immunol.* 2016:7.
38. Toussaint M, Jackson DJ, Swieboda D, et al. Host DNA released by NETosis promotes rhinovirus-induced type-2 allergic asthma exacerbation. *Nat Med.* 2017;23 (6):681– + .
39. Tang FSM, Hansbro PM, Burgess JK, Ammit AJ, Baines KJ, Oliver BG. A novel immunomodulatory function of neutrophils on rhinovirus-activated monocytes in vitro. *Thorax.* 2016;71(11):1039–1049.
40. Fraenkel DJ, Bardin PG, Sanderson G, Lampe F, Johnston SL, Holgate ST. Lower Airways Inflammation during rhinovirus colds in normal and in asthmatic subjects. *Am J Resp Crit Care.* 1995;151(3):879–886.

41. Calhoun WJ, Dick EC, Schwartz LB, Busse WW. A common cold virus, rhinovirus-16, potentiates airway inflammation after segmental antigen bronchoprovocation in allergic subjects. *J Clin Invest*. 1994;94(6):2200–2208.
42. Papadopoulos NG, Papi A, Meyer J, et al. Rhinovirus infection up-regulates eotaxin and eotaxin-2 expression in bronchial epithelial cells. *Clin Exp Allergy*. 2001;31 (7):1060–1066.
43. Sannohe S, Adachi T, Hamada K, et al. Upregulated response to chemokines in oxidative metabolism of eosinophils in asthma and allergic rhinitis. *Eur Respir J*. 2003;21 (6):925–931.
44. Grunberg K, Smits HH, Timmers MC, et al. Experimental rhinovirus 16 infection—effects on cell differentials and soluble markers in sputum in asthmatic subjects. *Am J Resp Crit Care*. 1997;156(2):609–616.
45. Jatakanon A, Lim S, Barnes PJ. Changes in exhaled nitric oxide and sputum eosinophils predict loss of asthma control. *Am J Resp Crit Care*. 1999;159(3):A121.
46. Avila PC, Abisheganaden JA, Wong H, et al. Effects of allergic inflammation of the nasal mucosa on the severity of rhinovirus 16 cold. *J Allergy Clin Immun*. 2000;105 (5):923–932.
47. Grissell TV, Powell H, Shafren DR, et al. Interleukin-10 gene expression in acute virus-induced asthma. *Am J Resp Crit Care*. 2005;172(4):433–439.
48. Takatsu K, Nakajima H. IL-5 and eosinophilia. *Curr Opin Immunol*. 2008;20 (3):288–294.
49. Message SD, Laza-Stanca V, Mallia P, et al. Rhinovirus-induced lower respiratory illness is increased in asthma and related to virus load and Th1/2 cytokine and IL-10 production. *Proc Natl Acad Sci USA*. 2008;105(36):13562–13567.
50. Hatchwell L, Collison A, Girkin J, et al. Toll-like receptor 7 governs interferon and inflammatory responses to rhinovirus and is suppressed by IL-5-induced lung eosinophilia. *Thorax*. 2015;70(9):854–861.
51. Handzel ZT, Busse WW, Sedgwick JB, et al. Eosinophils bind rhinovirus and activate virus-specific T cells. *J Immunol*. 1998;160(3):1279–1284.
52. Stone KD, Prussin C, Metcalfe DD. IgE, mast cells, basophils, and eosinophils. *J Allergy Clin Immun*. 2010;125(2):S73–S80.
53. Cromheecke JL, Nguyen KT, Huston DP. Emerging role of human basophil biology in health and disease. *Curr Allergy Asthm Rep*. 2014;14:1.
54. Nagata Y, Kamijuku H, Taniguchi M, Ziegler S, Seino K. Differential role of thymic stromal lymphopoietin in the induction of airway hyperreactivity and Th2 immune response in antigen-induced asthma with respect to natural killer T cell function. *Int Arch Allergy Immunol*. 2007;144(4):305–314.
55. Beale J, Jayaraman A, Jackson DJ, et al. Rhinovirus-induced IL-25 in asthma exacerbation drives type 2 immunity and allergic pulmonary inflammation. *Sci Transl Med*. 2014;6:256.
56. Akoto C, Davies DE, Swindle EJ. Mast cells are permissive for rhinovirus replication: potential implications for asthma exacerbations. *Clin Exp Allergy*. 2017;47 (3):351–360.
57. Chonmaitree T, Lettbrown MA, Tsong Y, Goldman AS, Baron S. Role of interferon in leukocyte histamine-release caused by common respiratory viruses. *J Infect Dis*. 1988;157(1):127–132.
58. Calhoun WJ, Swenson CA, Dick EC, Schwartz LB, Lemanske RF, Busse WW. Experimental rhinovirus-16 infection potentiates histamine-release after antigen bronchoprovocation in allergic subjects. *Am Rev Respir Dis*. 1991;144(6):1267–1273.
59. Hosoda M, Yamaya M, Suzuki T, et al. Effects of rhinovirus infection on histamine and cytokine production by cell lines from human mast cells and basophils. *J Immunol*. 2002;169(3):1482–1491.

60. Agrawal R, Wisniewski J, Yu MD, et al. Infection with human rhinovirus 16 promotes enhanced IgE responsiveness in basophils of atopic asthmatics. *Clin Exp Allergy*. 2014;44(10):1266–1273.
61. Wernersson S, Pejler G. Mast cell secretory granules: armed for battle. *Nat Rev Immunol*. 2014;14(7):478–494.
62. Seymour ML, Gilby N, Bardin PG, et al. Rhinovirus infection increases 5-lipoxygenase and cyclooxygenase-2 in bronchial biopsy specimens from nonatopic subjects. *J Infect Dis*. 2002;185(4):540–544.
63. Gern JE, Joseph B, Galagan DM, Borcherding WR, Dick EC. Rhinovirus inhibits antigen-specific T cell proliferation through an intercellular adhesion molecule-1-dependent mechanism. *J Infect Dis*. 1996;174(6):1143–1150.
64. Gern JE, Vrtis R, Kelly EAB, Dick EC, Busse WW. Rhinovirus produces nonspecific activation of lymphocytes through a monocyte-dependent mechanism. *J Immunol*. 1996;157(4):1605–1612.
65. Spurrell JCL, Wiehler S, Zaheer RS, Sanders SP, Proud D. Human airway epithelial cells produce IP-10 (CXCL10) in vitro and in vivo upon rhinovirus infection. *Am J Physiol-Lung C*. 2005;289(1):L85–L95.
66. Wark PAB, Bucchieri F, Johnston SL, et al. IFN-gamma-induced protein 10 is a novel biomarker of rhinovirus-induced asthma exacerbations. *J Allergy Clin Immun*. 2007;120(3):586–593.
67. Fehniger TA, Caligiuri MA. Interleukin 15: biology and relevance to human disease. *Blood*. 2001;97(1):14–32.
68. Jayaraman A, Jackson DJ, Message SD, et al. IL-15 complexes induce NK- and T-cell responses independent of type I IFN signaling during rhinovirus infection. *Mucosal Immunol*. 2014;7(5):1151–1164.
69. Moro K, Koyasu S. Innate production of Th2 cytokines by adipose tissue-associated c-Kit(+)Sca-1(+) lymphoid cells. *J Immunol*. 2010:184.
70. Neill DR, Wong SH, Bellosi A, et al. Nuocytes represent a new innate effector leukocyte that mediates type-2 immunity. *Nature*. 2010;464(7293):1367–U1369.
71. Chang YJ, Kim HY, Albacker LA, et al. Innate lymphoid cells mediate influenza-induced airway hyper-reactivity independently of adaptive immunity. *Nat Immunol*. 2011;12(7):631–U186.
72. Lloyd CM, Snelgrove RJ. Type 2 immunity: expanding our view. *Sci Immunol*. 2018;3:25.
73. Mjosberg J, Spits H. Type 2 innate lymphoid cells-new members of the "type 2 franchise" that mediate allergic airway inflammation. *Eur J Immunol*. 2012;42(5):1093–1096.
74. Hong JY, Bentley JK, Chung YT, et al. Neonatal rhinovirus induces mucous metaplasia and airways hyperresponsiveness through IL-25 and type 2 innate lymphoid cells. *J Allergy Clin Immun*. 2014;134(2):429– + .
75. Jackson DJ, Makrinioti H, Rana BMJ, et al. IL-33-dependent type 2 inflammation during rhinovirus-induced asthma exacerbations in vivo. *Am J Resp Crit Care*. 2014;190(12):1373–1382.
76. Duerr CU, McCarthy CD, Mindt BC, et al. Type I interferon restricts type 2 immunopathology through the regulation of group 2 innate lymphoid cells. *Nat Immunol*. 2016;17(1):65–75.
77. Levandowski RA, Ou DW, Jackson GG. Acute-phase decrease of lymphocyte-T subsets in rhinovirus infection. *J Infect Dis*. 1986;153(4):743–748.
78. Ilarraza R, Wu YQ, Skappak CD, Ajamian F, Proud D, Adamko DJ. Rhinovirus has the unique ability to directly activate human T cells in vitro. *J Allergy Clin Immun*. 2013;131(2):395–404.

79. Gern JE, Vrtis R, Grindle KA, Swenson C, Busse WW. Relationship of upper and lower airway cytokines to outcome of experimental rhinovirus infection. *Am J Resp Crit Care*. 2000;162(6):2226–2231.
80. Parry DE, Busse WW, Sukow KA, Dick CR, Swenson C, Gern JE. Rhinovirus-induced PBMC responses and outcome of experimental infection in allergic subjects. *J Allergy Clin Immun*. 2000;105(4):692–698.
81. Papadopoulos NG, Stanciu LA, Papi A, Holgate ST, Johnston SL. A defective type 1 response to rhinovirus in atopic asthma. *Thorax*. 2002;57(4):328–332.
82. Pritchard AL, White OJ, Burel JG, Upham JW. Innate interferons inhibit allergen and microbial specific T(H)2 responses. *Immunol Cell Biol*. 2012;90(10):974–977.
83. Pritchard AL, White OJ, Burel JG, Carroll ML, Phipps S, Upham JW. Asthma is associated with multiple alterations in anti-viral innate signalling pathways. *PLoS One*. 2014;9(9):e106501.
84. O'Shea JJ, Ma A, Lipsky P. Cytokines and autoimmunity. *Nat Rev Immunol*. 2002;2 (1):37–45.
85. Odemuyiwa SO, Ghahary A, Li YY, et al. Cutting edge: human eosinophils regulate T cell subset selection through indoleamine 2,3-dioxygenase. *J Immunol*. 2004;173 (10):5909–5913.
86. Bianco A, Sethi SK, Allen JT, Knight RA, Spiteri MA. Th2 cytokines exert a dominant influence on epithelial cell expression of the major group human rhinovirus receptor, ICAM-1. *Eur Respir J*. 1998;12(3):619–626.
87. Papi A, Johnston SL. Rhinovirus infection induces expression of its own receptor intercellular adhesion molecule 1 (ICAM-1) via increased NF-kappa B-mediated transcription. *J Biol Chem*. 1999;274(14):9707–9720.
88. Contoli M, Ito K, Padovani A, et al. Th2 cytokines impair innate immune responses to rhinovirus in respiratory epithelial cells. *Allergy*. 2015;70(8):910–920.
89. Divekar R, Kita H. Recent advances in epithelium-derived cytokines (IL-33, IL-25, and thymic stromal lymphopoietin) and allergic inflammation. *Curr Opin Allergy Clin Immunol*. 2015;15(1):98–103.
90. Perez GF, Rodriguez-Martinez CE, Nino G. Rhinovirus-induced airway disease: a model to understand the antiviral and Th2 epithelial immune dysregulation in childhood asthma. *J Invest Med*. 2015;63(6):792–795.
91. Ito T, Wang YH, Duramad O, et al. TSLP-activated dendritic cells induce an inflammatory T helper type 2 cell response through OX40 ligand. *J Exp Med*. 2005;202 (9):1213–1223.
92. Jurak LM, Xi Y, Landgraf M, Carroll ML, Murray L, Upham JW. Interleukin 33 selectively augments rhinovirus-induced type 2 immune responses in asthmatic but not healthy people. *Front Immunol*. 2018:9.
93. Miyauchi K. Helper T cell responses to respiratory viruses in the lung: development, virus suppression, and pathogenesis. *Viral Immunol*. 2017;30(6):421–430.
94. Stockinger B, Omenetti S. The dichotomous nature of T helper 17 cells. *Nat Rev Immunol*. 2017;17(9):535–544.
95. Ferretti S, Bonneau O, Dubois GR, Jones CE, Trifilieff A IL-17. produced by lymphocytes and neutrophils, is necessary for lipopolysaccharide-induced airway neutrophilia: IL-15 as a possible trigger. *J Immunol*. 2003;170(4):2106–2112.
96. Wiehler S, Proud D. Interleukin-17A modulates human airway epithelial responses to human rhinovirus infection. *Am J Physiol-Lung C*. 2007;293(2):L505–L515.
97. Choy DF, Hart KM, Borthwick LA, et al. T(H)2 and T(H)17 inflammatory pathways are reciprocally regulated in asthma. *Sci Transl Med*. 2015;7(301):301ra129.
98. Perez GF, Pancham K, Huseni S, et al. Rhinovirus-induced airway cytokines and respiratory morbidity in severely premature children. *Pediatr Allerg Immunol UK*. 2015;26(2):145–152.

99. Gong Y, Zhao C, Zhao P, et al. Role of IL-10-producing regulatory B cells in chronic hepatitis B virus infection. *Dig Dis Sci*. 2015;60(5):1308–1314.
100. Krishnamoorthy N, Khare A, Oriss TB, et al. Early infection with respiratory syncytial virus impairs regulatory T cell function and increases susceptibility to allergic asthma. *Nat Med*. 2012;18(10):1525–U1121.
101. Gaido CM, Granland C, Laing IA, et al. T-cell responses against rhinovirus species A and C in asthmatic and healthy children. *Immun Inflamm Dis*. 2018;6(1):143–153.
102. van Benten IJ, van Drunen CM, Koevoet JLM, et al. Reduced nasal IL-10 and enhanced TNF alpha responses during rhinovirus and RSV-Induced upper respiratory tract infection in atopic and non-atopic infants. *J Med Virol*. 2005;75 (2):348–357.
103. Thomas BJ, Lindsay M, Dagher H, et al. Transforming growth factor-beta enhances rhinovirus infection by diminishing early innate responses. *Am J Resp Cell Mol*. 2009;41(3):339–347.
104. Jartti T, Paul-Anttila M, Lehtinen P, et al. Systemic T-helper and T-regulatory cell type cytokine responses in rhinovirus vs. respiratory syncytial virus induced early wheezing: an observational study. *Resp Res*. 2009;10:85.
105. Roh DE, Park SH, Choi HJ, Kim YH. Comparison of cytokine expression profiles in infants with a rhinovirus induced lower respiratory tract infection with or without wheezing: a comparison with respiratory syncytial virus. *Korean J Pediatr*. 2017;60 (9):296–301.
106. Seyer M, Kirchberger S, Majdic O, et al. Human rhinoviruses induce IL-35-producing Treg via induction of B7-H1 (CD274) and sialoadhesin (CD169) on DC. *Eur J Immunol*. 2010;40(2):321–329.
107. Bielor C, Sopel N, Maier A, et al. Role of TGF-beta in anti-rhinovirus immune responses in asthmatic patients. *J Allergy Clin Immunol*. 2017;140(1):283–286.e10.
108. Gern JE, Busse WW. Association of rhinovirus infections with asthma. *Clin Microbiol Rev*. 1999;12(1):9– + .
109. Jacobs SE, Lamson DM, George St K, Walsh TJ. Human rhinoviruses. *Clin Microbiol Rev*. 2013;26(1):135–162.
110. Kennedy JL, Turner RB, Braciale T, Heymann PW, Borish L. Pathogenesis of rhinovirus infection. *Curr Opin Virol*. 2012;2(3):287–293.
111. Muehling LM, Mai DT, Kwok WW, Heymann PW, Pomes A, Woodfolk JA. Circulating memory CD4(+) T cells target conserved epitopes of rhinovirus capsid proteins and respond rapidly to experimental infection in humans. *J Immunol*. 2016;197(8):3214–3224.
112. Gaido CM, Stone S, Chopra A, Thomas WR, Le Souef PN, Hales BJ. Immunodominant T-cell epitopes in the VP1 capsid protein of rhinovirus species A and C. *J Virol*. 2016;90(23):10459–10471.
113. Mogabgab WJ, Pelon W. Problems in characterizing and identifying an apparently new virus found in association with mild respiratory disease in recruits. *Ann NY Acad Sci*. 1957;67(8):403–412.
114. Dick EC, Blumer CR, Evans AS. Epidemiology of infections with rhinovirus types 43 and 55 in a group of university of Wisconsin student families. *Am J Epidemiol*. 1967;86(2):386–400.
115. Hendley JO, Gwaltney Jr JM, Jordan Jr. WS. Rhinovirus infections in an industrial population. IV. Infections within families of employees during two fall peaks of respiratory illness. *Am J Epidemiol*. 1969;89(2):184–196.
116. Barclay WS, al-Nakib W, Higgins PG, Tyrrell DA. The time course of the humoral immune response to rhinovirus infection. *Epidemiol Infect*. 1989;103(3):659–669.
117. Message SD, Johnston SL. The immunology of virus infection in asthma. *Eur Respir J*. 2001;18(6):1013–1025.

118. Alper CM, Doyle WJ, Skoner DP, Buchman CA, Cohen S, Gwaltney JM. Prechallenge antibodies moderate disease expression in adults experimentally exposed to rhinovirus strain Hanks. *Clin Infect Dis.* 1998;27(1):119–128.
119. Iwasaki J, Smith WA, Khoo SK, et al. Comparison of rhinovirus antibody titers in children with asthma exacerbations and species-specific rhinovirus infection. *J Allergy Clin Immunol.* 2014;134(1):25–32.
120. van Kempen M, Bachert C, Van Cauwenberge P. An update on the pathophysiology of rhinovirus upper respiratory tract infections. *Rhinology.* 1999;37(3):97–103.
121. Glanville N, Johnston SL. Challenges in developing a cross-serotype rhinovirus vaccine. *Curr Opin Virol.* 2015;11:83–88.
122. Stenberg-Hammar K, Hedlin G, Soderhall C. Rhinovirus and preschool wheeze. *Pediatr Allergy Immunol.* 2017;28(6):513–520.
123. Edlmayr J, Niespodziana K, Popow-Kraupp T, et al. Antibodies induced with recombinant VP1 from human rhinovirus exhibit cross-neutralisation. *Eur Respir J.* 2011;37(1):44–52.
124. Niespodziana K, Napora K, Cabauatan C, et al. Misdirected antibody responses against an N-terminal epitope on human rhinovirus VP1 as explanation for recurrent RV infections. *FASEB J.* 2012;26(3):1001–1008.
125. McLean GR, Walton RP, Shetty S, et al. Rhinovirus infections and immunisation induce cross-serotype reactive antibodies to VP1. *Antiviral Res.* 2012;95(3):193–201.
126. Glanville N, McLean GR, Guy B, et al. Cross-serotype immunity induced by immunization with a conserved rhinovirus capsid protein. *PLoS Pathog.* 2013;9(9):e1003669.
127. Lee S, Nguyen MT, Currier MG, et al. A polyvalent inactivated rhinovirus vaccine is broadly immunogenic in rhesus macaques. *Nat Commun.* 2016;7:12838.

CHAPTER 5

Rhinoviruses and the onset of asthma

James E. Gern

[1]Departments of Pediatrics and Medicine, University of Wisconsin-Madison, Madison, WI, United States

5.1 INTRODUCTION

Asthma often begins in early childhood, and the first indication of asthma is usually episodic virus-induced wheezing. The clinical similarities between viral wheeze and the temporal progression from viral wheeze to asthma with recurrent wheezing triggered by multiple stimuli suggest a causal relationship. In fact, studies using palivizumab have established that early life wheezing illnesses caused by respiratory syncytial virus can initiate recurrent wheezing and probably nonallergic asthma during early childhood.[1–4] Mechanisms are poorly understood, but the severity of respiratory syncytial virus (RSV) illness correlates with the subsequent risk for developing asthma.[5] These findings suggest that virus-induced damage to the small airways during a key period of lung development could cause long-term disturbances in airway function and possible structure, and that these processes could be a root cause of childhood asthma. Notably, RSV infections are ubiquitous in early life; thus, it is not the infection per se but rather more severe RSV illnesses that confer increased risk for subsequent wheezing illnesses.

Rhinoviruses (RVs) were first described as common cold viruses, and in fact cause approximately half of all common colds. In addition to causing colds, RVs are also detected in acute wheezing illnesses in young children and in patients of all ages with chronic respiratory diseases such as asthma, cystic fibrosis, and chronic obstructive lung disease.[6] RV infection is the second leading cause for bronchiolitis during infancy.[7,8] Furthermore, children who wheeze with RVs are at especially high risk for developing asthma that is persistent, and perhaps accompanied by respiratory allergies. Both RV and RSV infections are ubiquitous in early childhood. Consequently (and similar to RSV), it is not RV infections that are associated with asthma, but rather more severe RV illnesses. This

Rhinovirus Infections
DOI: https://doi.org/10.1016/B978-0-12-816417-4.00005-6

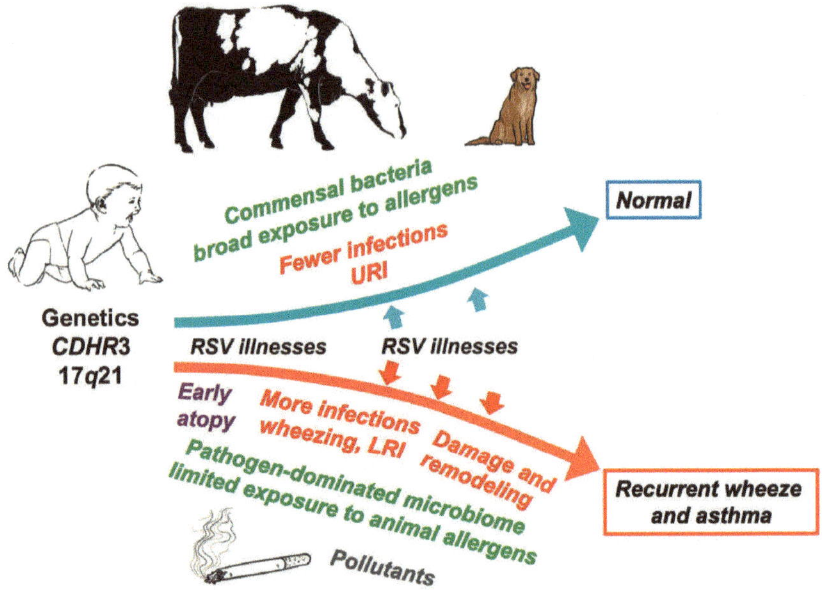

Figure 5.1 Personal and environmental factors associated with increased RV-induced wheezing and subsequent childhood asthma. *RV*, rhinovirus.

implies that limiting the severity of RV wheezing illnesses could also reduce the subsequent risk for developing asthma.

There are a number of factors related to the host, virus, and environment that promote more severe RV illnesses (Fig. 5.1). In the absence of clinically useful antiviral medications that can reduce the frequency and severity of RV infections, identification of cofactors for more severe RV illnesses could lead to new strategies for the prevention and treatment of RV infections. This review will focus on epidemiologic and experimental evidence linking RV illnesses in early life to childhood asthma. In addition, cofactors for more severe RV illnesses will be reviewed, and discussed in terms of their potential application to the prevention of early childhood asthma.

5.2 RHINOVIRUS INFECTIONS IN EARLY LIFE

RV has an especially large family tree,[9,10] and this is an important reason for the multitude of RV infections during childhood. RVs have three species (A, B, C) that are comprised of over 160 different types. Neutralizing epitopes of individual RV-A types ($n = 80$) and RV-B types

($n = 32$) are unique (RV-C types have yet to be tested), which means that infection with any given virus provides little protection against other RV types.[11,12] Since 20–30 RV types circulate in a community during peak seasons (generally spring and fall), and 70%–80% of the types change each season, children are generally susceptible to RVs circulating in the community year after year.[13,14] In fact, studies involving routine monitoring of young children have reported that ~40% of children are infected at any one point in time.[15] Even though most infections cause mild symptoms, the high rate of RV infections in the first few years of age leads to considerable morbidity. Toivonen et al. conducted a comprehensive study of RV infections in 923 children during the first 2 years of life, and collected samples during symptomatic periods and at routine intervals.[16] They documented an annual rate of 5.9 illnesses per child, and RV was responsible for 59% of these episodes (mean annual rate for RV illnesses 3.5 per child). This included rates of 47 per 100 children with RV-associated otitis media, and 61 per 100 children who were treated with an antibiotic during a RV infection. In addition, detection of RV is second only to RSV during episodes of bronchiolitis in the first year of life, and is the most commonly detected virus in wheezing children over the age of 2 years.[17,18] There are several differences between children who wheeze with RSV and those who wheeze with RV. As mentioned above, RSV wheezing tends to occur in younger children. By age 2 years ~90% of children are seropositive to RSV. Detectable antibody to RSV moderates the severity of, but does not prevent, subsequent RSV illnesses. Children who wheeze with RV are more likely to be atopic, as indicated by atopic dermatitis, allergic sensitization, and/or a parental history of allergy or asthma.[7,19–25]

5.3 THE ASSOCIATION BETWEEN RHINOVIRUS INFECTIONS IN EARLY LIFE AND ASTHMA

The association between RV wheezing in infancy and a high risk for subsequent development of asthma was first reported in a case–control study of children less than 2 years of age hospitalized for a wheezing illness in Finland.[19] Infants infected with RV were more likely to develop asthma by age 6 years compared with those infected with other respiratory viruses, including RSV and enteroviruses. These relationships were stable as the children aged, as documented by follow-up studies through age 18 years.[26–28] Additional studies conducted in Finland,[29–31] Rome,[32] the United States, and Japan[33] have since confirmed that infants hospitalized with RV are more likely to develop recurrent wheezing and asthma compared with infants who wheeze with other viruses.

These studies supported findings in two birth cohorts of children at increased risk for asthma based on parental history of allergies or asthma. In the Children's Asthma Study (CAS, Perth, Australia), asthma at age 5 years was positively related [odds ratio (OR) 4.1] to the combination of early allergic sensitization and wheezing illnesses with detection of RV or RSV.[34] In the Childhood Origins of Asthma study (COAST, Madison, WI, United States), the etiology of acute wheezing illnesses in early life was related to the risk for recurrent wheeze at age 3 years[35] and subsequent asthma.[36] The magnitude of the risk was related to the etiology of the wheezing illness: RV wheezing was associated with greater risk for asthma compared with RV wheezing (OR 9.8 vs 2.6). These relationships have been stable over time; at age 13 years, early life wheezing with RV (OR 3.3), but not RSV (OR 1.0), was associated with asthma.[37] As in the Australian study, RV wheezing and early sensitization had additive effects on the risk for subsequent asthma. The close association between the etiology of viral wheeze in early life and childhood asthma was further tested in the COAST study using a generalized additive logistic regression model, which allows for detection of nonlinear relationships between illnesses caused by specific viruses and the risk for asthma.[38] At each age that asthma was assessed (6, 8, 11, and 13 years), RV wheezing illnesses during infancy were more informative for asthma risk compared with non-RV wheezing illnesses.

In contrast to CAS, COAST, and the studies of hospitalized infants, a Danish cohort (Copenhagen Prospective Studies on Asthma in Childhood, or COPSAC) reported that wheezing episodes in early life predicted asthma independent of viral etiology.[39] The differences in findings could be related to the methods that were used for diagnostic virology. The recovery rates for viruses were lower in COPSAC (65%) compared with COAST (93%), and this is especially true for detection of RV (49% in COAST, 23% in COPSAC). RV-C illnesses are closely related to subsequent asthma,[40] and also require optimized primers in PCR-based assays for greatest sensitivity.[41] Incomplete detection of RV-C infections in COPSAC may have impaired the ability to detect relationships between viral etiology of wheezing illnesses and subsequent asthma.

As discussed previously, the close association between RV wheeze in early life and subsequent asthma has been demonstrated in high risk children, including those hospitalized with acute wheeze and those with increased risk for asthma based on parental history. In children participating in a population-based birth cohort study in the Netherlands, RV wheezing episodes in early life were associated with an increased risk for recurrent wheeze at age 3 years.[42] Further follow-up of this cohort should help to determine whether RV wheezing illnesses in low-risk infants are also a risk factor for subsequent asthma.

5.4 VIRAL FACTORS

The virulence of RV illnesses varies with species; in outpatient studies, RV-A and RV-C are more likely to cause symptomatic illnesses compared with RV-B, which is most often asymptomatic.[43] RV-C has also been associated with increased viral load,[44] an increased risk for viremia,[45–47] and greater lethargy compared with other RV species.[48] RV-C has been linked to greater severity of acute wheezing in children presented for emergency care,[49] and is detected more often than other RV species during intensive care unit admissions for acute wheeze.[50] Recurrent wheezing episodes are more likely to occur after acute infections with RV-C and/or RV-A.[51,52] In addition, many studies have found that RV-C is overrepresented in children with acute wheezing or other lower respiratory illnesses,[48,49,53–58] while others have reported similar contributions for RV-A and RV-C.[59–61] The reasons for the different findings in acute wheezing studies are speculative, but could include differences in study populations (e.g., age, asthma status), selection of controls for case–control studies, differences in circulating viruses, and viral diagnostics.[41,58,62]

Molecular and life cycle differences among the three viral species have could contribute to differences in virulence. RV-C growth is not inhibited at core body temperature (37°C), which could promote replication in small airways.[63] In addition, growth rates in differentiated cultures of airway epithelial cells and corresponding cytokine and chemokine responses are higher for RV-A and RV-C compared with RV-B.[64] The structure of RV-C has recently been demonstrated by cryo-electron microscopy, and the receptor binding platform is distinct from other RV species.[65] Accordingly, RV-A (ICAM-1 or (low density lipoprotein receptor) LDLR), RV-B (ICAM-1), and RV-C (CDHR3[66]) bind to different receptors. Studies are ongoing to link distinct characteristics of RV types or species to clinical outcomes.

5.5 HOST FACTORS

Host factors that influence the relationship between RV infections in early life and subsequent asthma include genetics, the early onset of respiratory allergies and other atopic features, and immunologic factors. Two genetic loci related to both RV wheeze and childhood asthma are 17q21 and *CDHR3*. In genome-wise association studies, 17q21 is the region that is most strongly and consistently associated with childhood asthma. This locus contains five genes that are in close linkage disequilibrium, and so which of the genes is functionally linked to asthma and the underlying mechanisms for this linkage is uncertain. The 17q21 variants were associated with the risk of early life wheezing illnesses with RV, but not RSV.[67] Interestingly, there was an interaction with respect to asthma; the associations of 17q21 variants with asthma were only significant for children who had had RV wheezing illnesses. The 17q21-associated asthma risk was independent of allergic sensitization, suggesting that 17q21 may regulate either the severity of the initial illness or else repair or remodeling of the airway following virus-induced damage.

Another genetic factor that influences the risk of RV wheezing and asthma is a single nucleotide polymorphism (rs6967330) in *CDHR3*. This allele has also been linked to early childhood asthma characterized by acute wheezing episodes. CDHR3 was subsequently identified as the RV-C receptor, and is expressed on ciliated airway epithelial cells.[66,68] The rs6967330$_{C\rightarrow T}$ is a functional variant that puts more CDHR3 on the surface of the cell. In the context of RV-C infection, this leads to greater viral binding and replication, and also increased frequency of infections and illnesses that are attributable to RV-C.[66,69] These findings suggest that the link between rs6967330$_{C\rightarrow T}$ and childhood asthma is mediated by increasing the number and severity of RV-C infections and illnesses. Thus, in genetically susceptible children, acute RV-C LRIs could damage and inflame the airways, possibly followed by airway remodeling and chronic obstruction.

Another important cofactor for the development of asthma following RV wheezing episodes is allergy. As discussed previously, early sensitization to respiratory allergies and RV wheezing both increase the risk for subsequent asthma, which raises the question as to whether

Table 5.1 Proposed mechanisms for allergic inflammation to impair antiviral responses.

Component	Effect on antiviral responses	References
IgE and $Fc_{\varepsilon}RI$	• ↑ IgE leads to ↑ FcεRI expression on monocytes and dendritic cells; ↑ FcεRI expression and especially cross-linking this receptor impairs interferon responses to viruses	[70,71]
Eosinophils	• Eosinophils suppress of TLR7 expression and interferon responses • Eosinophils secrete soluble TGFβ, which suppresses interferon responses	[72,73]
Type 2 cytokines	• IL-33 inhibits RV-induced IFN-λ responses	[74]
Transcription factors	• SOCS1 expression in bronchial epithelium leads to reduced RV-induced interferon responses • IL-13 and RV both induce FOXA3 expression with promotes goblet cell hyperplasia and reduces epithelial interferon responses	[75,76]

RV, rhinovirus; *TGF*, transforming growth factor-β.

allergies and RV wheezing are related. In the COAST birth cohort study, early allergic sensitization was a risk factor for RV wheezing, while viral wheezing episodes in early life did not increase the risk for allergic sensitization.[24] These findings suggest that allergic sensitization might increase the susceptibility to RV illnesses, perhaps through an immunologic mechanism. In fact, several experimental models have demonstrated potential mechanisms for allergic inflammation to inhibit antiviral responses and thereby increase the frequency and severity of RV illnesses (Table 5.1). An interaction between IgE-mediated allergic inflammation and antiviral mechanisms has also been demonstrated in clinical studies. Treatment of children with allergic asthma with omalizumab not only reduced virus-induced exacerbations, but also reduced the number of viral colds and the duration and magnitude of RV shedding, and increased RV-induced interferon-α (IFN-α) responses of PBMC and plasmacytoid dendritic cells ex vivo.[77–79] These findings provide evidence that IgE-mediated allergic inflammation increases susceptibility to RV infections and illnesses.

5.6 ENVIRONMENTAL FACTORS

Environmental factors in early life that reduce the risk for viral wheezing illnesses may also have parallel effects on the development of childhood asthma. Two environmental cofactors are airway bacteria and exposure to airborne pollutants.

In the Danish COPSAC birth cohort study, two key findings focused attention on the potential role of airway bacteria in asthma. First, infants who had one of three bacterial pathogens (*Streptococcus pneumoniae, Moraxella catarrhalis, Haemophilus influenzae*) cultured from the nasopharynx at 1 month of age were at increased risk for developing recurrent wheeze and early childhood asthma.[80] Second, analysis of nasal samples the first 3 years of life demonstrated that detection of either viral or bacterial pathogens were equally associated with wheezing illnesses (OR 2.8 for viruses and 2.9 for bacteria).[81] Following these studies, there have been a number of reports that acute viral wheezing illnesses are associated with increased detection of bacterial pathogens.[13,40,82–85] In addition, bacterial pathogens are more likely to be detected during infections with more virulent RV (e.g., RV-A and RV-C compared with RV-B).[83] Collectively, these studies strongly suggest that both respiratory viruses and the airway microbiome contribute to acute wheezing illnesses in childhood.

While wheezing illnesses have been linked to bacterial pathogens, studies of children with diverse home settings and environmental exposures suggest that certain commensal bacterial reduce the risk of acute wheezing illnesses and possibly asthma. For example, early life exposures that are inversely related to recurrent wheeze (which is generally associated with viral respiratory infection) include farm exposures and lifestyle,[86] and having a dog in the home.[87] Notably, there is an interaction between these environmental exposures, 17q21 genetics, and asthma; farm exposures and pets in the home lessen the risk of asthma only for children with high-risk genotypes.[86,88] Notably, children raised on dairy farms also have reduced rates of respiratory illnesses in early life.[89] For urban children living in economically disadvantaged neighborhoods, exposure to cat, mice, and cockroach proteins along with a rich home microbiome were related to reduced atopic wheeze[90] and later asthma.[91] These findings suggest that exposure to rich microbial sources and diverse sources of allergenic proteins in early life might reduce the frequency and severity of viral infections, and that this effect could reduce the subsequent risk for developing asthma.

Air pollution could also increase the susceptibility to RV infections. Many pollutants are sources of oxidant stress, which can inhibit RV-induced antiviral responses [92] and enhance proinflammatory cytokine to RV.[93] Clinical studies have identified links between secondhand smoke exposure and virus-induced wheezing illnesses during infancy,[94] and severity of RV-induced exacerbations in children with asthma.[95]

5.7 HOW COULD VIRAL WHEEZING ILLNESSES IN EARLY LIFE PROMOTE ASTHMA?

Interventional studies with palivizumab have established that more severe viral LRI contribute to the risk of recurrent wheeze and probably asthma.[1,4] Several mechanisms have been proposed to explain this association. First, virus-induced damage to the airway cells could lead to disordered repair (or remodeling) of the airways, or disruption of the normal sequence of events related to lung growth or development. In support of this theory, allergy is an important cofactor for the development of asthma following viral wheeze, suggesting the possibility that that allergic inflammation disrupts the repair process following viral infection. The relationship between the airway microbiome and childhood asthma may also depend on allergy. In the CAS cohort, airway colonization with bacterial pathogens in early childhood was found to be a risk factor for asthma, but only in sensitized children.[96] Nonsensitized children with pathogen-dominated airway microbiota were instead at increased risk for transient wheeze. Furthermore, RVs infections can induce vascular endothelial growth factor and transforming growth factor beta; these factors contribute to airway remodeling of blood vessels and subepithelial matrix proteins.[97,98] RV infections can also induce chemoattractants for airway smooth muscle cells.[99] Studies in animal models suggest that effects of RV infections on airway structure and function may be more pronounced for RV infections in early life.[100]

Another possibility is that repeated RV infections might alter the local mucosal environment to promote chronic inflammation, and possibly promote allergic sensitization. RV infections can induce "alarmins" such as IL-33 and thymic stromal lymphopoietin (TSLP) that can secondarily promote type 2 inflammation;[101] however, evidence for an increased risk for allergic sensitization following RV infection is lacking.[24]

5.8 THERAPEUTIC IMPLICATIONS AND CONCLUSION

Several approaches to antivirals with activity against RV are possible, including medications that directly target the virus, those that boost local antiviral immune responses, and development of a vaccine.[102,103] Since there are no antiviral medications that are specific for RV treatment or prevention, intervention studies to determine whether preventing RV infections in early life reduces that risk for developing asthma are lacking.

Even so, identifying cofactors for more severe RV wheezing illnesses presents other opportunities for prevention (Fig. 5.1). Natural exposures to animals on farms and in homes with pets and even pests suggest that exposures to diverse commensal microbes and perhaps a variety of allergenic proteins in early life could moderate the risk for both acute RV wheezing and subsequent asthma. As work proceeds to identify microbial communities or their metabolic products, candidates for interventional studies are likely to emerge. These studies will need to be guided with a thorough understanding of microbial and allergen exposures of healthy children.

While some bacteria may reduce the risk of wheezing illnesses, bacterial pathogens are likely to work in concert with viruses to increase the severity of respiratory illnesses. Two interventional studies with azithromycin have demonstrated short-term benefits during acute wheezing illnesses.[104,105] There are also concerns that using an antibiotic to reduce airway pathogens could also wipe out beneficial microbes, and studies are needed to determine whether treating acute wheezing illnesses with antibiotics is beneficial or detrimental in the long term.

Another approach to reducing viral wheezing illnesses is to reduce of allergic inflammation. In a randomized placebo-controlled study, prednisolone treatment of infants with RV wheezing during the acute illness did not prevent recurrent wheezing or asthma.[106] New biologics that inhibit type 2 inflammation, hopefully without the side effects of systemic corticosteroids, hold promise for reducing RV wheezing and asthma. Data demonstrating beneficial effects of omalizumab on susceptibility to RV illnesses[78] suggest that this approach may be successful.

In summary, RV infections are a source of considerable morbidity in the short term (wheezing illnesses), and possibly in the long term by promoting childhood asthma. Epidemiologic and experimental studies have provided solid data on risk factors for more severe RV illnesses. These studies raise hopes that interventions targeting these risk factors might be used to prevent early life wheezing illnesses and the subsequent development of childhood asthma.

CONFLICTS OF INTEREST DISCLOSURE

Dr. Gern has received grants from the National Institutes of Health/National Institutes of Allergy and Infectious Diseases (NIH/NIAID), consulting fees from PREP Biopharm Inc., Regeneron, Meissa Vaccines Inc. and MedImmune, outside the submitted work. In addition, Dr. Gern has a patent Methods of Propagating Rhinovirus C in Previously Unsusceptible Cell Lines, and a patent pending entitled Adapted Rhinovirus C.

REFERENCES

1. Simoes EA, Groothuis JR, Carbonell-Estrany X, et al. Palivizumab prophylaxis, respiratory syncytial virus, and subsequent recurrent wheezing. *J Pediatr.* 2007;151 (1):34–42. 42.e1.
2. Blanken MO, Rovers MM, Molenaar JM, et al. Respiratory syncytial virus and recurrent wheeze in healthy preterm infants. *N Engl J Med.* 2013;368(19):1791–1799.
3. Carroll KN, Gebretsadik T, Escobar GJ, et al. Respiratory syncytial virus immunoprophylaxis in high-risk infants and development of childhood asthma. *J Allergy Clin Immunol.* 2017;139(1):66–71.e3.
4. Mochizuki H, Kusuda S, Okada K, Yoshihara S, Furuya H, Simoes EA. Palivizumab prophylaxis in preterm infants and subsequent recurrent wheezing: 6 year follow up study. *Am J Respir Crit Care Med.* 2017;196:29–38.
5. Carroll KN, Wu P, Gebretsadik T, et al. The severity-dependent relationship of infant bronchiolitis on the risk and morbidity of early childhood asthma. *J Allergy Clin Immunol.* 2009;123(5):1055–1061. 1061.e1.
6. Kim CK, Callaway Z, Gern JE. Viral infections and associated factors that promote acute exacerbations of asthma. *Allergy Asthma Immunol Res.* 2018;10(1):12–17.
7. Jartti T, Lehtinen P, Vuorinen T, Ruuskanen O. Bronchiolitis: age and previous wheezing episodes are linked to viral etiology and atopic characteristics. *Pediatr Infect Dis J.* 2009;28(4):311–317.
8. Mansbach JM, Clark S, Teach SJ, et al. Children hospitalized with rhinovirus bronchiolitis have asthma-like characteristics. *J Pediatr.* 2016;172:202–204.e1.
9. Liggett SB, Bochkov YA, Pappas T, et al. Genome sequences of rhinovirus C isolates from Wisconsin pediatric respiratory studies. *Genome Announc.* 2014;2(2): e00203–e00214.
10. Palmenberg AC, Rathe JA, Liggett SB. Analysis of the complete genome sequences of human rhinovirus. *J Allergy Clin Immunol.* 2010;125(6):1190–1199.
11. Cooney MK, Kenny GE, Tam R, Fox JP. Cross relationships among 37 rhinoviruses demonstrated by virus neutralization with potent monotypic rabbit antisera. *Infect Immun.* 1973;7(3):335–340.
12. Mufson MA, Ludwig WM, James Jr HD, et al. Effect of neutralizing antibody on experimental rhinovirus infection. *JAMA.* 1963;186:578–584.
13. Kloepfer KM, Lee WM, Pappas TE, et al. Detection of pathogenic bacteria during rhinovirus infection is associated with increased respiratory symptoms and asthma exacerbations. *J Allergy Clin Immunol.* 2014;133(5):1301–1307.e1-3.
14. Kloepfer KM, Olenec JP, Lee WM, et al. Increased H1N1 infection rate in children with asthma. *Am J Respir Crit Care Med.* 2012;185(12):1275–1279.
15. Jartti T, Lee WM, Pappas T, Evans M, Lemanske Jr RF, Gern JE. Serial viral infections in infants with recurrent respiratory illnesses. *Eur Respir J.* 2008;32(2):314–320.

16. Toivonen L, Schuez-Havupalo L, Karppinen S, et al. Rhinovirus infections in the first 2 years of life. *Pediatrics*. 2016;138(3):e20161309.
17. Jartti T, Lehtinen P, Vuorinen T, et al. Respiratory picornaviruses and respiratory syncytial virus as causative agents of acute expiratory wheezing in children. *Emerg Infect Dis*. 2004;10(6):1095–1101.
18. Heymann PW, Carper HT, Murphy DD, et al. Viral infections in relation to age, atopy, and season of admission among children hospitalized for wheezing. *J Allergy Clin Immunol*. 2004;114(2):239–247.
19. Kotaniemi-Syrjanen A, Vainionpaa R, Reijonen TM, Waris M, Korhonen K, Korppi M. Rhinovirus-induced wheezing in infancy—the first sign of childhood asthma? *J Allergy Clin Immunol*. 2003;111(1):66–71.
20. Jartti T, Kuusipalo H, Vuorinen T, et al. Allergic sensitization is associated with rhinovirus-, but not other virus-, induced wheezing in children. *Pediatr Allergy Immunol*. 2010;21(7):1008–1014.
21. Miller EK, Williams JV, Gebretsadik T, et al. Host and viral factors associated with severity of human rhinovirus-associated infant respiratory tract illness. *J Allergy Clin Immunol*. 2011;127(4):883–891.
22. Carroll KN, Gebretsadik T, Minton P, et al. Influence of maternal asthma on the cause and severity of infant acute respiratory tract infections. *J Allergy Clin Immunol*. 2012;129(5):1236–1242.
23. Turunen R, Jartti T, Bochkov YA, Gern JE, Vuorinen T. Rhinovirus species and clinical characteristics in the first wheezing episode in children. *J Med Virol*. 2016;88 (12):2059–2068.
24. Jackson DJ, Evans MD, Gangnon RE, et al. Evidence for a causal relationship between allergic sensitization and rhinovirus wheezing in early life. *Am J Respir Crit Care Med*. 2012;185(3):281–285.
25. Nicolai A, Frassanito A, Nenna R, et al. Risk factors for virus-induced acute respiratory tract infections in children younger than 3 years and recurrent wheezing at 36 months follow-up after discharge. *Pediatr Infect Dis J*. 2017;36(2):179–183.
26. Hyvarinen MK, Kotaniemi-Syrjanen A, Reijonen TM, Korhonen K, Korppi MO. Teenage asthma after severe early childhood wheezing: an 11-year prospective follow-up. *Pediatr Pulmonol*. 2005;40(4):316–323.
27. Ruotsalainen M, Hyvarinen MK, Piippo-Savolainen E, Korppi M. Adolescent asthma after rhinovirus and respiratory syncytial virus bronchiolitis. *Pediatr Pulmonol*. 2013;48 (7):633–639.
28. Backman K, Ollikainen H, Piippo-Savolainen E, Nuolivirta K, Korppi M. Asthma and lung function in adulthood after a viral wheezing episode in early childhood. *Clin Exp Allergy*. 2018;48(2):138–146.
29. Lukkarinen M, Koistinen A, Turunen R, Lehtinen P, Vuorinen T, Jartti T. Rhinovirus-induced first wheezing episode predicts atopic but not nonatopic asthma at school age. *J Allergy Clin Immunol*. 2017;140(4):988–995.
30. Bergroth E, Aakula M, Korppi M, et al. Post-bronchiolitis use of asthma medication: a prospective 1-year follow-up study. *Pediatr Infect Dis J*. 2016;35(4):363–368.
31. Lehtinen P, Ruohola A, Vanto T, Vuorinen T, Ruuskanen O, Jartti T. Prednisolone reduces recurrent wheezing after a first wheezing episode associated with rhinovirus infection or eczema. *J Allergy Clin Immunol*. 2007;119(3):570–575.
32. Midulla F, Nicolai A, Ferrara M, et al. Recurrent wheezing 36 months after bronchiolitis is associated with rhinovirus infections and blood eosinophilia. *Acta Paediatr*. 2014;103(10):1094–1099.
33. Takeyama A, Hashimoto K, Sato M, et al. Clinical and epidemiologic factors related to subsequent wheezing after virus-induced lower respiratory tract infections in hospitalized pediatric patients younger than 3 years. *Eur J Pediatr*. 2014;173(7):959–966.

34. Kusel MM, de Klerk NH, Kebadze T, et al. Early-life respiratory viral infections, atopic sensitization, and risk of subsequent development of persistent asthma. *J Allergy Clin Immunol.* 2007;119(5):1105–1110.
35. Lemanske Jr RF, Jackson DJ, Gangnon RE, et al. Rhinovirus illnesses during infancy predict subsequent childhood wheezing. *J Allergy Clin Immunol.* 2005;116(3):571–577.
36. Jackson DJ, Gangnon RE, Evans MD, et al. Wheezing rhinovirus illnesses in early life predict asthma development in high-risk children. *Am J Respir Crit Care Med.* 2008;178(7):667–672.
37. Rubner FJ, Jackson DJ, Evans MD, et al. Early life rhinovirus wheezing, allergic sensitization, and asthma risk at adolescence. *J Allergy Clin Immunol.* 2017;139:501–507.
38. Anderson HM, Lemanske Jr RF, Evans MD, et al. Assessment of wheezing frequency and viral etiology on childhood and adolescent asthma risk. *J Allergy Clin Immunol.* 2017;139(2):692–694.
39. Bonnelykke K, Vissing NH, Sevelsted A, Johnston SL, Bisgaard H. Association between respiratory infections in early life and later asthma is independent of virus type. *J Allergy Clin Immunol.* 2015;136(1):81–86.e4.
40. Teo SM, Mok D, Pham K, et al. The infant nasopharyngeal microbiome impacts severity of lower respiratory infection and risk of asthma development. *Cell Host Microbe.* 2015;17(5):704–715.
41. Bochkov YA, Grindle K, Vang F, Evans MD, Gern JE. Improved molecular typing assay for rhinovirus species A, B, and C. *J Clin Microbiol.* 2014;52(7):2461–2471.
42. de Winter JJ, Bont L, Wilbrink B, van der Ent CK, Smit HA, Houben ML. Rhinovirus wheezing illness in infancy is associated with medically attended third year wheezing in low risk infants: results of a healthy birth cohort study. *Immun Inflamm Dis.* 2015;3(4):398–405.
43. Lee WM, Lemanske Jr RF, Evans MD, et al. Human rhinovirus species and season of infection determine illness severity. *Am J Respir Crit Care Med.* 2012;186(9):886–891.
44. Ng KT, Oong XY, Lim SH, et al. Viral load and sequence analysis reveal the symptom severity, diversity, and transmission clusters of rhinovirus infections. *Clin Infect Dis.* 2018;67(2):261–268.
45. Esposito S, Daleno C, Scala A, et al. Impact of rhinovirus nasopharyngeal viral load and viremia on severity of respiratory infections in children. *Eur J Clin Microbiol Infect Dis.* 2014;33(1):41–48.
46. Lu X, Schneider E, Jain S, et al. Rhinovirus viremia in patients hospitalized with community-acquired pneumonia. *J Infect Dis.* 2017;216(9):1104–1111.
47. Fuji N, Suzuki A, Lupisan S, et al. Detection of human rhinovirus C viral genome in blood among children with severe respiratory infections in the Philippines. *PLoS One.* 2011;6(11):e27247.
48. Lethbridge R, Prastanti F, Robertson C, et al. Prospective assessment of rhinovirus symptoms and species recurrence in children with and without an acute wheezing exacerbation. *Viral Immunol.* 2018;31(4):299–305.
49. Bizzintino J, Lee WM, Laing IA, et al. Association between human rhinovirus C and severity of acute asthma in children. *Eur Respir J.* 2011;37:1037–1042.
50. Cox DW, Khoo SK, Zhang G, et al. Rhinovirus is the most common virus and rhinovirus-C is the most common species in paediatric intensive care respiratory admissions. *Eur Respir J.* 2018;52(2):1800207.
51. Turunen R, Vuorinen T, Bochkov Y, Gern J, Jartti T. Clinical and virus surveillance after the first wheezing episode: special reference to rhinovirus A and C species. *Pediatr Infect Dis J.* 2017;36:539–544.
52. Halmo Hurdum S, Zhang G, Khoo SK, et al. Recurrent rhinovirus detections in children following a rhinovirus-induced wheezing exacerbation: a retrospective study. *Int J Pediatr Child Health.* 2015;3(1):10–18.

53. Mak RK, Tse LY, Lam WY, Wong GW, Chan PK, Leung TF. Clinical spectrum of human rhinovirus infections in hospitalized Hong Kong children. *Pediatr Infect Dis J.* 2011;30(9):749–753.
54. Miller EK, Edwards KM, Weinberg GA, et al. A novel group of rhinoviruses is associated with asthma hospitalizations. *J Allergy Clin Immunol.* 2009;123(1):98–104.
55. Pierangeli A, Ciccozzi M, Chiavelli S, et al. Molecular epidemiology and genetic diversity of human rhinovirus affecting hospitalized children in Rome. *Med Microbiol Immunol.* 2013;202(4):303–311.
56. Piralla A, Rovida F, Campanini G, et al. Clinical severity and molecular typing of human rhinovirus C strains during a fall outbreak affecting hospitalized patients. *J Clin Virol.* 2009;45(4):311–317.
57. Calvo C, Pozo F, Garcia-Garcia M, et al. Detection of new respiratory viruses in hospitalized infants with bronchiolitis: a three-year prospective study. *Acta Paediatr.* 2010;99:883–887.
58. Linder JE, Kraft DC, Mohamed Y, et al. Human rhinovirus C: age, season, and lower respiratory illness over the past 3 decades. *J Allergy Clin Immunol.* 2013;131(1):69–77.e1-6.
59. Iwane MK, Prill MM, Lu X, et al. Human rhinovirus species associated with hospitalizations for acute respiratory illness in young US children. *J Infect Dis.* 2011;204(11):1702–1710.
60. Calvo C, Casas I, Garcia-Garcia ML, et al. Role of rhinovirus C respiratory infections in sick and healthy children in Spain. *Pediatr Infect Dis J.* 2010;29(8):717–720.
61. Bruning AHL, Thomas XV, van der Linden L, et al. Clinical, virological and epidemiological characteristics of rhinovirus infections in early childhood: a comparison between non-hospitalised and hospitalised children. *J Clin Virol.* 2015;73:120–126.
62. Fawkner-Corbett DW, Khoo SK, Duarte CM, et al. Rhinovirus-C detection in children presenting with acute respiratory infection to hospital in Brazil. *J Med Virol.* 2016;88(1):58–63.
63. Ashraf S, Brockman-Schneider R, Bochkov YA, Pasic TR, Gern JE. Biological characteristics and propagation of human rhinovirus-C in differentiated sinus epithelial cells. *Virology.* 2013;436(1):143–149.
64. Nakagome K, Bochkov YA, Ashraf S, et al. Effects of rhinovirus species on viral replication and cytokine production. *J Allergy Clin Immunol.* 2014;134:332–341.
65. Liu Y, Hill MG, Klose T, et al. Atomic structure of a rhinovirus C, a virus species linked to severe childhood asthma. *Proc Natl Acad Sci USA.* 2016;113(32):8997–9002.
66. Bochkov YA, Watters K, Ashraf S, et al. Cadherin-related family member 3, a childhood asthma susceptibility gene product, mediates rhinovirus C binding and replication. *Proc Natl Acad Sci USA.* 2015;112(17):5485–5490.
67. Caliskan M, Bochkov YA, Kreiner-Moller E, et al. Rhinovirus wheezing illness and genetic risk of childhood-onset asthma. *N Engl J Med.* 2013;368(15):1398–1407.
68. Griggs TF, Bochkov YA, Basnet S, et al. Rhinovirus C targets ciliated airway epithelial cells. *Respir Res.* 2017;18(1):84.
69. Bonnelykke K, Coleman AT, Evans MD, et al. Cadherin-related family member 3 genetics and rhinovirus C respiratory illnesses. *Am J Respir Crit Care Med.* 2018;197(5):589–594.
70. Gill MA, Bajwa G, George TA, et al. Counterregulation between the FcepsilonRI pathway and antiviral responses in human plasmacytoid dendritic cells. *J Immunol.* 2010;184(11):5999–6006.
71. Durrani SR, Montville DJ, Pratt AS, et al. Innate immune responses to rhinovirus are reduced by the high-affinity IgE receptor in allergic asthmatic children. *J Allergy Clin Immunol.* 2012;130(2):489–495.
72. Hatchwell L, Collison A, Girkin J, et al. Toll-like receptor 7 governs interferon and inflammatory responses to rhinovirus and is suppressed by IL-5-induced lung eosinophilia. *Thorax.* 2015;70(9):854–861.

73. Mathur SK, Fichtinger PS, Kelly JT, Lee WM, Gern JE, Jarjour NN. Interaction between allergy and innate immunity: model for eosinophil regulation of epithelial cell interferon expression. *Ann Allergy Asthma Immunol.* 2013;111(1):25–31.
74. Werder RB, Zhang V, Lynch JP, et al. Chronic IL-33 expression predisposes to virus-induced asthma exacerbations by increasing type 2 inflammation and dampening antiviral immunity. *J Allergy Clin Immunol.* 2018;141(5):1607–1619.e9.
75. Gielen V, Sykes A, Zhu J, et al. Increased nuclear suppressor of cytokine signaling 1 in asthmatic bronchial epithelium suppresses rhinovirus induction of innate interferons. *J Allergy Clin Immunol.* 2015;136:177–188.e11.
76. Chen G, Korfhagen TR, Karp CL, et al. Foxa3 induces goblet cell metaplasia and inhibits innate antiviral immunity. *Am J Respir Crit Care Med.* 2014;189:301–313.
77. Teach SJ, Gill MA, Togias A, et al. Pre-seasonal treatment with either omalizumab an inhaled corticosteroid boost to prevent fall asthma exacerbations. *J Allergy Clin Immunol.* 2015;136:1476–1485.
78. Esquivel A, Busse WW, Calatroni A, et al. Effects of omalizumab on rhinovirus infections, illnesses, and exacerbations of asthma. *Am J Respir Crit Care Med.* 2017;196(8):985–992.
79. Gill MA, Liu AH, Calatroni A, et al. Enhanced plasmacytoid dendritic cell antiviral responses after omalizumab. *J Allergy Clin Immunol.* 2018;141(5):1735–1743.e9.
80. Bisgaard H, Hermansen MN, Buchvald F, et al. Childhood asthma after bacterial colonization of the airway in neonates. *N Engl J Med.* 2007;357(15):1487–1495.
81. Bisgaard H, Hermansen MN, Bonnelykke K, et al. Association of bacteria and viruses with wheezy episodes in young children: prospective birth cohort study. *BMJ.* 2010;341:c4978.
82. de Steenhuijsen Piters WA, Heinonen S, Hasrat R, et al. Nasopharyngeal microbiota, host transcriptome, and disease severity in children with respiratory syncytial virus infection. *Am J Respir Crit Care Med.* 2016;194(9):1104–1115.
83. Bashir H, Grindle K, Vrtis R, et al. Association of rhinovirus species with common cold and asthma symptoms and bacterial pathogens. *J Allergy Clin Immunol.* 2018;141 (2):822–824.e829.
84. Mansbach JM, Hasegawa K, Henke DM, et al. Respiratory syncytial virus and rhinovirus severe bronchiolitis are associated with distinct nasopharyngeal microbiota. *J Allergy Clin Immunol.* 2016;137(6):1909–1913.e4.
85. Hasegawa K, Mansbach JM, Ajami NJ, et al. Association of nasopharyngeal microbiota profiles with bronchiolitis severity in infants hospitalised for bronchiolitis. *Eur Respir J.* 2016;48(5):1329–1339.
86. Loss GJ, Depner M, Hose AJ, et al. The early development of wheeze. Environmental determinants and genetic susceptibility at 17q21. *Am J Respir Crit Care Med.* 2016;193(8):889–897.
87. Remes ST, Castro-Rodriguez JA, Holberg CJ, Martinez FD, Wright AL. Dog exposure in infancy decreases the subsequent risk of frequent wheeze but not of atopy. *J Allergy Clin Immunol.* 2001;108(4):509–515.
88. Stokholm J, Chawes BL, Vissing N, Bonnelykke K, Bisgaard H. Cat exposure in early life decreases asthma risk from the 17q21 high-risk variant. *J Allergy Clin Immunol.* 2018;141(5):1598–1606.
89. Ludka-Gaulke T, Ghera P, Waring SC, et al. Farm exposure in early childhood is associated with a lower risk of severe respiratory illnesses. *J Allergy Clin Immunol.* 2018;141(1):454–456.e4.
90. Lynch SV, Wood RA, Boushey H, et al. Effects of early-life exposure to allergens and bacteria on recurrent wheeze and atopy in urban children. *J Allergy Clin Immunol.* 2014;134(3):593–601.e12.

91. O'Connor GT, Lynch SV, Bloomberg GR, et al. Early-life home environment and risk of asthma among inner-city children. *J Allergy Clin Immunol.* 2018;141 (4):1468–1475.
92. Mihaylova VT, Kong Y, Fedorova O, et al. Regional differences in airway epithelial cells reveal tradeoff between defense against oxidative stress and defense against rhinovirus. *Cell Rep.* 2018;24(11):3000–3007.e3.
93. Spannhake EW, Reddy SP, Jacoby DB, Yu XY, Saatian B, Tian J. Synergism between rhinovirus infection and oxidant pollutant exposure enhances airway epithelial cell cytokine production. *Environ Health Perspect.* 2002;110(7):665–670.
94. Duff AL, Pomeranz ES, Gelber LE, et al. Risk factors for acute wheezing in infants in infants and children: viruses, passive smoke, and IgE antibodies to inhalant allergens. *Pediatrics.* 1993;92:535–540.
95. Chauhan AJ, Inskip HM, Linaker CH, et al. Personal exposure to nitrogen dioxide (NO_2) and the severity of virus-induced asthma in children. *Lancet.* 2003;361 (9373):1939–1944.
96. Aichbhaumik N, Zoratti EM, Strickler R, et al. Prenatal exposure to household pets influences fetal immunoglobulin E production. *Clin Exp Allergy.* 2008;38 (11):1787–1794.
97. Leigh R, Oyelusi W, Wiehler S, et al. Human rhinovirus infection enhances airway epithelial cell production of growth factors involved in airway remodeling. *J Allergy Clin Immunol.* 2008;121(5):1238–1245.
98. Tacon CE, Wiehler S, Holden NS, Newton R, Proud D, Leigh R. Human rhinovirus infection up-regulates MMP-9 production in airway epithelial cells via NF-{kappa}B. *Am J Respir Cell Mol Biol.* 2010;43(2):201–209.
99. Shariff S, Shelfoon C, Holden NS, et al. Human rhinovirus infection of epithelial cells modulates airway smooth muscle migration. *Am J Respir Cell Mol Biol.* 2017;56 (6):796–803.
100. Hong JY, Bentley JK, Chung Y, et al. Neonatal rhinovirus induces mucous metaplasia and airways hyperresponsiveness through IL-25 and type 2 innate lymphoid cells. *J Allergy Clin Immunol.* 2014;134(2):429–439.
101. Garcia-Garcia ML, Calvo C, Moreira A, et al. Thymic stromal lymphopoietin, IL-33, and periostin in hospitalized infants with viral bronchiolitis. *Medicine (Baltimore).* 2017;96(18):e6787.
102. Bauer L, Lyoo H, van der Schaar HM, Strating JR, van Kuppeveld FJ. Direct-acting antivirals and host-targeting strategies to combat enterovirus infections. *Curr Opin Virol.* 2017;24:1–8.
103. Papadopoulos NG, Megremis S, Kitsioulis NA, Vangelatou O, West P, Xepapadaki P. Promising approaches for the treatment and prevention of viral respiratory illnesses. *J Allergy Clin Immunol.* 2017;140(4):921–932.
104. Bacharier LB, Guilbert TW, Mauger DT, et al. Early administration of azithromycin and prevention of severe lower respiratory tract illnesses in preschool children with a history of such illnesses: a randomized clinical trial. *JAMA.* 2015;314 (19):2034–2044.
105. Stokholm J, Chawes BL, Vissing NH, et al. Azithromycin for episodes with asthma-like symptoms in young children aged 1-3 years: a randomised, double-blind, placebo-controlled trial. *Lancet Respir Med.* 2016;4(1):19–26.
106. Koistinen A, Lukkarinen M, Turunen R, et al. Prednisolone for the first rhinovirus-induced wheezing and 4-year asthma risk: a randomized trial. *Pediatr Allergy Immunol.* 2017;28(6):557–563.

CHAPTER 6

Exacerbations of chronic respiratory diseases

Michael R. Edwards[1,2], Andrew I. Ritchie[1,3] and Sebastian L. Johnston[1,2]

[1]Airway Disease Infection Section, National Heart and Lung Institute, Imperial College, London, United Kingdom
[2]MRC & Asthma UK Centre in Allergic Mechanisms of Asthma, London, United Kingdom
[3]Royal Brompton and Harefield NHS Trust, London, United Kingdom

6.1 INTRODUCTION

Human rhinovirus (RV) infection has emerged as an important trigger for exacerbations of chronic respiratory diseases including asthma, chronic obstructive pulmonary disease (COPD), cystic fibrosis (CF), and interstitial lung disease (ILD) including idiopathic pulmonary fibrosis (IPF). Epidemiological studies have played a central part in understanding the association of RV and other respiratory viruses in exacerbations of chronic respiratory diseases. This chapter offers a summary of the epidemiological literature that has linked RV infection with acute exacerbations, and Table 6.1 refers to the key references establishing the importance of RV infection in exacerbations of chronic respiratory diseases. Through a range of postulated mechanisms, RV infection elicits a complex host response that in part, may actually contribute to disease pathology and perpetuate symptoms rather than prevent virus infection and the associated symptoms. Whether or not RV infection directly triggers these events that add to, synergize with, or change the nature of the host response that drive disease, or whether the underlying disease makes certain individuals susceptible to RV infection is a pertinent question in the field. This chapter includes an in-depth commentary on the reasons why RV is the most common viral pathogen associated with exacerbations of these diseases; including the impact of current and emerging treatments on the outcome of RV infections and disease progression. Finally we explore the potential for new therapeutic opportunities to better manage chronic respiratory diseases whilst minimizing the acute and chronic disease promoting effects of RV infections.

Rhinovirus Infections
DOI: https://doi.org/10.1016/B978-0-12-816417-4.00006-8

Table 6.1 Key epidemiological evidence supporting the role of rhinovirus (RV) and other respiratory viruses as exacerbators of chronic respiratory diseases.

Reference	Disease	Insights
Johnston[1]	AE/Wheeze	Highlighted the importance of viruses, particularly RV to be associated with 85% of AE and wheeze events in children
Corne[2]	AE	Showed that while having the same frequency of RV infections, asthmatics had more severe and longer duration of symptoms versus nonasthmatics
Johnston[3]	AE	Identified viruses, particularly RV as vectors for the September asthma epidemic in the Northern Hemisphere
Bonnelykke[4]	AE/Wheeze	Demonstrated a link between CDHR3 variants and recurrent AE or wheeze in children
Seemungal[5]	COPD-E	Demonstrated association of RV with COPD-E and disease severity
Papi[6]	COPD-E	Showed that RV was the most readily virus detected in COPD-E versus stable COPD
Bafadhel[7]	COPD-E	Showed that RV associated COPD-E ere associated with certain biomarkers such as serum CXCL10
Collinson[8]	CF-E	Showed an association between picornaviruses, CF disease severity, and bacterial infections in CF-E
Wootton[9]	IPF-E	Showed that viruses, including RV, were associated with IPF-E and not stable disease

AE, Asthma exacerbations; *COPD*, chronic obstructive pulmonary disease; *IPF*, idiopathic pulmonary fibrosis; *CDHR3*, cadherin-related family member 3; *CF*, cystic fibrosis.

6.2 EXACERBATIONS OF ASTHMA

Asthma is a chronic disease of the conducting airways characterized by airway hyperreactivity that causes reversible airway smooth muscle (ASM) mediated bronchoconstriction. Another defining feature of asthma is airway inflammation that perpetuates airway hyperreactivity as well as mucus production, which together with airway inflammation cause airway narrowing, airway obstruction, and resistance to airflow and symptoms.[10] Asthma has been linked to numerous stimuli including airway infections, various allergens, exposure to airborne pollutants, exercise, stress, and nervous provocation.[10] The key feature of asthma is decreased lung function, measured as spontaneous changes in peak expiratory flow (PEF) or a decrease in forced expiratory volume in 1 second following provocation. Symptoms include nocturnal awakenings, dyspnea, wheeze, and tightening of the chest. The immediate effects of asthma are generally reversible by a short-acting β_2-agonist. Asthma is now considered a heterogeneous, complex disease and can present as a chronic, stable disease of different severity, including mild, moderate, or severe. Responses to treatments, such as glucocorticosteroids (GCs) and bronchodilators, may also vary, and be categorized as well, partial, or poorly controlled. The most common endophenotype is Th2/type 2 associated or allergic asthma, which is defined by a physician's diagnosis of asthma often associated with high serum Immunoglobulin E (IgE) levels, positive skin prick tests to aeroallergens, and/or confirmed diagnosis of an atopic disorder such as allergic rhinitis. Evidence of interleukin (IL)-5, IL-4, or IL-13 driven inflammation, or increased levels of eosinophils in blood or lower airway samples, or high levels of FeNO may also confirm Th2 or type-2 associated asthma. Other asthma endophenotypes may exist, including adult-onset asthma; steroid resistant, poorly steroid responsive, or neutrophilic asthma; fixed airflow limitation asthma; or more simply termed type 2 low asthma. However, these are not as well defined by genetic analysis or biomarkers as allergic asthma.[11,12] Currently, the exact nature or existence of these subgroups is vigorously debated.[13,14] Erroneous diagnosis, effects of treatments, and other confounders have led to a lack of consensus. The identification of correct asthma endophenotypes and implementation of precise medicine approaches represents an important ongoing challenge to the field.

Asthma exacerbations (AE) are defined when a stable individual with asthma suddenly becomes uncontrolled, experiences increased symptoms resulting in General practitioner (GP) consultation or hospitalization, and additional treatment including oral GCs. In the United States, there are approximately 1.6 M emergency room visits associated with asthma each year,[15,16] and approximately 5% of asthmatic individuals have at least one AE per year.[17] AE may have multiple predictors; however recent AE history is the strongest predictor for a future AE event. In children, the number of allergic sensitization triggers and race are other important predictors.[18] AE are more common in individuals with poor lung function, or who are classified as having severe or difficult to treat asthma, which may comprise up to 15% of all asthmatics.[18,19] AE may be life threatening, and results in significant morbidity worldwide including school/study absenteeism and sick leave. Recent economic studies have shown that individuals suffering from AE have increased both total healthcare and asthma specific healthcare costs, when compared with individuals who do not experience AE.[20] In unadjusted total costs, individuals that experience AE may accrue costs of >$9000 USD per year, compared with approximately $5000 USD per year for individuals who do not.

AE are triggered by many factors; however respiratory virus infections are the most common, associated with up to 85% of AE in both adults and children.[21–23] Historically, detection methods for viruses in respiratory clinical samples such as nasal secretions and throat swabs relied on in vitro cell culture based methods. This approach had limitations not least of which was the inability to propagate many clinically important viruses—including a significant number (approximately 40% based on current classification) of RVs we now know to belong to the RV-C species. In the mid-1980s nucleic acid based molecular methods become the method of choice, and in recent years, next generation sequencing[24] and DNA and protein based virus-chip methods[25,26] have become available with the added advantage of potentially discriminating different viruses of the same kind.

Viruses often detected in respiratory samples obtained during AE include respiratory syncytial viruses (RSVs), influenza viruses, adenoviruses, human RVs, coronaviruses, parainfluenza viruses (PIV), as well as the more recently identified metapneumoviruses and bocaviruses; however these are less common.[22,27–29] Among these respiratory viruses, RVs are the most common viruses detected and typically account for between 29% and 55% of virus-induced AE (Fig. 6.1) although this number can be as high as 87%.[22] Differences in detection rate involve volunteer demographics including age, geography, and method of detection.[22] Of the different RV genotypes, RV-A and RV-C are more commonly associated

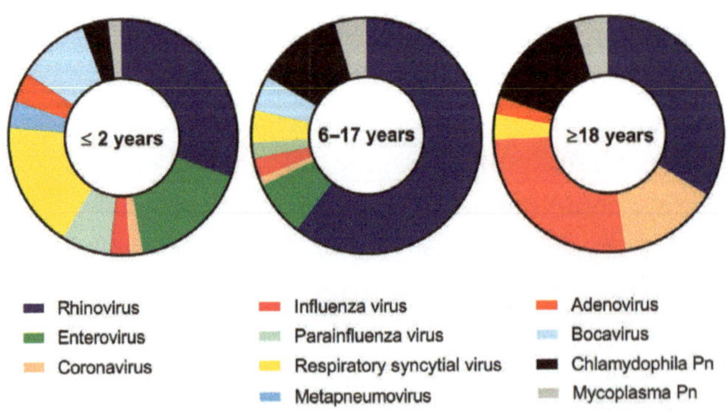

Figure 6.1 ***Association of different respiratory viruses and atypical bacteria in asthma by age expressed as a %.*** RV accounts for approximately 33% in children under 2 years of age, 55% in older children 6–17 years of age, and 29% of adults. Data are median % values and have been reproduced from published review articles.[22,23] *RV*, Rhinovirus.

with AE than RV-B. The RV-C group[30,31] shows unique sequences at the ICAM-1 and LDL receptor binding sites, initially suggesting they have a unique receptor,[32] which has since been identified as cadherin-related family member 3.[33] RV-C may contain a group of viruses that are associated with more severe AE events requiring hospitalization,[34] although how this occurs is unknown.[31] Epidemiological studies have shown that in the Northern Hemisphere, RV infections result in a marked increase in emergency room admissions due to AEs.[35] In fact this "asthma epidemic" has been shown to coincide with Labor Day in Canada,[3] the third week of September, approximately 2 weeks after children return to school, highlighting the important role school age children have as vectors of RV induced AEs.

While the strong association of RV infection and AE events from epidemiological studies investigating virus detection rates in hospitalized cases of AEs do not definitely prove RV causes AE, or infer mechanisms, they have been very important in highlighting the important connection between viral infections and asthma. More recently, human experimental infection models of RV-A in asthmatics[36–38] and mouse models of allergen exposure with RV infection[39–41] have also been established that thus definitively show the ability of RV infection to contribute to lower airway inflammation, and induce changes in lung function consistent with the symptomology observed in naturally occurring AEs.

6.3 EXACERBATIONS OF CHRONIC OBSTRUCTIVE PULMONARY DISEASE

COPD is progressive respiratory condition caused by exposure to cigarette smoke and in some parts of the world, biomass fuel burning. It is characterized by chronic inflammation of the small airways, accelerated loss of lung function and/or emphysematous disease associated with airflow limitation. It is a significant source of morbidity and in terms of mortality worldwide, COPD is currently the third leading cause of noncommunicable death.[42] The exact prevalence of COPD is difficult to accurately assess, indeed numbers varying between 200 and 600 million are still referenced. However, the recent update of the Global Burden of Disease has led to an agreed number of 328 million people having COPD worldwide, which includes 168 million men and 160 million women.[43]

The exact threshold for the duration/intensity of cigarette smoking that will result in COPD varies between individuals. In the absence of a predisposing factor (e.g., genetic, environmental, or occupational), smoking less than 10–15 pack-years of cigarettes is unlikely to result in COPD. Alternatively, the single best predictor for airflow obstruction is a history of more than 40 pack-years of smoking [positive likelihood ratio, 12 (95% CI, 2.7–50)].[44] As is the case with most chronic lung conditions, the disease course includes chronic respiratory symptoms such as wheeze and breathlessness, punctuated by periods of increased symptomatology, known as acute exacerbations.[45] Acute exacerbations (COPD-E) are significant events in COPD, as they accelerate disease progression, impair quality of life, and are the predominant cause of mortality. Additionally, they often necessitate unscheduled healthcare visits, treatment costs, and hospitalizations, which account for $18 billion in direct costs annually in the United States alone.[46,47]

Preventing COPD-E is a major unmet therapeutic goal. A crucial step toward this goal is the recognition that acute exacerbations are most commonly due to respiratory virus infection, specifically RV. It follows that the host immune response to virus infection may be impaired, and that a better understanding of how the immune response differs in COPD compared with smokers or healthy individuals has the potential to lead to the development of new therapies.

Historically, the main cause of COPD-E was considered to be bacterial infections and this belief is reflected in the continued extensive use of antibiotics in COPD exacerbations, despite weak evidence of their efficacy.[48,49] Earlier studies using diagnostic techniques such as culture detected viral infections in a minority of COPD exacerbations, resulting in the focus on bacteria. However, some RVs are difficult to culture in standard assays, and serology is not practical due to the presence of more than 100 serotyped strains and ~60 nonserotyped strains. Therefore, without polymerase chain reaction (PCR), these viruses cannot be easily identified. As a result, early studies using other diagnostic methods underestimated the prevalence of viruses being detected in 22%–64% of COPD exacerbations[5–7,49–63] with RVs the most prevalent.

As in asthma, ultimate proof of causation between RV infection and the induction COPD-E comes from a human model of experimental RV infection in COPD.[64–66] RV are identified in community treated mild exacerbations,[5] in hospitalized patients,[67,68] and in very severe exacerbations requiring intensive care.[62] Virus infection rates during acute exacerbations are unlikely to represent chronic infection or seasonal carriage, as viral PCRs are rarely positive when patients are sampled prior to[49,61,69] or after an exacerbation,[5,6,60] or in time-matched controls.[51,54,56,59] Whilst these studies support a causal relationship, they cannot exclude secondary causation (e.g., bacterial coinfection or air pollution). The novel experimental infection studies using RV in COPD subjects shows clear increases in lower respiratory tract symptoms following inoculation, along with airflow obstruction, and systemic and airway inflammation mimicking a COPD-E observed during naturally occurring infection. Indeed, in experimental infection studies, in over 90% of RV infected COPD subjects, COPD exacerbation criteria were met, suggesting that the virus detection frequencies reported in naturally occurring exacerbations (22%–64%) are actually a gross underestimate of the true frequency of viruses as precipitants of exacerbations. In the experimental infection model, RV RNA was present in the upper respiratory tract prior to progressing to the lower airway and causing COPD exacerbation onset, thereby excluding secondary causation.[64,65,70] Moreover these studies demonstrated a temporal relationship between viral replication in the airways, symptoms, and exacerbation onset, and viral clearance and exacerbation resolution. The virus load also correlated with markers of inflammation, and oxidative and nitrosative stress in the airways.[64,70] Therefore these studies have provided compelling evidence to the field that RV infection is directly linked to COPD-E.

6.4 EXACERBATIONS OF CYSTIC FIBROSIS

CF is a common Mendelian autosomal recessive genetic disorder that affects more than 70,000 individuals worldwide.[71] More than 2000 culpable gene variants have been identified. The most common, and widely studied, is the Phe508del, which is predominantly found in populations with northern European heritage;[72,73] whilst individuals with CF from other regions have a wider range of mutations with the Phe508del mutation being much less prevalent.

The mutation of this gene leads to defective function of the CF transmembrane conductance regulator (CFTR) protein. Pulmonary involvement is the most prominent manifestation of the disease due to tenacious mucus secretion, decreased mucociliary clearance, resulting in inflammation and recurrent infection.

Respiratory infections are the leading cause of acute exacerbations resulting in morbidity, decline in lung function, and hospitalizations. The predominant cause of infectious complications in CF is considered bacterial infection, with *Pseudomonas aeruginosa* the most common organism cultured. The role of a dysfunctional CTFR protein plays a key role in deficient innate immunity to *P. aeruginosa* infection due to resultant increased innate inflammation owing to disruptions in lipid metabolism, enhanced interactions of *P. aeruginosa* with epithelial cells, or a direct role for the CFTR in bacterial clearance subsequent to CFTR-mediated epithelial cell ingestion.[74] There is relatively little published work on the role of RV infections in CF exacerbations (CF-E), but recent studies suggest that viral infections have a significant impact. Thus the role of respiratory viruses is likely to have been significantly underestimated in the past. For example, prior to the utilization of modern PCR techniques, studies relying on serology, culture, and immunofluorescence implicated a viral cause in 10%–28% of exacerbations in CF patients.[75–78] However, PCR studies have increased this detection rate to 46% of exacerbating patients, compared with only 18% of patients in stable condition.[79] RV has been specifically identified in 13%–58% of CF patients with acute respiratory illness and was associated with worse respiratory symptoms, airway function deterioration, and greater incidence of secondary bacterial infection, compared with matched uninfected patients.[8,77,79,80] A number of different viral species have been detected in CF patients, with the most common being RVs, influenza viruses, and RSVs. Interestingly, the rates of viral infections in children with CF is not elevated in comparison to healthy children but importantly the severity of clinical illness associated with infection is demonstrably greater.[81] As is the case in other airways diseases, viral infections are associated with decline in lung function and more severe clinical illness in CF, indicating that they contribute to disease progression demonstrating the need for further research in this field.

6.5 EXACERBATIONS OF INTERSTITIAL LUNG DISEASE

The field of ILD includes a plethora of different, individual conditions that exhibit varying expected prognoses and disease courses. The most common ILD is IPF. IPF affects approximately 3 million people worldwide, with incidence increasing dramatically with age, it has a reported median survival of approximately 3 years from the time of diagnosis.[82] Age is the strongest demographic risk factor for IPF, suggesting that "accelerated" lung aging is a driving force for its development.[83] The pathogenesis of IPF development is an area of much debate. A sequence of three pathophysiologic stages has been described (see Table 6.2).[83]

Table 6.2 A summary of the pathophysiological stages proposed for idiopathic pulmonary fibrosis development.

Pathophysiologic stage

Predisposition	Epithelial cell dysfunction • Genetic factors • Exogenous exposures • Aging
Initiation	TGF-β activation Epithelial to mesenchymal transition Unfolded protein response Fibrocyte recruitment
Progression	Pathologic fibroblast differentiation Pathologic matrix remodeling Epigenetic changes

TGF, Transforming growth factor.

However, it is important to clarify that not all individuals in the predisposition stage will go on to exhibit clinically detectable disease. Similarly, not all patients with IPF develop all stages in a sequential order before established disease is detected.

As in asthma, COPD, and CF, the disease course of IPF commonly exhibit stable periods punctuated by acute exacerbations in symptoms, associated with a stepwise decline in lung function.[84] A proportion of these episodes may be the result of sequelae from infection that cannot be detected due to late presentation and limited microbiological detection methods, but studies of naturally occurring exacerbations have concluded that viruses only account for a small number of these events.[9,85,86]

Epidemiological evidence of infective etiology in IPF exacerbations (IPF-E) originates from work demonstrating that IPF-E was significantly more common in the winter and spring months and in patients who are taking immunosuppressive medications.[87–90] IPF increases the risk of pulmonary infections fourfold, although it is unclear how many of these are viral.[91] Moreover, community acquired pneumonia (both viral and bacterial) is associated with a higher mortality in ILD than otherwise healthy controls.[92]

The importance of viral infection in the disease progression of ILD is again indirectly supported by small trials of antiviral therapy. Two cases described by Tang et al. were positive for EBV on PCR of sputum samples, treated with valacyclovir, and went on to exhibit stabilization in at least one of the two cases.[93] A 14-day trial of ganciclovir in patients with severe IPF and positive EBV serology brought about improvements in 64% (9/14) patients in terms of three of FVC, shuttle walk test, DTPA scan, and steroid dose.[94] The small sample size, lack of a control group, and open-label design make it difficult to draw conclusions from this study. The optimal duration of therapy is also unclear, as 6/9 patients who responded and 3/5 patients who did not respond died within 12 months. Further studies are necessary to establish the true burden and consequences of RV, particularly given the paucity of effective treatments for IPF.

Chronic subclinical viral infection has been implicated in the pathogenesis of some ILDs, for example with the viruses EBV, CMV, HHV7 and HHV8, and hepatitis C. The hypothesis is that chronic and/or recurrent viral infection induces alveolar endothelial cell injury, in turn triggering the release of proinflammatory mediators, which bring about the increased disease pathology.

6.6 CHARACTERISTICS OF RHINOVIRUS THAT MAY PROMOTE EXACERBATIONS OF CHRONIC RESPIRATORY DISEASES

As RV is linked to exacerbation pathogenesis, a key question is therefore what properties of RV might contribute to disease mechanisms. Certainly the "commonness" and ease of transmission of the common cold virus; the fact there are any number of 160 + antigenically distinct viruses circulating at any one time is important. Furthermore, other factors that contribute to the predominance of RV in exacerbation of chronic respiratory diseases can be broadly described as follows: the ability of RV to (1) induce inflammation, nervous provocation, airway remodeling or repair processes, and airway mucus production; (2) synergize or add to ongoing, established inflammation (e.g., allergic or type-2 mediated immune pathways in asthma); (3) promote or contribute to secondary bacterial infections by degrading antimicrobial defenses (epithelial junctional proteins and airway barrier integrity and antimicrobial peptides); and (4) expose impaired or aberrant host antiviral innate immunity that is either intrinsic or caused by treatment regimes. It is likely that RV meets one or more of these criteria in a manner that makes it qualitatively different from other respiratory viruses. In the case of asthma, RV, or certain subtypes of RV, have been referred to as "asthmagenic viruses"[95] due to their ability to be

Table 6.3 Summary of characteristics of rhinovirus (RV) infection that may explain the high frequency of RV infection in exacerbations of chronic respiratory diseases.

Disease	Characteristic			
	Induce inflammation and repair, provoke nerves	**Interact with ongoing inflammation**	**Promote secondary bacterial infections**	**Benefit from treatment**
AE	X[a,b,c,d]	X[e,f]	X[g]	X[h]
COPD-E	X[a]	X[f]	X[g]	X[h]
CF-E	X[a]		X[g]	X[h]
IPF-E	X[c]			X[h]

AE, Asthma exacerbations; *COPD*, chronic obstructive pulmonary disease; *CF*, cystic fibrosis; *IPF*, idiopathic pulmonary fibrosis; *FGF*, fibroblast growth factor; *VEGF*, vascular endothelial growth factor; *TSLP*, thymic stromal lymphopoietin.
[a]Induce a broad range of inflammatory cytokines and chemokines including tumour necrosis factor (TNF), IL-1β, CXCL10, CCL5, granulocyte-macrophage colony-stimulating factor (GM-CSF), etc. and neutrophil chemokines CXCL8, CXCL5.
[b]Induce Th2 mediated allergic inflammation including proTh2 cytokines IL-33, IL-25, TSLP.
[c]Induce FGF, VEGF, collagen, perlecan, promoting remodeling and fibrosis.
[d]Evoke airway hyperresponsiveness or induce TRPV1 and TRPA1 receptors.
[e]Work in an additive manner with existing Th2 mediated inflammation enhancing IL-4 and IL-13 responses.
[f]Enhancing airway eosinophilia.
[g]Promote secondary bacterial infections, including proteobacteria including *Haemophilus influenzae* and *Moraxella catarrhalis* or *Pseudomonas aeruginosa* in CF.
[h]RV infections may be affected by the actions of GCs and β_2 agonists, which reduce antiviral immunity.

associated with AE more readily than other respiratory viruses. In one recent study, RV was described as exacerbating existing asthma, while the effects of influenza infection, which in terms of symptoms were quite different and more systemic, were described as asthma-augmented influenza infection,[96] supporting the idea that the nature of RV infection may be distinct from other respiratory viruses. In this section, the characteristics of RV infection that may explain its predominance in exacerbations of chronic respiratory diseases will be given due attention. Table 6.3 offers a summary.

1. *Induction of inflammation, nervous provocation, airway remodeling or repair processes, and airway mucus production.* A key feature of RV and other respiratory viruses in promoting exacerbations is their ability to induce either lung inflammation, nervous provocation, or in the case of asthma and IPF, lung injury, repair, or remodeling. Many of the innate receptors that recognize RV, including Toll-like receptors (TLRs),

RIG-I like helicases, and nucleotide-binding oligomerization domain (NOD)-like receptors, activate the NF-κB family of transcription factors, which induce over 100 proinflammatory and host response genes.[97] RV, RSV, and influenza induce a range of proinflammatory cytokines, chemokines, growth factors, and adhesion molecules and mucins,[98] thus contributing to lung inflammation.

The ability to promote lung inflammation is important in diseases where lung inflammation is key, and includes asthma, CF, and COPD. Mucins cause mucous plugging of the airway while cytokines facilitate cellular chemotaxis, activation, and proliferation of immune cells in the infected airway.[36,99,100] RV can increase airway hyperresponsiveness by increasing 5-lipoxygenase and the cyclooxygenase pathway enzyme COX-2.[101] While not as well characterized as asthma, inflammation is relevant in COPD, with a surprisingly low number of in vitro studies comparing the responses of cells from COPD patients to viral infection with those from healthy controls.[64,102,103] Similarly the few available studies examining naturally occurring virus-induced COPD exacerbations have demonstrated a role for inflammatory cells such as neutrophils and eosinophils[6] and proinflammatory mediators such as CXCL8/IL-8, CXCL-10/IP-10, and CCL-5/RANTES.[55,69,104] Experimental RV infection studies in COPD have indicated that RV induces airway neutrophilic inflammation and innate inflammatory meditators such as IL-1β, GM-CSF, CXCL8/IL-8, and TNF.[64,70,105] These, and other in vitro studies,[102] indicate an enhanced inflammatory response to virus infection in COPD subjects and this may be one mechanism whereby viruses induce exacerbations. The role of neutrophils is incompletely understood. While having the ability to damage tissues and contribute to airway inflammation, they undoubtedly have important roles in antibacterial defense, may produce type I IFNs,[106] and in mouse models, may have antiviral properties.[107] Of interest, a recent Phase 2a clinical trial of AZD5069, an inhibitor of the neutrophil chemokine receptor CXCR2, failed to affect severe AE rate, lung function of asthma control questionnaire (ACQ) score.[108] The mechanisms of viral-induced CF-E and increased clinical illness are not well characterized with inconsistent results from published studies. Whilst some groups have demonstrated increased production of proinflammatory cytokines and chemokines by airway epithelial cells obtained from CF patients compared with healthy controls,[81,109] others have failed to detect any differences.[110,111] It is unclear what leads to these contrasting observations.

RV infection has been shown to damage and compromise the integrity of the airway; through infecting airway epithelial cells and directly causing cell death[112] and/or epithelial cell shedding.[113,114] They can also affect epithelial permeability leading to increased airway inflammation and creating opportunities for increased secondary infections,[115] allergen uptake,[116] or exposure to environmental pollutants or irritants; all of which may trigger exacerbations of chronic respiratory diseases. In terms of nervous provocation, RV has recently been shown to upregulate the transient receptor potential (TRP) channel family members TRPA1 and TRPV1 in neuroblastoma cells[117] and may be important in the cough response[65] and sensitivity to methacholine[118] observed during infection.

RV can also induce mediators required for airway repair, which results in airway scarring or remodeling. These processes include angiogenesis, increased ASM proliferation and hypertrophy, and thicker basement membranes owing to collagen and fibronectin deposition as well as the generation of new lymphatic vessels that ultimately result in thicker airway walls and reduced lung function; the importance of these process is obvious in asthma. Whether or not RV infection directly initiates this process is unclear; however, RV can induce some of these proangiogenic mediators including fibroblast growth factor and vascular endothelial growth factor,[119–121] required for fibrosis in IPF, and the structural proteins perlecan and collagen[122] the building blocks of airway remodeling.

2. *Viruses synergize or add to ongoing, established inflammation.* Evidence for this property is greatest for RV infections in asthma, where a synergy between viral infections and allergen sensitization and exposure was first realized through epidemiological studies[123,124] and additional evidence for this interaction has been gained in both human challenge[36,37] and mouse models of RV.[39,40,125–127] In individuals with asthma, RV infection of the bronchial epithelium also likely has additive and synergistic effects on allergen sensitization and challenge as the epithelial derived pro-Th2 factors IL-33, IL-25, and thymic stromal lymphopoietin, which activate Th2 cells, DCs, and type 2 innate lymphoid cells, are induced readily by RV infection.[37,126,128] This now allows a direct link to how respiratory virus infection of the airway epithelium can augment existing Th2 pathways in asthma.

Whether or not increased RV disease burden in asthma is simply a result of a type 2 skewed environment or whether RV contributes to type 2 inflammation is a wider debate that is often discussed in asthma onset.[23,129] Concerning AE, RV is perhaps most easily understood as a direct contributor to established type 2 inflammation as discussed above, the results of which tip the fine balance between stable asthma and increased symptoms leading to AE. A study of cohabiting asthma and nonasthmatic pairs in the 1990s found that while the frequency of RV infection was not different between asthmatics and healthy volunteers, asthmatics had a more severe and longer duration of lower airway symptoms,[2] consistent with the concept that individuals with asthma may not be prone to greater number of RV infections, but deal with the infection in a fundamentally different manner from nonasthmatic individuals. Further studies using RV experimental challenge generally show that asthmatics and nonasthmatics have equal rates of successful infection,[36,37] while asthmatics generate greater lower airway symptom scores, which increase with decreased asthma control.[130] How underlying type 2 immunity and allergic inflammation may influence the course of RV infection is another matter, and will be discussed in detail below.

3. *To promote or contribute to secondary bacterial infections by degrading antimicrobial defenses.* The ability to promote secondary bacterial infection is most relevant for AE, or wheeze in children, COPD-E, and CF-E. As RV can disrupt the integrity of the airway epithelium, and reduce antibacterial cytokine responses[131] and degrade antimicrobial peptides,[132] this may promote secondary bacterial infection including enhanced dispersal,[115] further infiltration and colonization of bacterial species that may already be present at low level. In AE in children, RV can increase *Moraxella* abundance.[133,134] In CF, exacerbations associated with a respiratory viral infection are associated with increased bacterial load,[135] although not all studies report this association.[136]

   Viral and bacterial infections are common in COPD-E but when the role of dual virus—bacterial was previously considered studies appeared to indicate that it does not play a prominent role in COPD-E with low detection rates ranging from 6.5%[62] up to only 27%.[6,63,67] However, Papi et al. did observe that exacerbations with dual infection are more severe.[6] Then, in a study of experimental RV infection, secondary bacterial infections were detected in 60% of COPD patients with confirmed RV infection,[132] and increases in proteobacterial DNA sequences, including the potentially pathogenic *Haemophilus influenzae* have been reported.[66] Crucially viruses and bacteria were detected at separate time points with a pattern of virus infection occurring first and bacterial infection later. The implications were that studies of naturally occurring exacerbations that collect samples at a single time point are underestimating the true incidence of dual infection, and are likely missing secondary infections. This was confirmed in a more recent study of naturally occurring exacerbations in which patients were sampled at two time points during an exacerbation. When RV was initially detected but bacteria were absent, 73% had a bacteria detected when sampled again at day 14.[49]

There is also evidence that dual virus–bacterial infection contributes to COPD-E severity. In those COPD patients with *H. influenzae* detected at exacerbation, the presence of a symptomatic cold was associated with higher bacterial loads,[137] and in hospitalized patients with COPD-E, dual virus–bacterial infection correlated with impaired lung function and longer hospital stay.[6] During experimental RV infection, coinfection was associated with prolonged lower respiratory symptoms.[132] In summary, the role of dual virus–bacterial infections in COPD exacerbations is likely to have been overlooked and may be an important factor contributing to the severity of COPD exacerbations.

4. *To expose impaired or aberrant host antiviral innate immunity that is either intrinsic or caused by treatment regimes.* One example is that the above chronic respiratory diseases are associated in some way with impaired antiviral immunity. A striking example is the effects of type 2 mediated allergic inflammation in allergic asthma on the antiviral Interferon (IFN) response. Deficient type I and type III IFNs have been associated with upregulated type 2 pathways[138] that involve the presence of IgE,[139,140] skin prick test (SPTs),[141] or eosinophilic inflammation,[142] and it is believed this may be the root cause of impaired antiviral immunity, potentially resulting in higher virus loads,[37] and ultimately increased symptoms and worse lung function.

Several studies have observed impaired IFN in cells derived from people with asthma. Not all studies have observed this[143,144] and differences could be accounted for by different culture methods, virus stocks, asthma severity,[145] or unappreciated subtle differences in the asthma endophenotypes of the donors. In support of these ex vivo findings, virus load is higher in individuals with atopic asthma compared with nonasthmatics following experimental RV challenge;[37] and recently, bronchial epithelial IFN-α/β staining by immunohistochemistry was impaired in individuals with atopic asthma, and was inversely related to increased clinical symptoms and increased virus load following experimental RV challenge.[106]

Mechanisms associated with impaired IFN include the crosslinking of IgE on DCs derived from people with asthma prevents virus-induced IFN-α production;[139] BECs derived from people with asthma have been shown to have higher expression levels of suppressor of cytokine signaling-1, which when nuclear, prevents virus-induced IFN-β and IFN-λ production.[146] Macrophages may recognize respiratory viruses via TLR7, and TLR7 expression is deficient in macrophages,[147] and bronchial tissue[40] from people with asthma and may explain why their cells are also impaired in IFN production during viral infection.

Other than direct, cell culture based studies where IFN transcripts or protein are measured following ex vivo virus infection, an alternative method of perceiving impaired IFN in individuals with asthma is to use large transcriptomic studies. These studies have the advantages of often incorporating more donors than possible with culture based studies. A recent study investigated the gene expression patterns of bronchial brushings.[148] Genes strongly correlating with FeNO levels were used to identify clinically important clusters of asthma phenotypes. Although type I and III IFNs and their downstream genes did not form an identifiable cluster, a group of individuals with high Nos2 expression and genes associated with type 2 immunity were also found to have low expression levels of genes related to innate immunity and antimicrobial defense. The fact that the samples were taken during stable asthma (not at exacerbation) likely explains the lack of type I and III IFN and related genes found to be associated with any cluster of asthma phenotypes.

If the idea that type 2 inflammation may be related to impaired IFN is correct, then the association of increased severity of RV infection in AE could be explained by the type 2–dominated inflammatory environment providing a susceptibility factor for RV infection. In epidemiological studies of respiratory virus infections, the exact endophenotype of asthma is seldom studied, and there is no detailed data on IFN expression. Therefore, this idea remains a compelling hypothesis. In support, a recent study of over 300 children from a birth cohort study found that cluster analysis of ex vivo responses in peripheral blood mononuclear cell (PBMCs) at age 11 to different viruses and TLR ligands were related to asthma at ages through to age 16. One cluster, defined by very low type

1 IFN induction and high proinflammatory cytokine induction, had the strongest relationship with asthma risk, frequency of AE, and unscheduled visits to primary health services associated with asthma.[149] While impaired IFN in asthma is a controversial topic, and may be present in a yet to be properly classified endophenotype of type 2 asthma,[150] this phenomenon does bring about new avenues for therapy, including IFN therapy for asthma, as discussed below.

Impaired production of IFNs may increase susceptibility to virus infection in COPD. In the experimental RV infection studies in COPD, virus load was higher in the COPD subjects compared with non-COPD subjects.[64,70] This suggests that the mechanisms controlling viral replication are defective in COPD as all subjects were inoculated with the same intranasal virus dose. Despite this, it remains unclear whether production of IFN is impaired in all COPD patients. Ex vivo RV stimulated macrophages from COPD subjects produce less IFN compared with non-COPD subjects;[64] however, not all studies have supported this observation, with similar[103] and even increased[102] IFN production reported in cells obtained from COPD subjects. A recent study also observed impaired IFN-β induction in response to influenza virus infection. The failure to robustly induce IFN-β transcription was thought due to a lack of protein kinase R expression and lack of formation of the IFN-β enhanceosome at the IFN-β promoter.[151]

The presence of RV during CF exacerbations may also be attributed to impaired antiviral immunity. One possibility is the CF epithelium is intrinsically proinflammatory or alternatively antiviral innate immune responses in CF cells are inherently deficient. Zheng et al. support the premise that impaired innate host defense causes susceptibility to viral infections in patients with CF highlighting increased replication following PIV infection of cultured airway epithelial cells from CF donors with correction by subsequent administration of IFN-α.[81] Although IFN responses were not impaired, there was reduced induction of nitric oxide synthase 2 (NOS2) in CF airway epithelial cells. NOS2 is important as it is required for production of nitric oxide (NO), which is a known antiviral mediator. Therefore, impaired NO synthesis may be one mechanism of impaired antiviral host responses in CF. Holtzman et al. hypothesized "hypersusceptibility" to virus infection, via defective IFN pathways, is a unifying pathway in asthma, COPD, and now CF.[152] The overriding hypothesis is that chronic airway inflammation, regardless of origin, interferes with innate immunity causing a suboptimal or blunted antiviral response that leads to an imbalanced host response reinforcing the preexisting predominant inflammatory phenotype.

Viral exacerbations of IPF may also be a result of impaired antiviral immunity. A study of familial IPF has identified one way in which antiviral immunity can be affected in ILD. Genetic studies in six affected families identified a gene (ELMOD2).[153] Further work has demonstrated that ELMOD2 is involved in type I and III IFN induction in response to TLR3 activation, an important part of the innate immune response to viruses including RV.[154]

6.7 RHINOVIRUS–TREATMENT INTERACTIONS IN CHRONIC RESPIRATORY DISEASES

Medications used to treat patients with chronic respiratory diseases affect immunity and airway function and therefore have the potential to modify the response to RV and affect disease outcomes. Associations between RV infection and current therapy use may broadly be divided into two main groups; firstly RV or other respiratory viruses may benefit from the direct action of the therapy, for example through suppressing important host responses that have direct antiviral effects. Another is as a confounder; that is, therapy use may promote or coincide with some unknown risk or susceptibility factor (e.g., antibiotics). Surprisingly, little attention has been paid to the detrimental effects current treatments for chronic respiratory diseases may have on antiviral host responses. Each treatment and the effects they may have on immunity relevant to viral or bacterial infection in chronic respiratory diseases are discussed below.

GCs, β_2 agonists, and leukotriene receptor antagonists are examples of immunomodulatory or antiinflammatory medication widely used to treat chronic respiratory diseases, particularly asthma and COPD. As asthma is a chronic disease often requiring daily treatment with a range of immunomodulatory or immunosuppressive agents, such as GCs, this idea seems plausible. Hypogammaglobulinemia has been associated with GC use in asthmatics, although whether this contributes to increases in infection is not clear.[155] GCs, β_2 agonists, and PDE4 inhibitors can have potent suppressive effects on IFNs as seen in tissue culture experiments,[156,157] and recently Singanayagam et al. showed that in a mouse model, the GC fluticasone propionate could suppress IFN induction and antibacterial immunity, leading to increased virus-induced mucin expression and RV replication.

In IPF, the degree to which IPF-E is due to an impaired immune response arising from the ILD itself, as opposed to the immunosuppressive/antifibrotic treatments that many of these patients are given (e.g., GCs), is difficult to differentiate and research into the ILDs is further complicated by the heterogeneity of disorders, the short time frame available for treatment (3–5 years), and relatively imprecise diagnostic criteria.

Antibiotics may act as a confounder, and may be associated with respiratory infections. The hygiene hypothesis states that the increasing incidence of allergic diseases in the Western world is based on higher standards of personal cleanliness and has thus reduced the opportunity for cross infection of microorganisms.[158] Here antibiotics may alter the microbiome, and thus immune maturation and development.[158] Current international guidelines do not support the use of antibiotics for asthma, yet antibiotics are often prescribed as a general treatment for lower respiratory tract infections (LRTIs), and asthma despite evidence for a bacterial or viral etiology. Antibiotic exposure in the uterus has been associated with an increased risk of asthma in cohort analyses, and this association is more than tripled if antibiotics were used to treat respiratory infections rather than antibiotics used for either urinary tract or skin infections.[159] Early antibiotic use is also believed to increase asthma risk by 2–3 fold in 7–8 year olds.[160] Here antibiotics may drive changes in the microbiome, deplete potentially protective microorganisms including *Prevotella* spp., or contribute to airway dysbiosis[161,162] as seen in the gut[163] and affect immune development. The fact that this could alter immune responses to viral infections and thus predispose to AE rates seems plausible. In mice, long-term treatment with antibiotics negatively impacts on immunity to influenza.[164]

In a recent large cohort study, a significantly higher risk of physician-confirmed wheezing after antibiotic prescription and a twofold increase in severe wheeze or AE after antibiotic prescription was also observed. In children who wheezed, the risk of AE and admissions to hospital were also significantly increased in the 2 years after the first antibiotic prescription. Children who received antibiotics in infancy had significantly lower induction of cytokines from virus-infected PBMCs at age 11.[165] The conclusion was that an increased susceptibility to viral infections is associated with both early life antibiotic prescription and asthma risk, although the authors could not exclude reverse causation, in that antibiotic use or asthma influence PBMC responses to viruses later in life. Clearly this is an area of emerging interest and further large clinical studies are required to understand the interrelationships between RV and asthma and COPD treatments, and to determine whether certain treatments do benefit respiratory infections in asthma.

6.8 FUTURE DIRECTIONS FOR THERAPEUTIC APPROACHES

The predominance of RV infections in exacerbations of chronic respiratory diseases must therefore strengthen their claim as a direct target in these diseases. Unfortunately, the successful development of antivirals and vaccines for RV has been disappointing when compared with the development of successful antivirals for RSVs[166,167] and vaccines for influenza.[168] This next section will consider future directions of targeting RV directly, and will consider other therapy options available that may indirectly affect RV-induced host responses critical in exacerbations of chronic respiratory diseases.

Antivirals for RV, discussed elsewhere, have been seldom tested in individuals with chronic respiratory diseases. The 1990s and early 2000s saw a number of RV-specific antivirals tested in healthy volunteers; however none were considered efficacious enough to transcend the early clinical development pipeline.[169,170] Recently, a series of compounds that act on host *N*-myristoyltransferases were shown to have antiviral effects on in vitro models of RV infection by blocking virus capsid assembly, although efficacy was only observed when compounds were given during, or within, 3 hours of infection.[171] Challenges to the use of antivirals in chronic respiratory disease such as asthma and COPD are many, and have been discussed in a recent review article;[169] they include dosing, treatment timing relevant to infection, relationships between virus load and symptom onset and duration, and use of existing therapies.

IFN-β has potent effects on RV in vitro and in vivo[125,172] and IFN-β (SNG001) therapy was tested in asthma in 2014;[173] however it did not significantly alter lung function or ACQ scores in a study of naturally occurring infections. In a subgroup analysis, asthmatics with more severe disease did experience improvements in ACQ and morning PEF compared with placebo, suggesting that a pan antiviral may be useful in treating AE in more severe patients. A confounding factor for IFN-β therapy acting as an antiviral is the fact that IFN-β may provide additional therapeutic benefit by downregulating type 2 pathways.[150] This makes IFN therapy attractive as a therapy for AE, but complicates the interpretation of any positive effects; the fact that asthmatics may be defective in IFN induction[139,142,174,175] further makes IFN-therapy attractive, but may additionally complicate the interpretation of trials with IFNs. Additionally, IFN-inducing agents such as TLR9 agonists[176] may also hold promise for AE by downregulating type 2 immunity via inducing IFNs. The converse may also be true of anti-Th2 cytokine therapy, which does show impressive reductions in AE rate[12,177,178]; however it is unclear if they act by reducing ongoing type 2 inflammation alone, or by restoring a defective IFN-mediated antiviral response. Anti-IgE has been shown to affect AE rates in asthmatic children,[140,179] and a recent study of 300 volunteers showed that AE rate during anti-IgE therapy was directly related to ex vivo levels of IFN-α produced from RV infected PBMCs.[140] These mechanisms may also be relevant in COPD, as COPD patients may exhibit increased eosinophil infiltration during COPD-E.[6,7] Clearly, further careful research is needed in this area, and the use of anti-Th2 biologics and anti-IgE in established experimental RV challenge studies may also help discern their true mechanism of action and how asthmatics are benefiting.

Azithromycin, a macrolide antibiotic has shown promise in reducing rates of AE[180,181] and COPD-E,[182,183] and reducing the number of wheeze events[184] and LRTIs[185] in young children. The fact that some of these effects are observed in the absence of proven bacterial infection[184,185] is intriguing, suggesting that azithromycin may have additional benefits and is not just a mere antibiotic. In one study, the effects were most obvious for volunteers with proven RV infection[185] suggesting that RV may in some way be susceptible to azithromycin, although secondary bacterial infections that may have been a consequence of RV infection were not studied. Potential antiviral properties of azithromycin are supported by in vitro studies that show that azithromycin can boost RV-induced IFN responses, thus reducing viral load.[186–189] The augmentation of IFN is specific, and virus-induced inflammatory mediators are not affected,[186,187] although the mechanism remains to be properly elucidated, and the direct benefit of azithromycin on RV infection is yet to be proven in a clinical setting.

6.9 SUMMARY

Exacerbations of respiratory diseases represent a major unmet need in respiratory medicine. The past 20 years have undoubtedly put respiratory viral infections center stage as triggers for exacerbations, and their perceived importance will only increase as diagnostic tests improve. Many challenges still remain in elucidating mechanisms of disease immunopathology, identifying at-risk populations, optimizing treatment, and identifying future therapeutic targets. For basic research, an important directive will be to continue to translate findings from in vitro or ex vivo culture systems and animal models, into human models. The different respiratory diseases are also at different stages in terms of addressing unmet therapeutic needs; a subpopulation of asthmatics experiencing AE will benefit from additional anti-type 2 cytokine therapy, while a better understanding of basic disease mechanisms is needed for other subgroups. These new disease modifying approaches are yet to be applied to COPD-E, and there are reasons for cautious optimism that COPD-E may benefit from approaches successfully applied to asthma. CF and the ILDs represent different challenges, more basic research is required to better understand the role of viruses, underlying mechanisms of disease and other triggers of exacerbations; these two diseases may also be candidate diseases for direct antiviral therapy.

REFERENCES

1. Johnston SL, Pattemore PK, Sanderson G, et al. Community study of role of viral infections in exacerbations of asthma in 9-11 year old children. *BMJ*. 1995;310 (6989):1225−1229.
2. Corne JM, Marshall C, Smith S, et al. Frequency, severity, and duration of rhinovirus infections in asthmatic and non-asthmatic individuals: a longitudinal cohort study. *Lancet*. 2002;359(9309):831−834.
3. Johnston NW, Johnston SL, Duncan JM, et al. The September epidemic of asthma exacerbations in children: a search for etiology. *J Allergy Clin Immunol*. 2005;115 (1):132−138.
4. Bonnelykke K, Sleiman P, Nielsen K, et al. A genome-wide association study identifies CDHR3 as a susceptibility locus for early childhood asthma with severe exacerbations. *Nat Genet*. 2013;46(1):51−55.
5. Seemungal T, Harper-Owen R, Bhowmik A, et al. Respiratory viruses, symptoms, and inflammatory markers in acute exacerbations and stable chronic obstructive pulmonary disease. *Am J Respir Crit Care Med*. 2001;164(9):1618−1623.
6. Papi A, Bellettato CM, Braccioni F, et al. Infections and airway inflammation in chronic obstructive pulmonary disease severe exacerbations. *Am J Respir Crit Care Med*. 2006;173:1114−1121.
7. Bafadhel M, McKenna S, Terry S, et al. Acute exacerbations of COPD: identification of biological clusters and their biomarkers. *Am J Respir Crit Care Med*. 2011;184:662−671.
8. Collinson J, Nicholson KG, Cancio E, et al. Effects of upper respiratory tract infections in patients with cystic fibrosis. *Thorax*. 1996;51(11):1115−1122.
9. Wootton SC, Kim DS, Kondoh Y, et al. Viral infection in acute exacerbation of idiopathic pulmonary fibrosis. *Am J Respir Crit Care Med*. 2011;183(12):1698−1702.

10. Bateman ED, Hurd SS, Barnes PJ, et al. Global strategy for asthma management and prevention: GINA executive summary. *Eur Respir J*. 2008;31(1):143–178.
11. Woodruff PG, Boushey HA, Dolganov GM, et al. Genome-wide profiling identifies epithelial cell genes associated with asthma and with treatment response to corticosteroids. *Proc Natl Acad Sci USA*. 2007;104(40):15858–15863.
12. Nair P, Pizzichini MM, Kjarsgaard M, et al. Mepolizumab for prednisone-dependent asthma with sputum eosinophilia. *N Engl J Med*. 2009;360(10):985–993.
13. Lotvall J, Akdis CA, Bacharier LB, et al. Asthma endotypes: a new approach to classification of disease entities within the asthma syndrome. *J Allergy Clin Immunol*. 2011;127(2):355–360.
14. Fahy JV. Type 2 inflammation in asthma—present in most, absent in many. *Nat Rev Immunol*. 2014;15(1):57–65.
15. GINA. Global initiative for asthma. In: *Global Strategy for Asthma Management and Prevention*; 2016. Available from: www.ginasthma.org. Accessed 19.06.18.
16. Rui P, Kang K, Albert M. *National Hospital Ambulatory Medical Care Survey: 2013 Emergency Department Summary Tables*; 2018. <https://www.cdc.gov/nchs/data/ahcd/nhamcs_emergency/2013_ed_web_tables.pdf>.
17. Slejko JF, Ghushchyan VH, Sucher B, et al. Asthma control in the United States, 2008-2010: indicators of poor asthma control. *J Allergy Clin Immunol*. 2014;133 (6):1579–1587.
18. Chipps BE, Zeiger RS, Borish L, et al. Key findings and clinical implications from the epidemiology and natural history of asthma: outcomes and treatment regimens (TENOR) study. *J Allergy Clin Immunol*. 2012;130(2):332–342.e10.
19. Tough SC, Hessel PA, Ruff M, Green FH, Mitchell I, Butt JC. Features that distinguish those who die from asthma from community controls with asthma. *J Asthma*. 1998;35(8):657–665.
20. Ivanova JI, Bergman R, Birnbaum HG, Colice GL, Silverman RA, McLaurin K. Effect of asthma exacerbations on health care costs among asthmatic patients with moderate and severe persistent asthma. *J Allergy Clin Immunol*. 2012;129(5):1229–1235.
21. Gern JE. The ABCs of rhinoviruses, wheezing, and asthma. *J Virol*. 2010;84 (15):7418–7426.
22. Papadopoulos NG, Christodoulou I, Rohde G, et al. Viruses and bacteria in acute asthma exacerbations—a GA(2) LEN-DARE systematic review. *Allergy*. 2010;66 (4):458–468.
23. Edwards MR, Bartlett NW, Hussell T, Openshaw P, Johnston SL. The microbiology of asthma. *Nat Rev Microbiol*. 2012;10(7):459–471.
24. Lysholm F, Wetterbom A, Lindau C, et al. Characterization of the viral microbiome in patients with severe lower respiratory tract infections, using metagenomic sequencing. *PLoS One*. 2012;7(2):e30875.
25. Niespodziana K, Stenberg-Hammar K, Megremis S, et al. PreDicta chip-based high resolution diagnosis of rhinovirus-induced wheeze. *Nat Commun*. 2018;9(1):2382.
26. Kistler A, Avila PC, Rouskin S, et al. Pan-viral screening of respiratory tract infections in adults with and without asthma reveals unexpected human coronavirus and human rhinovirus diversity. *J Infect Dis*. 2007;196(6):817–825.
27. Khetsuriani N, Kazerouni NN, Erdman DD, et al. Prevalence of viral respiratory tract infections in children with asthma. *J Allergy Clin Immunol*. 2007;119(2):314–321.
28. Arden KE, Chang AB, Lambert SB, Nissen MD, Sloots TP, Mackay IM. Newly identified respiratory viruses in children with asthma exacerbation not requiring admission to hospital. *J Med Virol*. 2010;82(8):1458–1461.
29. Antunes H, Rodrigues H, Silva N, et al. Etiology of bronchiolitis in a hospitalized pediatric population: prospective multicenter study. *J Clin Virol*. 2010;48 (2):134–136.

30. Palmenberg AC, Spiro D, Kuzmickas R, et al. Sequencing and analyses of all known human rhinovirus genomes reveal structure and evolution. *Science*. 2009;324 (5923):55–59.
31. Lau SK, Yip CC, Tsoi HW, et al. Clinical features and complete genome characterization of a distinct human rhinovirus (HRV) genetic cluster, probably representing a previously undetected HRV species, HRV-C, associated with acute respiratory illness in children. *J Clin Microbiol*. 2007;45(11):3655–3664.
32. Bochkov YA, Palmenberg AC, Lee WM, et al. Molecular modeling, organ culture and reverse genetics for a newly identified human rhinovirus C. *Nat Med*. 2011;17 (5):627–632.
33. Bochkov YA, Watters K, Ashraf S, et al. Cadherin-related family member 3, a childhood asthma susceptibility gene product, mediates rhinovirus C binding and replication. *Proc Natl Acad Sci USA*. 2015;112(17):5485–5490.
34. Bizzintino J, Lee WM, Laing IA, et al. Association between human rhinovirus C and severity of acute asthma in children. *Eur Respir J*. 2010;37(5):1037–1042.
35. Busti AJ, Bain AM, Hall 2nd RG, et al. Effects of atazanavir/ritonavir or fosamprenavir/ritonavir on the pharmacokinetics of rosuvastatin. *J Cardiovasc Pharmacol*. 2008;51 (6):605–610.
36. Message SD, Laza-Stanca V, Mallia P, et al. Rhinovirus-induced lower respiratory illness is increased in asthma and related to virus load and Th1/2 cytokine and IL-10 production. *Proc Natl Acad Sci USA*. 2008;105(36):13562–13567.
37. Jackson DJ, Makrinioti H, Rana BM, et al. IL-33-dependent type 2 inflammation during rhinovirus-induced asthma exacerbations in vivo. *Am J Respir Crit Care Med*. 2014;190(12):1373–1382.
38. Grunberg K, Smits HH, Timmers MC, et al. Experimental rhinovirus 16 infection. Effects on cell differentials and soluble markers in sputum in asthmatic subjects. *Am J Respir Crit Care Med*. 1997;156(2 Pt 1):609–616.
39. Bartlett NW, Walton RP, Edwards MR, et al. Mouse models of rhinovirus-induced disease and exacerbation of allergic airway inflammation. *Nat Med*. 2008;14 (2):199–204.
40. Hatchwell L, Collison A, Girkin J, et al. Toll-like receptor 7 governs interferon and inflammatory responses to rhinovirus and is suppressed by IL-5-induced lung eosinophilia. *Thorax*. 2015;70(9):854–861.
41. Toussaint M, Jackson DJ, Swieboda D, et al. Host DNA released by NETosis promotes rhinovirus-induced type-2 allergic asthma exacerbation. *Nat Med*. 2017;23(6):681–691.
42. Wang H, Naghavi M, Allen C, et al. Global, regional, and national life expectancy, all-cause mortality, and cause-specific mortality for 249 causes of death, 1980–2015: a systematic analysis for the Global Burden of Disease Study 2015. *Lancet*. 2016;388 (10053):1459–1544.
43. Vos T, Flaxman AD, Naghavi M, et al. Years lived with disability (YLDs) for 1160 sequelae of 289 diseases and injuries 1990-2010: a systematic analysis for the Global Burden of Disease Study 2010. *Lancet (London, England)*. 2012;380(9859):2163–2196.
44. Qaseem A, Wilt TJ, Weinberger SE, et al. Diagnosis and management of stable chronic obstructive pulmonary disease: a clinical practice guideline update from the American College of Physicians, American College of Chest Physicians, American Thoracic Society, and European Respiratory Society. *Ann Intern Med*. 2011;155 (3):179–191.
45. Beasley V, Joshi PV, Singanayagam A, Molyneaux PL, Johnston SL, Mallia P. Lung microbiology and exacerbations in COPD. *Int J Chron Obstruct Pulmon Dis*. 2012;7:555–569.
46. Donaldson GC, Seemungal TA, Bhowmik A, Wedzicha JA. Relationship between exacerbation frequency and lung function decline in chronic obstructive pulmonary disease. *Thorax*. 2002;57(10):847–852.

47. Anzueto A, Sethi S, Martinez FJ. Exacerbations of chronic obstructive pulmonary disease. *Proc Am Thorac Soc.* 2007;4(7):554–564.
48. Vollenweider DJ, Jarrett H, Steurer-Stey CA, Garcia-Aymerich J, Puhan MA. Antibiotics for exacerbations of chronic obstructive pulmonary disease. *Cochrane Database Syst Rev.* 2012;12:CD010257.
49. George SN, Garcha DS, Mackay AJ, et al. Human rhinovirus infection during naturally occurring COPD exacerbations. *Eur Respir J.* 2014;44(1):87–96.
50. Ko FW, Ip M, Chan PK, et al. Viral etiology of acute exacerbations of COPD in Hong Kong. *Chest.* 2007;132(3):900–908.
51. Hutchinson AF, Ghimire AK, Thompson MA, et al. A community-based, time-matched, case-control study of respiratory viruses and exacerbations of COPD. *Respir Med.* 2007;101(12):2472–2481.
52. Camargo Jr CA, Ginde AA, Clark S, Cartwright CP, Falsey AR, Niewoehner DE. Viral pathogens in acute exacerbations of chronic obstructive pulmonary disease. *Intern Emerg Med.* 2008;3(4):355–359.
53. Bozinovski S, Hutchinson A, Thompson M, et al. Serum amyloid a is a biomarker of acute exacerbations of chronic obstructive pulmonary disease. *Am J Respir Crit Care Med.* 2008;177(3):269–278.
54. Almansa R, Sanchez-Garcia M, Herrero A, et al. Host response cytokine signatures in viral and nonviral acute exacerbations of chronic obstructive pulmonary disease. *J Interferon Cytokine Res.* 2011;31(5):409–413.
55. Pant S, Walters EH, Griffiths A, Wood-Baker R, Johns DP, Reid DW. Airway inflammation and anti-protease defences rapidly improve during treatment of an acute exacerbation of COPD. *Respirology.* 2009;14(4):495–503.
56. Rohde G, Wiethege A, Borg I, et al. Respiratory viruses in exacerbations of chronic obstructive pulmonary disease requiring hospitalisation: a case-control study. *Thorax.* 2003;58(1):37–42.
57. Tan WC, Xiang X, Qiu D, Ng TP, Lam SF, Hegele RG. Epidemiology of respiratory viruses in patients hospitalized with near-fatal asthma, acute exacerbations of asthma, or chronic obstructive pulmonary disease. *Am J Med.* 2003;115(4):272–277.
58. Qiu Y, Zhu J, Bandi V, et al. Biopsy neutrophilia, neutrophil chemokine and receptor gene expression in severe exacerbations of chronic obstructive pulmonary disease. *Am J Respir Crit Care Med.* 2003;168(8):968–975.
59. McManus TE, Marley AM, Baxter N, et al. Respiratory viral infection in exacerbations of COPD. *Respir Med.* 2008;102(11):1575–1580.
60. Kherad O, Kaiser L, Bridevaux PO, et al. Upper-respiratory viral infection, biomarkers, and COPD exacerbations. *Chest.* 2010;138(4):896–904.
61. Singh M, Lee SH, Porter P, et al. Human rhinovirus proteinase 2A induces TH1 and TH2 immunity in patients with chronic obstructive pulmonary disease. *J Allergy Clin Immunol.* 2010;125(6):1369–1378.e2.
62. Cameron RJ, de Wit D, Welsh TN, Ferguson J, Grissell TV, Rye PJ. Virus infection in exacerbations of chronic obstructive pulmonary disease requiring ventilation. *Intensive Care Med.* 2006;32(7):1022–1029.
63. Perotin JM, Dury S, Renois F, et al. Detection of multiple viral and bacterial infections in acute exacerbation of chronic obstructive pulmonary disease: a pilot prospective study. *J Med Virol.* 2013;85(5):866–873.
64. Mallia P, Message SD, Gielen V, et al. Experimental rhinovirus infection as a human model of chronic obstructive pulmonary disease exacerbation. *Am J Respir Crit Care Med.* 2010;183(6):734–742.
65. Mallia P, Message SD, Kebadze T, Parker HL, Kon OM, Johnston SL. An experimental model of rhinovirus induced chronic obstructive pulmonary disease exacerbations: a pilot study. *Respir Res.* 2006;7:116.

66. Molyneaux PL, Mallia P, Cox MJ, et al. Outgrowth of the bacterial airway microbiome after rhinovirus exacerbation of chronic obstructive pulmonary disease. *Am J Respir Crit Care Med.* 2013;188(10):1224–1231.
67. Biancardi E, Fennell M, Rawlinson W, Thomas PS. Viruses are frequently present as the infecting agent in acute exacerbations of chronic obstructive pulmonary disease in patients presenting to hospital. *Intern Med J.* 2016;46(10):1160–1165.
68. Sanz I, Tamames S, Rojo S, et al. Viral etiology of chronic obstructive pulmonary disease exacerbations during the A/H1N1pdm09 pandemic and postpandemic period. *Adv Virol.* 2015;2015:560679.
69. Bafadhel M, McKenna S, Terry S, et al. Acute exacerbations of chronic obstructive pulmonary disease: identification of biologic clusters and their biomarkers. *Am J Respir Crit Care Med.* 2011;184(6):662–671.
70. Footitt J, Mallia P, Durham AL, et al. Oxidative and nitrosative stress and histone deacetylase-2 activity in exacerbations of COPD. *Chest.* 2016;149(1):62–73.
71. Cuppens H, Boulyjenkov V, Calafell F, et al. *The Molecular Genetic Epidemiology of Cystic Fibrosis: Report of a Joint Meeting of WHO/ECFTN/ICF(M)A/ECFS*; 2004.
72. Burgel PR, Bellis G, Olesen HV, et al. Future trends in cystic fibrosis demography in 34 European countries. *Eur Respir J.* 2015;46(1):133–141.
73. Parkins MD, Parkins VM, Rendall JC, Elborn S. Changing epidemiology and clinical issues arising in an ageing cystic fibrosis population. *Ther Adv Respir Dis.* 2011;5 (2):105–119.
74. Goldberg JB, Pier G. The role of the CFTR in susceptibility to *Pseudomonas aeruginosa* infections in cystic fibrosis. *Trends Microbiol.* 2000;8(11):514–520.
75. Wang EE, Prober CG, Manson B, Corey M, Levison H. Association of respiratory viral infections with pulmonary deterioration in patients with cystic fibrosis. *N Engl J Med.* 1984;311(26):1653–1658.
76. Ong EL, Ellis ME, Webb AK, et al. Infective respiratory exacerbations in young adults with cystic fibrosis: role of viruses and atypical microorganisms. *Thorax.* 1989;44(9):739–742.
77. Smyth AR, Smyth RL, Tong CY, Hart CA, Heaf DP. Effect of respiratory virus infections including rhinovirus on clinical status in cystic fibrosis. *Arch Dis Child.* 1995;73(2):117–120.
78. Ramsey BW, Gore EJ, Smith AL, Cooney MK, Redding GJ, Foy H. The effect of respiratory viral infections on patients with cystic fibrosis. *Am J Dis Child.* 1989;143 (6):662–668.
79. Wat D, Gelder C, Hibbitts S, et al. The role of respiratory viruses in cystic fibrosis. *J Cyst Fibros.* 2008;7(4):320–328.
80. van Ewijk BE, van der Zalm MM, Wolfs TF, et al. Prevalence and impact of respiratory viral infections in young children with cystic fibrosis: prospective cohort study. *Pediatrics.* 2008;122(6):1171–1176.
81. Zheng S, De BP, Choudhary S, et al. Impaired innate host defense causes susceptibility to respiratory virus infections in cystic fibrosis. *Immunity.* 2003;18(5):619–630.
82. Ley B, Ryerson CJ, Vittinghoff E, et al. A multidimensional index and staging system for idiopathic pulmonary fibrosis. *Ann Intern Med.* 2012;156(10):684–691.
83. Wolters PJ, Collard HR, Jones KD. Pathogenesis of idiopathic pulmonary fibrosis. *Annu Rev Pathol.* 2014;9:157–179.
84. Collard HR, Ryerson CJ, Corte TJ, et al. Acute exacerbation of idiopathic pulmonary fibrosis. An international working group report. *Am J Respir Crit Care Med.* 2016;194(3):265–275.
85. Huie TJ, Olson AL, Cosgrove GP, et al. A detailed evaluation of acute respiratory decline in patients with fibrotic lung disease: aetiology and outcomes. *Respirology.* 2010;15(6):909–917.

86. Ushiki A, Yamazaki Y, Hama M, Yasuo M, Hanaoka M, Kubo K. Viral infections in patients with an acute exacerbation of idiopathic interstitial pneumonia. *Respir Invest.* 2014;52(1):65–70.
87. Collard HR, Yow E, Richeldi L, Anstrom KJ, Glazer C. Suspected acute exacerbation of idiopathic pulmonary fibrosis as an outcome measure in clinical trials. *Respir Res.* 2013;14(1):73.
88. Johannson KA, Vittinghoff E, Lee K, et al. Acute exacerbation of idiopathic pulmonary fibrosis associated with air pollution exposure. *Eur Respir J.* 2014;43 (4):1124–1131.
89. Simon-Blancal V, Freynet O, Nunes H, et al. Acute exacerbation of idiopathic pulmonary fibrosis: outcome and prognostic factors. *Respiration.* 2012;83(1):28–35.
90. Petrosyan F, Culver DA, Reddy AJ. Role of bronchoalveolar lavage in the diagnosis of acute exacerbations of idiopathic pulmonary fibrosis: a retrospective study. *BMC Pulm Med.* 2015;15:70.
91. Collard HR, Ward AJ, Lanes S, Cortney Hayflinger D, Rosenberg DM, Hunsche E. Burden of illness in idiopathic pulmonary fibrosis. *J Med Econ.* 2012;15(5):829–835.
92. Dusemund F, Chronis J, Baty F, Albrich WC, Brutsche MH. The outcome of community-acquired pneumonia in patients with chronic lung disease: a case-control study. *Swiss Med Wkly.* 2014;144:w14013.
93. Tang YW, Johnson JE, Browning PJ, et al. Herpesvirus DNA is consistently detected in lungs of patients with idiopathic pulmonary fibrosis. *J Clin Microbiol.* 2003;41(6):2633–2640.
94. Egan JJ, Adamali HI, Lok SS, Stewart JP, Woodcock AA. Ganciclovir antiviral therapy in advanced idiopathic pulmonary fibrosis: an open pilot study. *Pulm Med.* 2011;2011:240805.
95. Stone Jr CA, Miller EK. Understanding the association of human rhinovirus with asthma. *Clin Vaccine Immunol.* 2016;23(1):6–10.
96. Guibas GV, Tsolia M, Christodoulou I, Stripeli F, Sakkou Z, Papadopoulos NG. Distinction between rhinovirus-induced acute asthma and asthma-augmented influenza infection. *Clin Exp Allergy.* 2018;48(5):536–543.
97. O'Neill LA, Bowie AG. The family of five: TIR-domain-containing adaptors in Toll-like receptor signalling. *Nat Rev Immunol.* 2007;7(5):353–364.
98. Edwards MR, Bartlett NW, Clarke D, Birrell M, Belvisi M, Johnston SL. Targeting the NF-kappaB pathway in asthma and chronic obstructive pulmonary disease. *Pharmacol Ther.* 2009;121(1):1–13.
99. Pizzichini MM, Pizzichini E, Efthimiadis A, et al. Asthma and natural colds. Inflammatory indices in induced sputum: a feasibility study. *Am J Respir Crit Care Med.* 1998;158(4):1178–1184.
100. Jarjour NN, Gern JE, Kelly EA, Swenson CA, Dick CR, Busse WW. The effect of an experimental rhinovirus 16 infection on bronchial lavage neutrophils. *J Allergy Clin Immunol.* 2000;105(6 Pt 1):1169–1177.
101. Seymour ML, Gilby N, Bardin PG, et al. Rhinovirus infection increases 5-lipoxygenase and cyclooxygenase-2 in bronchial biopsy specimens from nonatopic subjects. *J Infect Dis.* 2002;185(4):540–544.
102. Schneider D, Ganesan S, Comstock AT, et al. Increased cytokine response of rhinovirus-infected airway epithelial cells in chronic obstructive pulmonary disease. *Am J Respir Crit Care Med.* 2010;182(3):332–340.
103. Baines KJ, Hsu AC, Tooze M, Gunawardhana LP, Gibson PG, Wark PA. Novel immune genes associated with excessive inflammatory and antiviral responses to rhinovirus in COPD. *Respir Res.* 2013;14:15.
104. Rohde G, Borg I, Wiethege A, et al. Inflammatory response in acute viral exacerbations of COPD. *Infection.* 2008;36(5):427–433.

105. Mallia P, Message SD, Contoli M, et al. Neutrophil adhesion molecules in experimental rhinovirus infection in COPD. *Respir Res*. 2013;14(1):72.
106. Zhu J, Message SD, Mallia P, et al. Bronchial mucosal IFN-alpha/beta and pattern recognition receptor expression in patients with experimental rhinovirus-induced asthma exacerbations. *J Allergy Clin Immunol*. 2018;143:114–125.e4.
107. Tate MD, Ioannidis LJ, Croker B, Brown LE, Brooks AG, Reading PC. The role of neutrophils during mild and severe influenza virus infections of mice. *PLoS One*. 2011;6(3):e17618.
108. O'Byrne PM, Metev H, Puu M, et al. Efficacy and safety of a CXCR2 antagonist, AZD5069, in patients with uncontrolled persistent asthma: a randomised, double-blind, placebo-controlled trial. *Lancet Respir Med*. 2016;4(10):797–806.
109. Sutanto EN, Kicic A, Foo CJ, et al. Innate inflammatory responses of pediatric cystic fibrosis airway epithelial cells: effects of nonviral and viral stimulation. *Am J Respir Cell Mol Biol*. 2011;44(6):761–767.
110. Black HR, Yankaskas JR, Johnson LG, Noah TL. Interleukin-8 production by cystic fibrosis nasal epithelial cells after tumor necrosis factor-alpha and respiratory syncytial virus stimulation. *Am J Respir Cell Mol Biol*. 1998;19(2):210–215.
111. Kieninger E, Vareille M, Kopf BS, et al. Lack of an exaggerated inflammatory response on virus infection in cystic fibrosis. *Eur Respir J*. 2012;39(2):297–304.
112. Bossios A, Psarras S, Gourgiotis D, et al. Rhinovirus infection induces cytotoxicity and delays wound healing in bronchial epithelial cells. *Respir Res*. 2005;6:114.
113. Reed SE, Boyde A. Organ cultures of respiratory epithelium infected with rhinovirus or parainfluenza virus studied in a scanning electron microscope. *Infect Immun*. 1972;6(1):68–76.
114. Turner RB, Hendley JO, Gwaltney Jr. JM. Shedding of infected ciliated epithelial cells in rhinovirus colds. *J Infect Dis*. 1982;145(6):849–853.
115. Bassetti S, Bischoff WE, Walter M, et al. Dispersal of *Staphylococcus aureus* into the air associated with a rhinovirus infection. *Infect Control Hosp Epidemiol*. 2005;26 (2):196–203.
116. Marsland BJ, Scanga CB, Kopf M, Le Gros G. Allergic airway inflammation is exacerbated during acute influenza infection and correlates with increased allergen presentation and recruitment of allergen-specific T-helper type 2 cells. *Clin Exp Allergy*. 2004;34(8):1299–1306.
117. Abdullah H, Heaney LG, Cosby SL, McGarvey LP. Rhinovirus upregulates transient receptor potential channels in a human neuronal cell line: implications for respiratory virus-induced cough reflex sensitivity. *Thorax*. 2014;69(1):46–54.
118. Cheung D, Dick EC, Timmers MC, de Klerk EP, Spaan WJ, Sterk PJ. Rhinovirus inhalation causes long-lasting excessive airway narrowing in response to methacholine in asthmatic subjects in vivo. *Am J Respir Crit Care Med*. 1995;152(5 Pt 1):1490–1496.
119. Baluk P, Tammela T, Ator E, et al. Pathogenesis of persistent lymphatic vessel hyperplasia in chronic airway inflammation. *J Clin Invest*. 2005;115(2):247–257.
120. Volonaki E, Psarras S, Xepapadaki P, Psomali D, Gourgiotis D, Papadopoulos NG. Synergistic effects of fluticasone propionate and salmeterol on inhibiting rhinovirus-induced epithelial production of remodelling-associated growth factors. *Clin Exp Allergy*. 2006;36(10):1268–1273.
121. Tourdot S, Mathie S, Hussell T, et al. Respiratory syncytial virus infection provokes airway remodelling in allergen-exposed mice in absence of prior allergen sensitization. *Clin Exp Allergy*. 2008;38(6):1016–1024.
122. Kuo C, Lim S, King NJ, et al. Rhinovirus infection induces extracellular matrix protein deposition by primary bronchial epithelial cells and lung fibroblasts. *Respirology*. 2011;16(2):367–377.

123. Green RM, Custovic A, Sanderson G, Hunter J, Johnston SL, Woodcock A. Synergism between allergens and viruses and risk of hospital admission with asthma: case-control study. *BMJ*. 2002;324(7340):763.
124. Murray CS, Poletti G, Kebadze T, et al. Study of modifiable risk factors for asthma exacerbations: virus infection and allergen exposure increase the risk of asthma hospital admissions in children. *Thorax*. 2006;61(5):376–382.
125. Bartlett NW, Slater L, Glanville N, et al. Defining critical roles for NF-kappaB p65 and type I interferon in innate immunity to rhinovirus. *EMBO Mol Med*. 2012;4 (12):1244–1260.
126. Beale J, Jayaraman A, Jackson DJ, et al. Rhinovirus-induced IL-25 in asthma exacerbation drives type 2 immunity and allergic pulmonary inflammation. *Sci Transl Med*. 2014;6(256):256ra134.
127. Collison A, Hatchwell L, Verrills N, et al. The E3 ubiquitin ligase midline 1 promotes allergen and rhinovirus-induced asthma by inhibiting protein phosphatase 2A activity. *Nat Med*. 2013;19(2):232–237.
128. Kato A, Favoreto Jr S, Avila PC, Schleimer RP. TLR3- and Th2 cytokine-dependent production of thymic stromal lymphopoietin in human airway epithelial cells. *J Immunol*. 2007;179(2):1080–1087.
129. Jackson DJ, Evans MD, Gangnon RE, et al. Evidence for a causal relationship between allergic sensitization and rhinovirus wheezing in early life. *Am J Respir Crit Care Med*. 2011;185(3):281–285.
130. Jackson DJ, Trujillo-Torralbo MB, Del-Rosario J, et al. The influence of asthma control on the severity of virus-induced asthma exacerbations. *J Allergy Clin Immunol*. 2015;136(2):497–500.e3.
131. Oliver BG, Lim S, Wark P, et al. Rhinovirus exposure impairs immune responses to bacterial products in human alveolar macrophages. *Thorax*. 2008;63(6):519–525.
132. Mallia P, Footitt J, Sotero R, et al. Rhinovirus infection induces degradation of antimicrobial peptides and secondary bacterial infection in chronic obstructive pulmonary disease. *Am J Respir Crit Care Med*. 2012;186(11):1117–1124.
133. Kloepfer KM, Sarsani VK, Poroyko V, et al. Community-acquired rhinovirus infection is associated with changes in the airway microbiome. *J Allergy Clin Immunol*. 2017;140(1):312–315.e8.
134. Kloepfer KM, Lee WM, Pappas TE, et al. Detection of pathogenic bacteria during rhinovirus infection is associated with increased respiratory symptoms and asthma exacerbations. *J Allergy Clin Immunol*. 2014;133(5):1301–1307. 1307.e1–3.
135. Wark PA, Tooze M, Cheese L, et al. Viral infections trigger exacerbations of cystic fibrosis in adults and children. *Eur Respir J*. 2012;40(2):510–512.
136. Chin M, De Zoysa M, Slinger R, et al. Acute effects of viral respiratory tract infections on sputum bacterial density during CF pulmonary exacerbations. *J Cyst Fibros*. 2015;14(4):482–489.
137. Wilkinson TM, Hurst JR, Perera WR, Wilks M, Donaldson GC, Wedzicha JA. Effect of interactions between lower airway bacterial and rhinoviral infection in exacerbations of COPD. *Chest*. 2006;129(2):317–324.
138. Baraldo S, Contoli M, Bazzan E, et al. Deficient antiviral immune responses in childhood: distinct roles of atopy and asthma. *J Allergy Clin Immunol*. 2012;130 (6):1307–1314.
139. Gill MA, Bajwa G, George TA, et al. Counterregulation between the FcepsilonRI pathway and antiviral responses in human plasmacytoid dendritic cells. *J Immunol*. 2010;184(11):5999–6006.
140. Teach SJ, Gill MA, Togias A, et al. Preseasonal treatment with either omalizumab or an inhaled corticosteroid boost to prevent fall asthma exacerbations. *J Allergy Clin Immunol*. 2015;136(6):1476–1485.

141. Sykes A, Edwards MR, Macintyre J, et al. Rhinovirus 16-induced IFN-alpha and IFN-beta are deficient in bronchoalveolar lavage cells in asthmatic patients. *J Allergy Clin Immunol.* 2012;129(6):1506−1514.e6.
142. Contoli M, Message SD, Laza-Stanca V, et al. Role of deficient type III interferon-lambda production in asthma exacerbations. *Nat Med.* 2006;12(9):1023−1026.
143. Patel DA, You Y, Huang G, et al. Interferon response and respiratory virus control are preserved in bronchial epithelial cells in asthma. *J Allergy Clin Immunol.* 2014.
144. Lopez-Souza N, Favoreto S, Wong H, et al. In vitro susceptibility to rhinovirus infection is greater for bronchial than for nasal airway epithelial cells in human subjects. *J Allergy Clin Immunol.* 2009;123(6):1384−1390.e2.
145. Sykes A, Macintyre J, Edwards MR, et al. Rhinovirus-induced interferon production is not deficient in well controlled asthma. *Thorax.* 2014;69(3):240−246.
146. Gielen V, Sykes A, Zhu J, et al. Increased nuclear suppressor of cytokine signaling 1 in asthmatic bronchial epithelium suppresses rhinovirus induction of innate interferons. *J Allergy Clin Immunol.* 2015;136(1):177−188.e11.
147. Rupani H, Martinez-Nunez RT, Dennison P, et al. Toll-like receptor 7 is reduced in severe asthma and linked to an altered microRNA profile. *Am J Respir Crit Care Med.* 2016;194(1):26−37.
148. Modena BD, Tedrow JR, Milosevic J, et al. Gene expression in relation to exhaled nitric oxide identifies novel asthma phenotypes with unique biomolecular pathways. *Am J Respir Crit Care Med.* 2014;190(12):1363−1372.
149. Custovic A, Belgrave D, Lin L, et al. Cytokine responses to rhinovirus and development of asthma, allergic sensitization, and respiratory infections during childhood. *Am J Respir Crit Care Med.* 2018;197(10):1265−1274.
150. Edwards MR, Strong K, Cameron A, Walton RP, Jackson DJ, Johnston SL. Viral infections in allergy and immunology: how allergic inflammation influences viral infections and illness. *J Allergy Clin Immunol.* 2017;140(4):909−920.
151. Hsu AC, Parsons K, Moheimani F, et al. Impaired antiviral stress granule and IFN-beta enhanceosome formation enhances susceptibility to influenza infection in chronic obstructive pulmonary disease epithelium. *Am J Respir Cell Mol Biol.* 2016;55(1):117−127.
152. Holtzman M, Patel D, Kim HJ, You Y, Zhang Y. Hypersusceptibility to respiratory viruses as a shared mechanism for asthma, chronic obstructive pulmonary disease, and cystic fibrosis. *Am J Respir Cell Mol Biol.* 2011;44(6):739−742.
153. Hodgson U, Pulkkinen V, Dixon M, et al. ELMOD2 is a candidate gene for familial idiopathic pulmonary fibrosis. *Am J Hum Genet.* 2006;79(1):149−154.
154. Pulkkinen V, Bruce S, Rintahaka J, et al. ELMOD2, a candidate gene for idiopathic pulmonary fibrosis, regulates antiviral responses. *FASEB J.* 2010;24(4):1167−1177.
155. Hamilos DL, Young RM, Peter JB, Agopian MS, Ikle DN, Barka N. Hypogammaglobulinemia in asthmatic patients. *Ann Allergy.* 1992;68(6):472−481.
156. Edwards MR, Facchinetti F, Civelli M, Villetti G, Johnston SL. Anti-inflammatory effects of the novel inhaled phosphodiesterase type 4 inhibitor CHF6001 on virus-inducible cytokines. *Pharmacol Res Perspect.* 2016;4(1):e00202.
157. Thomas BJ, Porritt RA, Hertzog PJ, Bardin PG, Tate MD. Glucocorticosteroids enhance replication of respiratory viruses: effect of adjuvant interferon. *Sci Rep.* 2014;4:7176.
158. Strachan DP. Family size, infection and atopy: the first decade of the "hygiene hypothesis". *Thorax.* 2000;55(Suppl 1):S2−10.
159. Ortqvist AK, Lundholm C, Kieler H, et al. Antibiotics in fetal and early life and subsequent childhood asthma: nationwide population based study with sibling analysis. *BMJ.* 2014;349:g6979.

160. Droste JH, Wieringa MH, Weyler JJ, Nelen VJ, Vermeire PA, Van Bever HP. Does the use of antibiotics in early childhood increase the risk of asthma and allergic disease? *Clin Exp Allergy*. 2000;30(11):1547–1553.
161. Huang YJ, Nariya S, Harris JM, et al. The airway microbiome in patients with severe asthma: Associations with disease features and severity. *J Allergy Clin Immunol*. 2015;136(4):874–884.
162. Chung KF. Potential role of the lung microbiome in shaping asthma phenotypes. *Ann Am Thorac Soc*. 2017;14(Supplement_5):S326–S331.
163. Kim YG, Udayanga KG, Totsuka N, Weinberg JB, Nunez G, Shibuya A. Gut dysbiosis promotes M2 macrophage polarization and allergic airway inflammation via fungi-induced PGE(2). *Cell Host Microbe*. 2014;15(1):95–102.
164. Ichinohe T, Pang IK, Kumamoto Y, et al. Microbiota regulates immune defense against respiratory tract influenza A virus infection. *Proc Natl Acad Sci USA*. 2011;108(13):5354–5359.
165. Semic-Jusufagic A, Belgrave D, Pickles A, et al. Assessing the association of early life antibiotic prescription with asthma exacerbations, impaired antiviral immunity, and genetic variants in 17q21: a population-based birth cohort study. *Lancet Respir Med*. 2014;2(8):621–630.
166. DeVincenzo JP, McClure MW, Symons JA, et al. Activity of oral ALS-008176 in a respiratory syncytial virus challenge study. *N Engl J Med*. 2015;373(21):2048–2058.
167. DeVincenzo JP, Whitley RJ, Mackman RL, et al. Oral GS-5806 activity in a respiratory syncytial virus challenge study. *N Engl J Med*. 2014;371(8):711–722.
168. Turner PJ, Southern J, Andrews NJ, Miller E, Erlewyn-Lajeunesse M. Safety of live attenuated influenza vaccine in atopic children with egg allergy. *J Allergy Clin Immunol*. 2015;136(2):376–381.
169. Edwards MR, Walton RP, Jackson DJ, et al. The potential of anti-infectives and immunomodulators as therapies for asthma and asthma exacerbations. *Allergy*. 2018;73(1):50–63.
170. Papadopoulos NG, Megremis S, Kitsioulis NA, Vangelatou O, West P, Xepapadaki P. Promising approaches for the treatment and prevention of viral respiratory illnesses. *J Allergy Clin Immunol*. 2017;140(4):921–932.
171. Mousnier A, Bell AS, Swieboda DP, et al. Fragment-derived inhibitors of human N-myristoyltransferase block capsid assembly and replication of the common cold virus. *Nat Chem*. 2018;10(6):599–606.
172. Cakebread JA, Xu Y, Grainge C, et al. Exogenous IFN-beta has antiviral and anti-inflammatory properties in primary bronchial epithelial cells from asthmatic subjects exposed to rhinovirus. *J Allergy Clin Immunol*. 2011;127(5):1148–1154.e9.
173. Djukanovic R, Harrison T, Johnston SL, et al. The effect of inhaled IFN-beta on worsening of asthma symptoms caused by viral infections. A randomized trial. *Am J Respir Crit Care Med*. 2014;190(2):145–154.
174. Wark PA, Johnston SL, Bucchieri F, et al. Asthmatic bronchial epithelial cells have a deficient innate immune response to infection with rhinovirus. *J Exp Med*. 2005;201(6):937–947.
175. Edwards MR, Regamey N, Vareille M, et al. Impaired innate interferon induction in severe therapy resistant atopic asthmatic children. *Mucosal Immunol*. 2012;6(4):797–806.
176. Jackson S, Candia AF, Delaney S, et al. First-in-human study with the inhaled TLR9 oligonucleotide agonist AZD1419 results in interferon responses in the lung, and is safe and well-tolerated. *Clin Pharmacol Ther*. 2017;104:335–345.
177. Castro M, Wenzel SE, Bleecker ER, et al. Benralizumab, an anti-interleukin 5 receptor alpha monoclonal antibody, versus placebo for uncontrolled eosinophilic

asthma: a phase 2b randomised dose-ranging study. *Lancet Respir Med.* 2014;2 (11):879–890.
178. Pavord ID, Korn S, Howarth P, et al. Mepolizumab for severe eosinophilic asthma (DREAM): a multicentre, double-blind, placebo-controlled trial. *Lancet.* 2012;380 (9842):651–659.
179. Busse WW, Morgan WJ, Gergen PJ, et al. Randomized trial of omalizumab (anti-IgE) for asthma in inner-city children. *N Engl J Med.* 2011;364(11):1005–1015.
180. Brusselle GG, Vanderstichele C, Jordens P, et al. Azithromycin for prevention of exacerbations in severe asthma (AZISAST): a multicentre randomised double-blind placebo-controlled trial. *Thorax.* 2013;68(4):322–329.
181. Gibson PG, Yang IA, Upham JW, et al. Effect of azithromycin on asthma exacerbations and quality of life in adults with persistent uncontrolled asthma (AMAZES): a randomised, double-blind, placebo-controlled trial. *Lancet.* 2017;390 (10095):659–668.
182. Albert RK, Connett J, Bailey WC, et al. Azithromycin for prevention of exacerbations of COPD. *N Engl J Med.* 2011;365(8):689–698.
183. Uzun S, Djamin RS, Kluytmans JA, et al. Azithromycin maintenance treatment in patients with frequent exacerbations of chronic obstructive pulmonary disease (COLUMBUS): a randomised, double-blind, placebo-controlled trial. *Lancet Respir Med.* 2014;2(5):361–368.
184. Stokholm J, Chawes BL, Vissing NH, et al. Azithromycin for episodes with asthma-like symptoms in young children aged 1-3 years: a randomised, double-blind, placebo-controlled trial. *Lancet Respir Med.* 2016;4(1):19–26.
185. Bacharier LB, Guilbert TW, Mauger DT, et al. Early administration of azithromycin and prevention of severe lower respiratory tract illnesses in preschool children with a history of such illnesses: a randomized clinical trial. *JAMA.* 2015;314 (19):2034–2044.
186. Gielen V, Johnston SL, Edwards MR. Azithromycin induces anti-viral responses in bronchial epithelial cells. *Eur Respir J.* 2010;36(3):646–654.
187. Porter JD, Watson J, Roberts LR, et al. Identification of novel macrolides with anti-bacterial, anti-inflammatory and type I and III IFN-augmenting activity in airway epithelium. *J Antimicrob Chemother.* 2016;71(10):2767–2781.
188. Schogler A, Kopf BS, Edwards MR, et al. Novel antiviral properties of azithromycin in cystic fibrosis airway epithelial cells. *Eur Respir J.* 2014;45(2):428–439.
189. Menzel M, Akbarshahi H, Bjermer L, Uller L. Azithromycin induces anti-viral effects in cultured bronchial epithelial cells from COPD patients. *Sci Rep.* 2016;6:28698.

CHAPTER 7

The interplay of the host, virus, and the environment

Peter Wark[1], Teresa Williams[2,3] and Prabuddha Pathinayake[2,4]
[1]Centre for Healthy Lungs, HMRI, University of Newcastle, New Lambton, NSW, Australia
[2]Priority Research Centre for Healthy Lungs, Hunter Medical Research Institute, The University of Newcastle, Callaghan, NSW, Australia
[3]School of Biomedical Sciences and Pharmacy, University of Newcastle, Callaghan, NSW, Australia
[4]School of Medicine and Public Health, University of Newcastle, Callaghan, NSW, Australia

7.1 INTRODUCTION

The development of infection relies upon a complex interaction between the virus and the host. The pathogenicity of the virus is contingent upon its opportunity to be exposed to the host and its ability to infect and replicate in the host's cells. Countering this within the host are the innate and adaptive immune response. In both cases the environment may influence this interaction, tilting it in favor of either the host or the virus, a concept described in the 1960s by George McNew.[1] In this model, six factors determined the severity of the disease in the host, including the inherent susceptibility of host, virulence of pathogen, prevalence of pathogen, seasonal development, duration of the infection period, and the severity of environment collectively. Later, this model was adapted to study various public health issues and describe and predict the pathogenicity of human diseases.

Human rhinoviruses (RVs) are the major causative agent of common cold and are highly diverse in nature, and are likely to have coevolved with humans over millennia. Generally, they infect the upper respiratory tract and lead to relatively minor illnesses and in many cases few if any symptoms. However, in exceptional circumstances they can lead to more severe disease. The best-described and most widely evident complications have been their association with worsened asthma. They have also been shown to worsen symptoms and airway inflammation in people with non-allergic airways disease: those with chronic obstructive pulmonary disease (COPD) and cystic fibrosis (CF). However, RV infection may also lead to more severe lower respiratory tract infections in those without chronic lung disease such as bronchiolitis and pneumonia; this is seen in those who are immunocompromised[2] as well as at the extremes of age.[3,4]

Rhinovirus Infections
DOI: https://doi.org/10.1016/B978-0-12-816417-4.00007-X

We will explore aspects of this complex interplay between the host and virus in this chapter.

7.2 SEVERE INFECTION WITH RHINOVIRUS

RV infections are in general innocuous and in many individuals low level infection of the upper respiratory tract is even asymptomatic, with a relationship seen between greater viral load and the presence of symptoms.[5]

The host adaptive immune response is likely to play an important role in preventing severe RV infection. Symptomatic infections are more likely in young children, with their still developing adaptive immune systems compared with their parents.[6] In support of this observation is the role RV plays in bronchiolitis, the most common cause for admission in infants under the age of 1, and after respiratory syncytial virus (RSV), RV is the most common cause.[7] Though children hospitalized with bronchiolitis and RV tend to be less unwell than those with RSV, they are more likely to be diagnosed with asthma later when of school age.[8] More severe disease outbreaks though can be seen. A case–control study from a Vietnamese orphanage described severe acute respiratory distress with a very high mortality in 12 infants and attributed this to RV-C,[9] implying that the combination of age and poor living conditions combined to increase susceptibility to RV.

At the other extreme of age, the elderly are also particularly susceptible to RV infection. One Italian study demonstrated that RV was the most frequently detected pathogen in elderly patients presenting to the emergency room with an acute respiratory illness.[10] While in a cohort of elderly patients admitted to hospital over 12 months with a respiratory illness, those with RV detected, compared with those with influenza, were more likely to be in residential care, at greater risk of pneumonia, and in fact had a longer length of stay.[11] Similarly in an English community study, RV was an important cause of morbidity in the elderly and those at risk were more likely to have chronic medical illnesses or were smokers.[12]

Apart from the very young and the elderly those at risk of severe respiratory infection from RV tend to be those with substantial immunosuppression.[13] In these patients not only can infection be more severe, but the individuals also fail to clear the virus highlighting the important role of the immune response in eliminating replicating viruses.[14]

In the immunocompetent host, RV infection may still be an important cause of morbidity through its association with pneumonia. RVs have been detected in 4%–45% of children and 2%–17% of adults with community acquired pneumonia (CAP).[15] In many of these cases RV is found in association with other viruses and bacteria and it remains unclear whether RV is the primary pathogen causing pneumonia or whether it predisposes to infection with other pathogens, especially bacteria such as *Streptococcus pneumoniae*. One New Zealand study assessed the presence viruses by polymerase chain reaction (PCR) in 304 immunocompetent adults admitted to hospital with CAP.[16] The most frequently isolated virus was RV in 13%, which occurred throughout the year, followed by Influenza, which had a predominance of being isolated during late winter. *S. pneumoniae* was the most frequent bacteria isolated in 31%. While the most frequent dual infections were with RV and *S. pneumoniae* in 11% followed by Influenza and *S. pneumoniae*. Using multiple logistic regression the investigators also showed that increasing age, male sex, and mixed RV/pneumococcal infection were independently associated with more severe clinical disease. Though it was the presence of mixed RV/pneumococcal disease that had the strongest association [as measured by the pneumonia severity index; odds ratio 11.52 (1.09–121.89)]. RV infection has also been shown to increase acquisition and transmission of *S. pneumoniae*.[17] Epidemiological studies have shown a link between times of high RV activity and the incidence of invasive pneumococcal disease in children.[18] A biologically plausible explanation of this association is that RV infection of the airway epithelial cells can upregulate transcription factors and increase *S. pneumoniae* adhesion to the epithelium, at least partially via increased expression of platelet activating factor–receptor, with increased pneumococcal adhesion required as the first step toward acquisition and the development of invasive pneumococcal disease.[19]

An evolving area of research is the airway environment created by the microbiome. Recently, RV infection has been shown to influence and be influenced by the airway microbiome. In a prospective study of 32 healthy infants, the nasopharyngeal microbiota were monitored over the first year of life. Asymptomatic RV infection had no impact, however, more frequent symptomatic infections led to lower microbial biodiversity.[20] In children RV infection similarly leads to a change in the airway microbiota, with an increase in *Moraxella* species seen.[21] In those with higher RV replication an association was also seen with increased *Haemophilus influenzae* and *S. pneumoniae*.[21] Interestingly preexisting bacterial microbiota also predicted the severity of RV infection symptoms, with asymptomatic infections being associated with higher levels of *Corynebacterium* species. In a small study in adults, RV infection was associated with an increase in *H. influenzae* and other proteobacteriae that lasted

42 days, though this was only in those with existing COPD, with no change seen in the healthy controls.[22]

The interaction between the airway microbiome and viral infections is clearly important in influencing the immune responses to infection in a bidirectional way but we are only just starting to study the complexity of this relationship. The important early life findings that symptomatic RV infections are associated with the development of later childhood asthma underscores this importance as well as the complex relationship that exists between airway inflammation associated with asthma and susceptibility to RV infection.[23]

7.3 ALLERGIC AIRWAY INFLAMMATION AND RHINOVIRUS INFECTION

The equilibrium between antiviral and inflammatory responses is fundamental in resolving RV infections, as discussed in previous chapters. Dysregulation of this equilibrium, particularly by allergens in atopic individuals, may influence the antiviral response and skew it toward active type 2 inflammation. RV infections in people with asthma are both more prolonged and characterized by more severe symptoms than in people without asthma.[24–26]

Allergy arises from inflammation induced by an IgE-mediated immune response to small doses of harmless environmental factors or allergens such as pollen, house dust mites, or animal dander.[27] Antigen presenting cells process allergens and present them to CD4 + T cells inducing differentiation into Th2 cells, releasing proinflammatory mediators IL-4 and IL-13, which drive B cell class switching to produce allergen specific-IgE, which binds to the high affinity receptors FcεRI located on mast cells and basophils and cross-links the receptor and sensitizes the cell.[28,29] Once sensitized, subsequent reexposure to the allergen will lead to allergic inflammation, which can be divided into two components: the early and late phase.

In the early phase, tissue resident IgE coated mast cells encounter an allergen and form an IgE-antigen complex, which results in degranulation, releasing histamines, prostaglandins, leukotrienes, chemokines, and cytokines.[30,31] This results in smooth muscle contraction (airway narrowing), increased mucus production, and chemokine release to recruit more immune cells.[30,32] In the late phase, typically 8–12 hours later, chemokines secreted in the early phase have recruited leukocytes, mainly eosinophils, T lymphocytes, basophils, and neutrophils,[33] which infiltrate the affected area and cause additional degranulation or secretion of inflammatory cytokines, such as IL-4, -5, and -13. Proinflammatory cytokines recruit more leukocytes and cause B cell class switching to produce more IgE, priming immune cells for future responses and further sensitizing them.

Allergen sensitization and subsequent exposure coupled with RV infections appear to work synergistically to increase the severity and risk of an allergic airway disease (AAD) exacerbation, particularly in early life.[34–37] Noting the interactions between virus and allergic responses, Calhoun et al. published that there was increased histamine and eosinophil recruitment throughout the late phase during a cold in atopic participants.[38] Further, the effects of inflammation persisted up to 1 month in atopic individuals, but were absent in nonatopic participants. Lemanske et al. additionally posited a "two-hit" hypothesis noting the interaction between allergen and virus appeared bidirectional in that viral infections could influence the development of allergy[39,40] and that allergy could influence the lower airway response to viral infections.[41,42] Investigations into this bidirectional interaction in AAD revealed deficiencies in interferon (IFN) production and an increased type 2 inflammatory response to viral infection, altering the Th1/Th2 axis in response to RV infection.

7.3.1 Abnormal antiviral signaling in the airways of those with allergic asthma

In 2005, Wark et al. identified that when human primary bronchial epithelial cells (pBECs) cultured from the airways of subjects with asthma and atopy were infected with RV, they exhibited an impaired innate immune response.[43] Asthmatic airway epithelial cells infected with RV-A16 expressed delayed and significantly reduced type 1 IFN-β, reduced apoptosis, and increased RV replication. Administration of exogenous IFN-β induced apoptosis and reduced viral replication, indicating that deficient IFN-β production may contribute to severity of viral infections in asthmatics. As type 3 IFN-λ responses are induced by type 1 IFNs, investigations into IFN-λ in asthmatics BECs and macrophages in response to RV infection revealed deficient type 3 IFN production in atopic individuals. Contoli et al. noted that exacerbation severity and cold symptom scores were inversely proportional to IFN-λ generation.[44] Since these early findings, investigations to understand the mechanism of dysregulated IFN production in BECs in asthma have led to many candidate pathways (Table 7.1).

Table 7.1 Studies of asthmatic cells infected with rhinovirus (RV) that report interferon deficiency.

Asthmatic subtype studied (vs healthy control)	Cell type studied	RV strain	Interferon studied	Additional markers modified in asthmatics	Reference
Atopic asthma	PBMC	RV-A16, RV-A30	↓ IFN-γ	↑ IL-4	[45]
Atopic asthma	PBMC	RV-A16	↓ IFN-β, α	↓ IRF1,7, STAT1, TLR7/8, NFkB	[46]
Severe asthma	alveolar macrophages	RV-A16	↓ IFN-β, α	↓ TLR7	[47]
Atopic asthma	HBECs, alveolar macrophages	RV-A16	↓ IFN-λ	–	[44]
Atopic asthma	HBECs	RV-A16	↓ IFN-β	↓ Caspase 3/7 ↑ CXCL-10	[43,48]
Atopic asthma, nonatopic asthma, nonatopic nonasthmatic	HBECs	RV-A16	↓ IFN-β, λ	↑ IL-4	[49]
Asthma	HBECs	RV-A1	↓ IFN-β	↑ TGFβ1, via SOCS1, 3	[50]
Atopic asthma	HBECs	RV-A16, RV-A1	↓ IFN-β, λ	↓ TLR3	[51]
Asthma	Eosinophils, BEAS-2B	RV-A16, RV-A1	↓ IFN-β, λ	↑ TGFβ1	[52]
Asthma	HBECs	RV-A1	↓ IFN-λ	↑ IL-6, CXCL8	[53]
Atopic asthma	HBECs	RV-A1	↓ IFN-β, λ	↑ SOCS1	[54]
Atopic asthma	BECs, AECs	RV-A1, RV-B14	↓ IFN-β	↑ LDLR, IL-6, CXCL8, CCL5, CXCL10, IL-1β	[55]

BEC, Bronchial epithelial cell; *IFN*, interferon; *IRF*, interferon regulatory factor; *PBMC*, peripheral blood monocyte; *SOCS*, suppressor of cytokine signaling; *TGF*, transforming growth factor; *TLR*, Toll-like receptor.

With evidence of deficient IFN in vitro in both pBECs and peripheral blood monocytes (PBMCs), a clinical trial was designed with administration of IFN-β during the onset of a "cold." Unfortunately, the trial did not achieve its primary outcome of attenuated asthma symptoms by viral infections, though it appeared to have a greater effect in those with poorly controlled asthma, and the investigators did notice a boosted innate immune response and enhanced peak expiratory flow.[56] As this trial required presentation to clinic within 24 hours of developing flu or cold symptoms, the time point for administration of treatment may have affected the efficacy in asthmatic patients. Future prospective studies catered toward IFN-β treatment in atopic asthmatics with pretreatment during seasonally peaks are needed to address the full effectiveness of IFN-β administration.

Not all studies have found IFN deficiencies in asthmatic cells.[57,58] These findings may have to do with participant numbers, age, severity, or asthma control. A study in HBECs infected with RV concluded that there was no difference in induction of type 1 or 3 IFNs in well-controlled atopic asthmatics.[59] These findings stress the importance of other environmental, epigenetic factors and the effect of persisting type 2 inflammation in individuals with AAD during RV exacerbations, which could impair the antiviral response.

AAD is driven by type 2 inflammation, which has immune suppressive effects on antiviral immunity; one such activator of inflammation is transforming growth factor beta (TGFβ), which has effects on multiple cell types. Pretreatment of fibroblasts with TGFβ1 can transdifferentiate fibroblasts into myofibroblasts augmenting a proinflammatory response. In fibroblasts from asthmatics or fibroblasts pretreated with TGFβ1, RV infection exhibited a blunted type 1 IFN as well as increased viral replication.[60] Xatzipsalti et al. noted that PBMCs from atopic asthmatics had increased TGFβ1, which resulted in significantly decreased IL-6, CXCL8 (IL-8), and CCL5 (RANTES) when cocultured with an epithelial cell line (BEAS-2B) cells infected with RV.[61] The addition of exogenous TGFβ in BECs resulted in increased RV replication and a reduction in IFN-β and IFN-λ.[50] Neutralizing pBECs against TGFβ increased IFN-β production and decreased RV replication. Bedke et al. posited that this mechanism could be through suppressor of cytokine signaling (SOCS) 1 or 3. Following this finding, Gielen et al. showed that SOCS1 was increased in asthmatics and that overexpression of SOCS1 inhibited IFN-β activation and subsequent activation of the type 3 IFN-λ.

Infection with RV or administration of IL-4 or IL-13 could induce SOCS1, and its expression was correlated with total serum IgE. The TGFβ-SOCS signaling cascade helps illustrate how type 2 inflammation can influence IFN expression at multiple points through multiple mechanisms.[54]

7.4 EVIDENCE OF IMPAIRED SYSTEMIC IMMUNE RESPONSES TO VIRUS INFECTION WITH ALLERGIC AIRWAYS DISEASE

In 2010, Olenec et al. published a connection between RV infection and allergen sensitization in nasal lavage collected during April and September. Atopy was associated with more symptomatic illness, however, did not influence the amount of RV present, as verified by qPCR.[62] Studies done in both pBECs and PBMCs from atopic asthmatics negatively correlated both serum IgE levels and its receptor FcεRI with RV induced IFN-α and IFN-λ, demonstrating a potential mechanism, which corroborated previous findings.[49,63]

Toll-like receptors (TLRs) are fundamental in detecting pathogen-associated molecular patterns. TLR7 in particular is important in activating an immune response against single stranded RNAs present during RV infection in immune cells, such as plamacytoid dendritic cells (pDCs). Studies investigating TLR7 expression in eosinophilic and atopic asthmatics found decreased gene expression[64] and protein levels by immunohistochemistry[65] compared with control subjects. This decrease was associated with decreased airway IFN-λ messenger RNA (mRNA) expression. In TLR7$^{-/-}$ mice, infection with RV-A1 had a marked increase in RV replication and eosinophil recruitment, treatment with exogenous IFN-β, or transfer of functional pDCs reduced the exaggerated response. With indications of impaired systemic immunity, precision medicine targeting specific markers of type 2 inflammation were investigated.

Monoclonal antibodies targeting IgE (omalizumab) were the first biological agent that demonstrated improved outcomes for patients with poorly controlled asthma despite maximal therapy with inhaled corticosteroids and long-acting beta agonists. In particular, administration was shown to significantly reduce acute exacerbations of asthma.[66] In a randomized double-blind study of omalizumab (or placebo) therapy, in addition to regular asthma control in adolescents from the inner city, anti-IgE therapy reduced the number of exacerbations and days of illness.[67]

However, treatment did not affect the number of viruses obtained, supporting previous findings that allergy alters the response to virus and not prevalence.

Expanding upon the patient demographics in the previous trial, the preventative omalizumab or step-up therapy for severe fall exacerbations (PROSE) study targeted children going back to school in the fall. Teach et al. determined that PBMCs collected before randomization had significantly suppressed IFN-α responses to RV infection, when in the presence of an IgE cross-linking antibody.[68] During the intervention phase, post-randomization, collected PBMCs underwent the same testing with IgE cross-linking and RV infection. Only the participants with omalizumab treatment had demonstrated significantly increased IFN-α production compared with placebo. Additional analysis of PBMCs by Gill et al. from the PROSE study detected that omalizumab treatment reduced FceRIα expression on pDCs, which resulted in an increase in IFN-α production.[69] Interestingly, these findings were only associated with RV infection, as omalizumab treatment did not affect pDCs infected with influenza. Investigation of RV prevalence collected from weekly nasal mucus samples in the PROSE study determined that RV was present in 57% of exacerbations, indicating omalizumab treatment could be effective against the majority of viral infections.[70] In the RV detected samples, treatment with omalizumab decreased duration of RV infections, peak viral shedding, and frequency of RV illness. Omalizumab trials have supported the use of monoclonal antibodies against IgE as a promising therapeutic against allergy and RV induced exacerbations.

7.5 VIRUS INFECTION WORSENS TYPE 2 AIRWAY INFLAMMATION

During a viral infection, a robust type 1 antiviral response is generated to resolve infection; however, in individuals with AAD, this is not always the case. As discussed previously, atopic individuals can have a lagging or deficient immune response, and there is evidence to indicate that increased type 2 inflammation may be the underlying issue. A trio of upstream type 2 inflammatory cytokines IL-25, IL-33, and Thymic stromal lymphopoietin (TSLP) have been linked as important epithelial derived mediators that are upregulated during RV infection and asthma exacerbations.

In 2014, an experimental human RV infection model discovered that IL-25 induction was greater in those with asthma compared with healthy controls. Baseline expression of IL-25 was also higher at baseline in asthmatics and was correlated with atopic status. Utilizing a mouse model of allergic inflammation during RV infection, IL-25 receptor blockade distinctly suppressed type 2 cytokines such as IL-4, IL-13, and IL-33. An increase in mucus production and reduction in recruitment of immune cells such as eosinophils, neutrophils, Th2 cells, and type 2 innate lymphoid cells (ILC2s) were also reported. Both human and mouse models provide strong evidence that targeting upstream markers such as IL-25 could reduce the type 2 signaling cascade.[71]

Similarly, IL-33 was increased in collected nasal and bronchial fluid from asthmatic compared with healthy individuals using the same human RV infection model. RV induced IL-33 expression in asthmatics could be correlated with exacerbation severity as determined by viral load and levels of IL-5 and IL-13 during infection.[24] The Addition of conditioned media from RV infected pBECs strongly induced IL-4, IL-5, and IL-13 in cultured T cells and IL-5 and IL-13 in ILC2s. Similarly to IL-25, in vitro receptor blockade of IL-33 suppressed induction of type 2 cytokines from cultured immune cells. In another study, neutralization of IL-33 with an antibody promoted IFN-β and IFN-λ release from asthmatic pBECs and in a mouse model against both major and minor group strains of RV.[72]

Unlike the quick danger response of IL-33 and IL-25, TSLP is primarily involved in maturation of immune cells toward a type 2 response. In vitro models in pBECs have found TSLP increases during RV infection.[73,74] A study using double-stranded RNA as a mimic for viral infection found TSLP is further increased synergistically when IL-4 is present. In individuals with COPD, RV infection of pBECs was also found to increase TSLP through TLR3 signaling, which was verified with dsRNA.[75] However, the effect of RV infection on TSLP levels has not been thoroughly investigated in humans yet.

Downstream of epithelial derived IL-25, IL-33 and TSLP are the hallmark cytokines of type 2 inflammation IL-4, IL-5, and IL-13. A study by Jackson et al. demonstrated that in a human RV-induced exacerbation model, in vivo infection of asthmatics induced IL-4, IL-5, and IL-13, which was absent from the healthy controls.[24] Increased IL-13 and IL-4 were additionally reported in BECs and nasal epithelial cells (NECs) brushing from individuals with asthma and allergic rhinitis respectively.

Induction of type 2 cytokines impaired IFN-β and IFN-λ production, TLR3, and IFN regulatory factor 3 (IRF3), all critical components of innate immune signaling. In asthmatic BECs, decreased TLR3 was associated with an increase in RV RNA in atopic individuals compared with nonatopic. Additionally, IL-13 or virus induced IFN-β can activate Foxa3, a molecule required for mucin production and goblet cell differentiation. From tissue sections and cultured asthmatics BECs, foxa3 was overexpressed in both asthma and COPD and expression inhibited IFN stimulated genes such as TLR3, IRF3, MDA5, and RIG-I,[76] further indicating the role type 2 inflammation has on impaired antiviral responses.

Additional monoclonal antibodies that target other aspects of the type 2 immune response have now been developed and are now either available for use in severe asthma or have advanced through phase 3 clinical trials such as anti-IL5 (mepolizumab, benralizumab),[77,78] anti-IL-4/13 receptor (dupilumab),[79] and the anti-TSLP (tezipelumab).[80] These agents have demonstrated efficacy in patients with severe asthma, in particular through reducing the frequency of acute exacerbations. Clearly, they reduce associated type 2 immune responses in the airways, though whether this influences responses to virus infections directly or indirectly remains to be determined.

7.6 CHRONIC INFLAMMATORY AIRWAYS DISEASES NOT ASSOCIATED WITH ALLERGIC AND TYPE 2 INFLAMMATION ARE ALSO SUSCEPTIBLE TO RHINOVIRUS INFECTION: CYSTIC FIBROSIS

Susceptibility to RV infection is also associated with non−type 2 chronic airway inflammations, especially in people with CF and COPD.

CF is a genetic disorder that leads to alteration of the CF Transmembrane Regulator protein, a chloride channel crucial for regulating ion and water composition in the airway mucus, resulting in desiccated secretions, bacterial colonization and intense neutrophilic airway inflammation the release of oxidants and proteases. These changes result in progressive airway damage resulting in airflow obstruction and bronchiectasis, which correlate with lung function deterioration and respiratory exacerbations.[81] Infants under the age of 1 year with CF do not experience more frequent symptomatic infections with RV.[20] However, as children with CF get older and develop more airways disease, RV infection becomes associated with pulmonary exacerbations[82] and RV is detected more frequently and persists for longer than in healthy children.[83] CF children also have increased risk of virus acquisition, replication, and impaired innate immune responses to RV.[83,84] Pulmonary exacerbations associated with RV in children are also more severe, with the children less likely to recover lung function.[85] In adults with CF, viral infections, including RV, are also associated with pulmonary exacerbations that require treatment, though there is mixed evidence that they influence the underlying infection with bacterial pathogens such as *Pseudomonas*.[86,87]

A reason for increased susceptibility to RV in CF, unlike asthma, is not so apparent. CF airway epithelial cells do not demonstrate impaired type 1 IFN responses to RV but do demonstrate a heightened inflammatory response with increased RV replication.[88] In one study in vitro *Pseudomonas* exposure impaired type 1 IFN production to RV in CF BECs.[89] However, this was not replicated in a more recent publication, where type 1 IFN responses were intact and not impaired by *Pseudomonas* virulence proteins, though RV replication was nonetheless higher in CF cells.[90] An alternate reason for susceptibility to RV infection must therefore exist. Vitamin D is known to play an important role in host response to infection[91,92] and as a fat soluble vitamin is often deficient in CF. Vitamin D deficiency has been shown to increase susceptibility to RV in CF patients.[93] In vitro pretreatment with vitamin D significantly reduces the RV-A16 virus load in CF pBECs by increasing the expression of antimicrobial peptide LL-37. Similarly, the level of RV-A16 load in bronchial lavage of CF patients negatively correlates with LL-37 protein level presents demonstrating a potentially important role of vitamin D induced LL-37 in controlling RV infections.[94]

7.7 CHRONIC OBSTRUCTIVE PULMONARY DISEASE AND SUSCEPTIBILITY TO RHINOVIRUS INFECTION

Patients with COPD are also vulnerable to various respiratory virus infections, including RV. COPD is a disorder of progressive airflow limitation characterized by emphysema and/or chronic bronchitis. Airway inflammation in COPD has been associated with increased numbers of alveolar macrophages, neutrophils, T lymphocytes (predominantly TC1, TH1, and TH17 cells), and ILCs recruited from the circulation.[95] Excessive oxidative stress is thought to be a major driving force in COPD-related inflammation resulting in tissue damage, impaired antiprotease defenses, DNA damage, structural and immune cell senescence, as well as systemic inflammation.[95,96]

RV infection is commonly associated with acute exacerbations of COPD. In a cohort of exacerbating patients with COPD, RV accounted for 43% of the virus infections or was present in 12% of all the exacerbations.[97] At the time of exacerbation an increase in RV load was seen to temporally associate with worsened symptoms, while those with a history of more frequent exacerbations appear to be particularly susceptible to the effects of RV infection.[98] The reason for this susceptibility as in CF remains unclear. Paradoxically COPD airway epithelial cells have been shown to have increased baseline expression of inflammatory cytokines and type 1/3 IFNs.[99] In an elegant study of experimental RV infection in volunteers with mild COPD compared with matched controls, those with COPD were shown to have an increase in lower respiratory tract symptoms, airway and systemic inflammation in response to infection, a clinical picture consistent with an exacerbation. In addition, virus load correlated with airway inflammatory markers, virus load was higher than in controls and in vitro type 1/3 IFN production from cells obtained at bronchiolar lavage in subjects with COPD was impaired.[100] Experimental RV infection also enhanced recruitment of neutrophils and T cells to the lungs, demonstrating a potential for RV infection alone to worsen airway inflammation in COPD, at least in the short term.[101,102] Despite these findings a mechanism(s) to account for this susceptibility remains elusive. RV infection is also known to increase oxidative and nitrosative stress markers in the lungs of COPD, but not in healthy subjects.[103] This enhanced oxidative stress impaired monocyte/macrophage responses to RV and this correlated with RV load. Perhaps it is in the setting of this preexistent proinflammatory environment that RV infection worsens COPD.

7.8 CELLULAR OXIDATIVE STRESS AND INCREASED SUSCEPTIBILITY TO RHINOVIRUS INFECTION

Oxidative stress is a result of imbalance between free radical production and the antioxidant capacity of the cell. That can result from inhibition of antioxidant enzyme activity or by increased production of oxidants. Both reactive oxygen species (ROS) and reactive nitrogen species are known as prooxidants, and an upregulation of either of these species can result in cellular oxidative stress. Accumulation of prooxidants causes oxidative damage to proteins, lipids, and DNA, which leads various inflammatory responses and cytotoxic events.[104,105]

Notably, oxidative stress can increase ICAM-1 expression, which is the binding site for major group RV on epithelial cells,[107] via IL-6/AKT/STAT3/NF-κB (nuclear factor kappa-light-chain-enhancer of activated B cells) signaling activation.[106] A lung epithelial cell line (A549 cells) exposed to particulate matters (PMs) significantly increased intracellular ROS level and ICAM-1 expression. Treating these cells with antioxidants attenuated the PM induced ROS and ICAM-1 expression suggesting a direct role of ROS in regulating ICAM-1 receptor expression.[106] Similarly, exposure to cigarette smoke extract (CSE) upregulated ICAM-1 expression in pBECs[107] and this might also due to the effect of CSE induced intracellular prooxidants.[108,109] In smokers and patients with chronic airflow limitation, ICAM-1 expression in both large and small airways is upregulated.[110] In these patients ICAM-1 expression is abnormally increased particularly in goblet cells and submucosal glands of the large airways.[110] This overexpression of ICAM-1 might increase the susceptibility to infection by major group RVs and aggravate virus induced asthma exacerbations. Additionally, RV infection itself upregulates cellular ICAM-1 expression regardless of the virus strain in airway epithelial cells via increased NF-κB mediated transcription.[111] This facilitates further virus attachment and replication of major group RVs. Of note, RV upregulates cellular ROS accumulation and ROS could activate NF-κB transcription. Treatment with antioxidants abolished RV induced NF-κB activation and IL-8 production.[112] Therefore, ICAM-1 upregulation during RV infection might regulates through ROS induced NF-κB activity, however mechanisms remains further elucidated.

CSE plays a crucial role in the pathogenesis of COPD. Exposure to CSE decreases innate antiviral responses of epithelial cells to RV infection.[113–115] Preexposure of airway epithelial cells to CSE significantly reduced the RV-A16 or poly I:C induced IFN-β, CCL-5, and CXCL-10 expression. It also reduced the activation of STAT-1 and JNK signaling and increased RV-A16 replication.[114] Similarly, pBECs from COPD patients cultured at the air–liquid interfaces and exposed to whole cigarette smoke demonstrated impaired induction of the antiviral genes, IFN-λ1, OAS1, and MX1, to RV-A16.[113] However, CSE has wide ranging effects on cellular signaling, mitochondrial metabolism, redox pathways, chemokine signaling, and the expression of cellular pattern recognition receptors.[115] Therefore, it's hard to conclude that increased

RV pathogenesis by CSE is just due to impaired antiviral signaling and remains worthy of further investigation.

Oxidative stress has the potential to disrupt epithelial barrier function, an important host defense against all pathogens by attenuating epithelial tight junctions via TRPM2-phospholipase Cγ1 (PLCγ1)-protein kinase Cα (PKCα) signaling pathway.[116] Under increased oxidative stress conditions, airway epithelial cells open up the transient receptor potential melastatin (TRPM) 2 channel and this allows $Ca2+$ influx, which consequently phosphorylates PLCγ1 and activates PKCα. PKCα activation results in reduction of zona occluding-1 (ZO-1) and claudin-2 causing opening of tight junctions in the airway epithelium.[116] While this effect has not been seen in COPD, it has with asthmatic epithelial cells. The loss of epithelial integrity may lead to various opportunistic infections by facilitating entry of viruses or bacteria to penetrate into the deeper layers within the airways resulting invasive infections.[117,118] In cell cultures obtained from a pediatric cohort, asthmatic individuals showed significantly reduced tight junction protein expressions concomitant with lower epithelial resistance compared with the nonasthmatic group. When these differentiated cultures were infected with RV-A1, both asthmatic and nonasthmatic cells demonstrated reduced membrane tight junction protein expression though nonasthmatic cells recovered after 48 hours while in asthmatic cells the effect was sustained for longer.[119] This suggests that the asthmatic epithelium is intrinsically more permeable and RV infection further reduces the integrity. In another in vitro study, RV infection resulted in barrier dysfunction in polarized airway epithelial cells by increasing cellular oxidative stress. The dsRNA produced during RV replication interacts with Rac-1 dependent nicotinamide adenine dinucleotide phosphate (NADPH) oxidase 1 resulting in overproduction of cellular prooxidants. Consequently, this enhanced ROS acts on dissociation of ZO-1 leading to disruption of epithelial barrier function.[120] It is unclear why asthmatic epithelium possesses a reduced epithelial integrity and is more vulnerable to RV induced barrier dysfunction. Possibly, inherent chronic oxidative stress in asthmatic airways might weaken the tight junction of airway epithelial cells. Therefore, it is suggested that this compromised barrier function would facilitate the movement of pathogens across the epithelium and their interaction with the basolateral receptors, therefore potentially enhancing airway inflammation and exacerbation episodes in asthma.[119]

Exposure to oxidative stress,[121] allergens,[122] and CSE[123] all could induce ER stress and an unfolded protein response (UPR) in the airways.[124] UPR is an adaptive mechanism induced by cells to regulate the cellular protein homeostasis. Enhanced ER stress shuts off global cellular cap-dependent protein translation, which inhibits various cellular protein syntheses and shifts the cellular protein translation process into the internal ribosome-entry site (IRES) dependent pathway.[125] Most picornaviruses use this highly structured IRES sequence located in the 5′ untranslated region of their mRNA to translate viral proteins during replication. Therefore, this ER stress dependent switching of protein translation could facilitate virus replication while impairing self-defense.[125,126] ER stress and UPR activation could interfere with cellular antiviral defense in a number of ways.[127] For instance, virus induced UPR may lead to degradation of IFN-α/β receptor alpha chain (IFNAR1) and inhibit type 1 IFNs induced antiviral protein transcription in favor of virus replication.[128] In contrast, treating with tunicamycin a potent UPR-inducing agent substantially upregulated lipopolysaccharide or a synthetic RNA poly I:C induced type 1 IFNs however, tunicamycin alone did not affect IFNs expression.[129,130] Therefore, UPR regulated antiviral mechanisms remains unclear. That might be specific to the type of stimulant and the longevity of ER stress.

There is no direct evidence to demonstrate if ER stress alters RV replication or its pathogenicity. However, the ER resident protein ORMDL3 is closely associated with development of childhood asthma[131] and is capable of dysregulating UPR[132] and influencing RV replication. Increased expression of ORMDL3 in transgenic mice demonstrated reduced RV viral load and increased antiviral responses.[133] Moreover, upregulated ORMDL3 in human leukocytes infected with RV resulted increased expression of ER chaperons and type 1 IFNs.[134]

The chronic type 2 inflammatory environment in asthmatic airways facilitates RV infection and virus induced asthma exacerbations. Of note, Bhakta et al. recently demonstrated that increased type 2 airway inflammation is associated with poor lung function and increased epithelial ER stress/UPR. The activation of IRE1-XBP1 signaling cascade of UPR was significantly correlated with airway type 2 inflammation in mild asthma patients suggesting potential role of ER stress in regulating type 2 inflammation.[135] Additionally, non−type 2, IFN driven inflammation characterized by increased IFN-stimulated gene expression in the lungs was also associated with poor lung function and increased ER stress/UPR in the airway epithelium.[135] Therefore, ongoing chronic

ER stress in asthmatic epithelium may perpetually upregulate airway inflammation and act as a common downstream pathway linking both type 2 and non−type 2 inflammation. This dysregulated inflammatory environment with heightened ER stress in asthmatic epithelium might enhance the susceptibility to RV infection.

Airway remodeling refers to alterations in structural cells and tissues characterized by airway wall thickening, fibrosis in the subepithelial region, myocyte hypertrophy and hyperplasia, myofibroblast hyperplasia, and mucous metaplasia, all key features of asthma. Notably, IL-13 which is a major type 2 cytokine has the ability to induce epithelial remodeling leading to mucous metaplasia, enhanced tissue hyaluronic acid accumulation, and subepithelial fibrosis.[136,137] Mucous metaplasia induced by IL-13 as a result of increased goblet cells in the epithelium increases the susceptibility of airway epithelium to RV infections. Upon infection, goblet cells were preferentially infected by RV and IL-13 treatment further increased the RV infection by increasing the goblet cell number in the epithelium. This increased susceptibility however was not related to RV binding receptor expression, but was suggested to be due to IL-13 induced structural changes and loss of epithelial complexity.[138] The mechanisms behind IL-13 induced epithelial metaplasia are not as clear. Interestingly, the ER stress transducer IRE1-β has previously shown to be involved in regulation of epithelial mucus production in airways via UPR activation.[139] As IL-13 upregulates ER stress in airway epithelium,[135] IL-13 induced mucous metaplasia and loss of epithelial integrity might be regulated via ER-UPR activation leading increased susceptibility to RV infections. Therefore, mitigating ER stress would be a potential future therapeutic approach for controlling asthma and RV induced exacerbations, though further investigations are required.

7.9 CONCLUSIONS

RV is an important human pathogen. While in most individuals it causes minor symptoms it does play a more sinister and destructive role in causing respiratory disease. For the very young and old and those with immune compromise RV is anything but an innocuous pathogen, potentially causing life-threatening disease. In the larger numbers of people with chronic respiratory disease, especially asthma and COPD, RV is an important trigger to acute worsening of their chronic

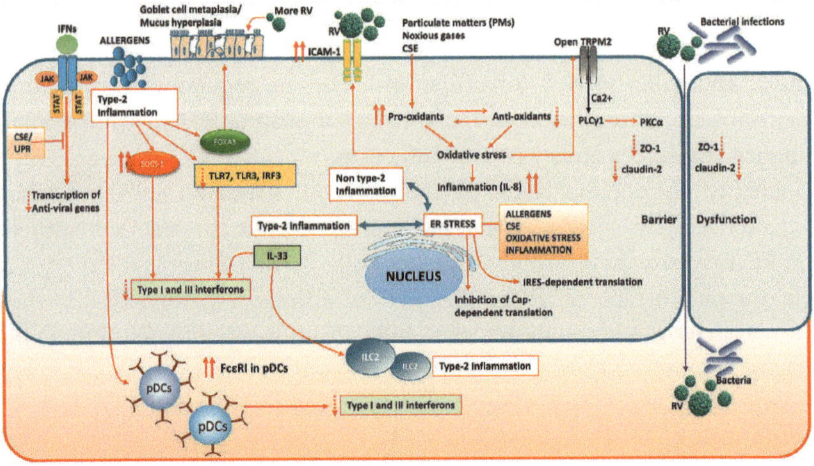

Figure 7.1 ***Mechanisms of rhinovirus interactions with the host and the environment.*** Accumulation of prooxidants and impaired antioxidant defense results in increased cellular oxidative stress in airway epithelium. This upregulates ICAM-1 expression via NF-kb transcription activation, which facilitates binding of major group RVs. RV infection itself upregulates ICAM-1 expression. Increased oxidative stress induces epithelial barrier dysfunction and impairs epithelial integrity. This facilitates movement of pathogen including viruses across the epithelium. Persistent allergen exposure creates a type 2 inflammatory environment. Increased type 2 cytokines (IL-4, IL-13) upregulate SOCS1 expression and impair type 1 and 3 IFN induction. A type 2 environment also reduces the expression of TLR3, TLR7, and IRF3, which are important regulators of innate antiviral response. In addition, allergen exposure increases the expression of FcεRI receptors in pDCs, which ultimately impairs type 1 and 3 IFN production. Type 2 cytokine IL-13 upregulates FOXA3 expression, which induces goblet cell metaplasia and mucus hypersecretion. Increased goblet cells in airway epithelium facilitate RV binding by reducing epithelial integrity. RV infection drastically increases epithelial derived cytokines IL-33, IL-25, and TSLP. IL-33 down regulates type 1 and 3 IFN production and all of them collectively enhances type 2 inflammation. Increased oxidative stress and type 2 inflammation increases protein misfolding, ER stress, and activates UPR. Elevated UPR individually associates with type 2 inflammation and non–type 2 inflammation in the airways working as a common downstream regulator of airway inflammation. *CSE,* cigarette smoke extract; *ER,* endoplasmic reticulum; *IFN,* Interferon; *IRF,* interferon regulatory factor; *pDC,* plamacytoid dendritic cells; *SOCS,* suppressor of cytokine signaling; *TLR,* Toll-like receptor; *UPR,* unfolded protein response.

inflammatory airways disease. In this setting the disease state compromise the host's innate and acquired immune responses in numerous and subtle ways (Fig. 7.1). An understanding of these mechanisms will be essential in reducing the burden of acute exacerbations in both asthma and COPD.

REFERENCES

1. McNEW GL. The nature, origin, and evolution of parasitism. In: Horsfall JG, Dimond AE, eds. *Plant Pathology: An Advanced Treatise*. New York: Academic Press; 1960:19–69.
2. Pappas DE, Hendley JO, Hayden FG, Winther B. Symptom profile of common colds in school-aged children. *Pediatr Infect Dis J*. 2008;27(1):8–11.
3. Gern JE. The ABCs of rhinoviruses, wheezing, and asthma. *J Virol*. 2010;84 (15):7418–7426.
4. Lee WM, Lemanske Jr. RF, Evans MD, et al. Human rhinovirus species and season of infection determine illness severity. *Am J Respir Crit Care Med*. 2012;186 (9):886–891.
5. Granados A, Peci A, McGeer A, Gubbay JB. Influenza and rhinovirus viral load and disease severity in upper respiratory tract infections. *J Clin Virol*. 2017;86:14–19.
6. Peltola V, Waris M, Osterback R, Susi P, Ruuskanen O, Hyypia T. Rhinovirus transmission within families with children: incidence of symptomatic and asymptomatic infections. *J Infect Dis*. 2008;197(3):382–389.
7. Mansbach JM, Piedra PA, Teach SJ, et al. Prospective multicenter study of viral etiology and hospital length of stay in children with severe bronchiolitis. *Arch Pediatr Adolesc Med*. 2012;166(8):700–706.
8. Mansbach JM, Clark S, Teach SJ, et al. Children hospitalized with rhinovirus bronchiolitis have asthma-like characteristics. *J Pediatr*. 2016;172:202–204.e1.
9. Hai Le T, Bich VT, Ngai Le K, et al. Fatal respiratory infections associated with rhinovirus outbreak, Vietnam. *Emerg Infect Dis*. 2012;18(11):1886–1888.
10. Pierangeli A, Scagnolari C, Selvaggi C, et al. Rhinovirus frequently detected in elderly adults attending an emergency department. *J Med Virol*. 2011;83 (11):2043–2047.
11. Hung IF, Zhang AJ, To KK, et al. Unexpectedly higher morbidity and mortality of hospitalized elderly patients associated with rhinovirus compared with influenza virus respiratory tract infection. *Int J Mol Sci*. 2017;18(2).
12. Nicholson KG, Kent J, Hammersley V, Cancio E. Risk factors for lower respiratory complications of rhinovirus infections in elderly people living in the community: prospective cohort study. *BMJ*. 1996;313(7065):1119–1123.
13. Kraft CS, Jacob JT, Sears MH, Burd EM, Caliendo AM, Lyon GM. Severity of human rhinovirus infection in immunocompromised adults is similar to that of 2009 H1N1 influenza. *J Clin Microbiol*. 2012;50(3):1061–1063.
14. Engelmann I, Dewilde A, Lazrek M, et al. In vivo persistence of human rhinoviruses in immunosuppressed patients. *PLoS One*. 2017;12(2):e0170774.
15. Pavia AT. What is the role of respiratory viruses in community-acquired pneumonia?: What is the best therapy for influenza and other viral causes of community-acquired pneumonia? *Infect Dis Clin North Am*. 2013;27(1):157–175.
16. Jennings LC, Anderson TP, Beynon KA, et al. Incidence and characteristics of viral community-acquired pneumonia in adults. *Thorax*. 2008;63(1):42–48.
17. Karppinen S, Terasjarvi J, Auranen K, et al. Acquisition and transmission of *Streptococcus pneumoniae* are facilitated during rhinovirus infection in families with children. *Am J Respir Crit Care Med*. 2017;196(9):1172–1180.
18. Peltola V, Heikkinen T, Ruuskanen O, et al. Temporal association between rhinovirus circulation in the community and invasive pneumococcal disease in children. *Pediatr Infect Dis J*. 2011;30(6):456–461.
19. Ishizuka S, Yamaya M, Suzuki T, et al. Effects of rhinovirus infection on the adherence of *Streptococcus pneumoniae* to cultured human airway epithelial cells. *J Infect Dis*. 2003;188(12):1928–1939.

20. Korten I, Kieninger E, Klenja S, et al. Respiratory viruses in healthy infants and infants with cystic fibrosis: a prospective cohort study. *Thorax*. 2018;73(1):13–20.
21. Kloepfer KM, Sarsani VK, Poroyko V, et al. Community-acquired rhinovirus infection is associated with changes in the airway microbiome. *J Allergy Clin Immunol*. 2017;140(1):312–315.e8.
22. Molyneaux PL, Mallia P, Cox MJ, et al. Outgrowth of the bacterial airway microbiome after rhinovirus exacerbation of chronic obstructive pulmonary disease. *Am J Respir Crit Care Med*. 2013;188(10):1224–1231.
23. Leino A, Lukkarinen M, Turunen R, et al. Pulmonary function and bronchial reactivity 4 years after the first virus-induced wheezing. *Allergy*. 2018.
24. Jackson DJ, Makrinioti H, Rana BMJ, et al. IL-33-dependent type 2 inflammation during rhinovirus-induced asthma exacerbations in vivo. *Am J Respir Crit Care Med*. 2014;190(12):1373–1382.
25. Corne JM, Marshall C, Smith S, et al. Frequency, severity, and duration of rhinovirus infections in asthmatic and non-asthmatic individuals: a longitudinal cohort study. *Lancet*. 2002;359(9309):831–834.
26. Message SD, Laza-Stanca V, Mallia P, et al. Rhinovirus-induced lower respiratory illness is increased in asthma and related to virus load and Th1/2 cytokine and IL-10 production. *Proc Natl Acad Sci USA*. 2008;105(36):13562–13567.
27. Skoner DP. Allergic rhinitis: definition, epidemiology, pathophysiology, detection, and diagnosis. *J Allergy Clin Immunol*. 2001;108(1 Suppl):S2–S8.
28. Turner H, Kinet JP. Signalling through the high-affinity IgE receptor Fc epsilonRI. *Nature*. 1999;402(6760 Suppl):B24–B30.
29. Pawankar R. Mast cells as orchestrators of the allergic reaction: the IgE-IgE receptor mast cell network. *Curr Opin Allergy Clin Immunol*. 2001;1(1):3–6.
30. Williams CM, Galli SJ. The diverse potential effector and immunoregulatory roles of mast cells in allergic disease. *J Allergy Clin Immunol*. 2000;105(5):847–859.
31. Carroll NG, Mutavdzic S, James AL. Distribution and degranulation of airway mast cells in normal and asthmatic subjects. *Eur Respir J*. 2002;19(5):879–885.
32. Oettgen HC, Geha RS. IgE in asthma and atopy: cellular and molecular connections. *J Clin Invest*. 1999;104(7):829–835.
33. Naclerio RM, Proud D, Togias AG, et al. Inflammatory mediators in late antigen-induced rhinitis. *N Eng J Med*. 1985;313(2):65–70.
34. Jackson DJ, Gangnon RE, Evans MD, et al. Wheezing rhinovirus illnesses in early life predict asthma development in high-risk children. *Am J Respir Crit Care Med*. 2008;178(7):667–672.
35. Soto-Quiros M, Avila L, Platts-Mills TA, et al. High titers of IgE antibody to dust mite allergen and risk for wheezing among asthmatic children infected with rhinovirus. *J Allergy Clin Immunol*. 2012;129(6):1499–1505.e5.
36. Murray CS, Poletti G, Kebadze T, et al. Study of modifiable risk factors for asthma exacerbations: virus infection and allergen exposure increase the risk of asthma hospital admissions in children. *Thorax*. 2006;61(5):376–382.
37. Green RM, Custovic A, Sanderson G, Hunter J, Johnston SL, Woodcock A. Synergism between allergens and viruses and risk of hospital admission with asthma: case-control study. *BMJ*. 2002;324(7340):763.
38. Calhoun WJ, Dick EC, Schwartz LB, Busse WW. A common cold virus, rhinovirus 16, potentiates airway inflammation after segmental antigen bronchoprovocation in allergic subjects. *J Clin Invest*. 1994;94(6):2200–2208.
39. Frick OL. Effect of respiratory and other virus infections on IgE immunoregulation. *J Allergy Clin Immunol*. 1986;78(5 Pt 2):1013–1018.

40. Lemanske Jr. RF, Dick EC, Swenson CA, Vrtis RF, Busse WW. Rhinovirus upper respiratory infection increases airway hyperreactivity and late asthmatic reactions. *J Clin Invest.* 1989;83(1):1–10.
41. Martinez FD. Viral infections and the development of asthma. *Am J Respir Crit Care Med.* 1995;151(5):1644–1647. discussion 1647–1648.
42. Bardin PG, Fraenkel DJ, Sanderson G, et al. Amplified rhinovirus colds in atopic subjects. *Clin Exp Allergy.* 1994;24(5):457–464.
43. Wark PAB, Johnston SL, Bucchieri F, et al. Asthmatic bronchial epithelial cells have a deficient innate immune response to infection with rhinovirus. *J Exp Med.* 2005;201 (6):937–947.
44. Contoli M, Message SD, Laza-Stanca V, et al. Role of deficient type III interferon-lambda production in asthma exacerbations. *Nat Med.* 2006;12(9):1023–1026.
45. Papadopoulos NG, Stanciu LA, Papi A, Holgate ST, Johnston SL. A defective type 1 response to rhinovirus in atopic asthma. *Thorax.* 2002;57(4):328–332.
46. Pritchard AL, White OJ, Burel JG, Carroll ML, Phipps S, Upham JW. Asthma is associated with multiple alterations in anti-viral innate signalling pathways. *PLoS One (Electronic Resource).* 2014;9(9):e106501.
47. Rupani H, Martinez-Nunez RT, Dennison P, et al. Toll-like receptor 7 is reduced in severe asthma and linked to an altered microRNA profile. *Am J Respir Crit Care Med.* 2016;194(1):26–37.
48. Wark PAB, Bucchieri F, Johnston SL, et al. IFN-gamma-induced protein 10 is a novel biomarker of rhinovirus-induced asthma exacerbations. *J Allergy Clin Immunol.* 2007;120(3):586–593.
49. Baraldo S, Contoli M, Bazzan E, et al. Deficient antiviral immune responses in childhood: distinct roles of atopy and asthma. *J Allergy Clin Immunol.* 2012;130 (6):1307–1314.
50. Bedke N, Sammut D, Green B, et al. Transforming growth factor-beta promotes rhinovirus replication in bronchial epithelial cells by suppressing the innate immune response. *PLoS One (Electronic Resource).* 2012;7(9):e44580.
51. Edwards MR, Regamey N, Vareille M, et al. Impaired innate interferon induction in severe therapy resistant atopic asthmatic children. *Mucosal Immunol.* 2013;6 (4):797–806.
52. Mathur SK, Fichtinger PS, Kelly JT, Lee W-M, Gern JE, Jarjour NN. Interaction between allergy and innate immunity: model for eosinophil regulation of epithelial cell interferon expression. *Ann Allergy Asthma Immunol.* 2013;111(1):25–31.
53. Parsons KS, Hsu AC, Wark PA. TLR3 and MDA5 signalling, although not expression, is impaired in asthmatic epithelial cells in response to rhinovirus infection. *Clin Exp Allergy.* 2014;44(1):91–101.
54. Gielen V, Sykes A, Zhu J, et al. Increased nuclear suppressor of cytokine signaling 1 in asthmatic bronchial epithelium suppresses rhinovirus induction of innate interferons. *J Allergy Clin Immunol.* 2015;136(1):177–188.e1.
55. Kicic A, Stevens PT, Sutanto EN, et al. Impaired airway epithelial cell responses from children with asthma to rhinoviral infection. *Clin Exp Allergy.* 2016;46 (11):1441–1455.
56. Djukanovic R, Harrison T, Johnston SL, et al. The effect of inhaled IFN-beta on worsening of asthma symptoms caused by viral infections. A randomized trial. *Am J Respir Crit Care Med.* 2014;190(2):145–154.
57. Lopez-Souza N, Favoreto S, Wong H, et al. In vitro susceptibility to rhinovirus infection is greater for bronchial than for nasal airway epithelial cells in human subjects. *J Allergy Clin Immunol.* 2009;123(6):1384–1390.e2.

58. Bochkov YA, Hanson KM, Keles S, Brockman-Schneider RA, Jarjour NN, Gern JE. Rhinovirus-induced modulation of gene expression in bronchial epithelial cells from subjects with asthma. *Mucosal Immunol.* 2010;3(1):69–80.
59. Sykes A, Macintyre J, Edwards MR, et al. Rhinovirus-induced interferon production is not deficient in well controlled asthma. *Thorax.* 2014;69(3):240–246.
60. Thomas BJ, Lindsay M, Dagher H, et al. Transforming growth factor-beta enhances rhinovirus infection by diminishing early innate responses. *Am J Respir Cell Mol Biol.* 2009;41(3):339–347.
61. Xatzipsalti M, Psarros F, Konstantinou G, et al. Modulation of the epithelial inflammatory response to rhinovirus in an atopic environment. *Clin Exp Allergy.* 2008;38(3):466–472.
62. Olenec JP, Kim WK, Lee WM, et al. Weekly monitoring of children with asthma for infections and illness during common cold seasons. *J Allergy Clin Immunol.* 2010;125(5):1001–1006.e1.
63. Durrani SR, Montville DJ, Pratt AS, et al. Innate immune responses to rhinovirus are reduced by the high-affinity IgE receptor in allergic asthmatic children. *J Allergy Clin Immunol.* 2012;130(2):489–495.
64. Hatchwell L, Collison A, Girkin J, et al. Toll-like receptor 7 governs interferon and inflammatory responses to rhinovirus and is suppressed by IL-5-induced lung eosinophilia. *Thorax.* 2015;70(9):854–861.
65. Shikhagaie MM, Andersson CK, Mori M, et al. Mapping of TLR5 and TLR7 in central and distal human airways and identification of reduced TLR expression in severe asthma. *Clin Exp Allergy.* 2014;44(2):184–196.
66. Normansell RWS, Milan SJ, Walters E, Nair P. Omalizumab for asthma in adults and children. *Cochrane Database Syst Rev.* 2014;(1):CD003559. Available from: https://doi.org/10.1002/14651858.CD003559.pub4.
67. Busse WW, Morgan WJ, Gergen PJ, et al. Randomized trial of omalizumab (anti-IgE) for asthma in inner-city children. *N Eng J Med.* 2011;364(11):1005–1015.
68. Teach SJ, Gill MA, Togias A, et al. Preseasonal treatment with either omalizumab or an inhaled corticosteroid boost to prevent fall asthma exacerbations. *J Allergy Clin Immunol.* 2015;136(6):1476–1485.
69. Gill MA, Liu AH, Calatroni A, et al. Enhanced plasmacytoid dendritic cell antiviral responses after omalizumab. *J Allergy Clin Immunol.* 2018;141(5):1735–1743.e9.
70. Esquivel A, Busse WW, Calatroni A, et al. Effects of omalizumab on rhinovirus infections, illnesses, and exacerbations of asthma. *Am J Respir Crit Care Med.* 2017;196(8):985–992.
71. Beale J, Jayaraman A, Jackson DJ, et al. Rhinovirus-induced IL-25 in asthma exacerbation drives type 2 immunity and allergic pulmonary inflammation. *Sci Transl Med.* 2014;6(256):256ra134.
72. Werder RB, Zhang V, Lynch JP, et al. Chronic IL-33 expression predisposes to virus-induced asthma exacerbations by increasing type 2 inflammation and dampening antiviral immunity. *J Allergy Clin Immunol.* 2018;141(5):1607–1619.e9.
73. Kato A, Favoreto Jr. S, Avila PC, Schleimer RP. TLR3- and Th2 cytokine-dependent production of thymic stromal lymphopoietin in human airway epithelial cells. *J Immunol.* 2007;179(2):1080–1087.
74. Mehta AK, Duan W, Doerner AM, et al. Rhinovirus infection interferes with induction of tolerance to aeroantigens through OX40 ligand, thymic stromal lymphopoietin, and IL-33. *J Allergy Clin Immunol.* 2016;137(1):278–288.e6.
75. Calven J, Yudina Y, Hallgren O, et al. Viral stimuli trigger exaggerated thymic stromal lymphopoietin expression by chronic obstructive pulmonary disease epithelium: role of endosomal TLR3 and cytosolic RIG-I-like helicases. *J Innate Immun.* 2012;4(1):86–99.

76. Chen G, Korfhagen TR, Karp CL, et al. Foxa3 induces goblet cell metaplasia and inhibits innate antiviral immunity. *Am J Respir Crit Care Med.* 2014;189(3):301–313.
77. Castro M, Zangrilli J, Wechsler ME, et al. Reslizumab for inadequately controlled asthma with elevated blood eosinophil counts: results from two multicentre, parallel, double-blind, randomised, placebo-controlled, phase 3 trials. *Lancet Respir Med.* 2015;3(5):355–366.
78. Ortega HG, Liu MC, Pavord ID, et al. Mepolizumab treatment in patients with severe eosinophilic asthma. *N Engl J Med.* 2014;371(13):1198–1207.
79. Wenzel S, Ford L, Pearlman D, et al. Dupilumab in persistent asthma with elevated eosinophil levels. *N Engl J Med.* 2013;368(26):2455–2466.
80. Corren J, Parnes JR, Wang L, et al. Tezepelumab in adults with uncontrolled asthma. *N Eng J Med.* 2017;377(10):936–946.
81. Cantin AM, Hartl D, Konstan MW, Chmiel JF. Inflammation in cystic fibrosis lung disease: pathogenesis and therapy. *J Cyst Fibro.* 2015;14(4):419–430.
82. Stelzer-Braid S, Liu N, Doumit M, et al. Association of rhinovirus with exacerbations in young children affected by cystic fibrosis: preliminary data. *J Med Virol.* 2017;89 (8):1494–1497.
83. Dijkema JS, van Ewijk BE, Wilbrink B, Wolfs TF, Kimpen JL, van der Ent CK. Frequency and duration of rhinovirus infections in children with cystic fibrosis and healthy controls: a longitudinal cohort study. *Pediatr Infect Dis J.* 2016;35(4):379–383.
84. Zheng S, Xu W, Bose S, Banerjee AK, Haque SJ, Erzurum SC. Impaired nitric oxide synthase-2 signaling pathway in cystic fibrosis airway epithelium. *Am J Physiol Lung Cell Mol Physiol.* 2004;287(2):L374–L381.
85. Cousin M, Molinari N, Foulongne V, et al. Rhinovirus-associated pulmonary exacerbations show a lack of FEV1 improvement in children with cystic fibrosis. *Influenza Other Respir Viruses.* 2016;10(2):109–112.
86. Wark PA, Tooze M, Cheese L, et al. Viral infections trigger exacerbations of cystic fibrosis in adults and children. *Eur Respir J.* 2012;40(2):510–512.
87. Chin M, De Zoysa M, Slinger R, et al. Acute effects of viral respiratory tract infections on sputum bacterial density during CF pulmonary exacerbations. *J Cyst Fibros.* 2015;14(4):482–489.
88. Sutanto EN, Kicic A, Foo CJ, et al. Innate inflammatory responses of pediatric cystic fibrosis airway epithelial cells: effects of nonviral and viral stimulation. *Am J Respir Cell Mol Biol.* 2011;44(6):761–767.
89. Chattoraj SS, Ganesan S, Faris A, Comstock A, Lee WM, Sajjan US. *Pseudomonas aeruginosa* suppresses interferon response to rhinovirus infection in cystic fibrosis but not in normal bronchial epithelial cells. *Infect Immun.* 2011;79(10):4131–4145.
90. Dauletbaev N, Das M, Cammisano M, et al. Rhinovirus load is high despite preserved interferon-beta response in cystic fibrosis bronchial epithelial cells. *PLoS One.* 2015;10(11):e0143129.
91. Hansdottir S, Monick MM, Hinde SL, Lovan N, Look DC, Hunninghake GW. Respiratory epithelial cells convert inactive vitamin D to its active form: potential effects on host defense. *J Immunol (Baltimore, MD: 1950).* 2008;181(10):7090–7099.
92. Di Rosa M, Malaguarnera M, Nicoletti F, Malaguarnera L. Vitamin D3: a helpful immuno-modulator. *Immunology.* 2011;134(2):123–139.
93. Schögler A, Muster RJ, Kieninger E, et al. Vitamin D represses rhinovirus replication in cystic fibrosis cells by inducing LL-37. *Eur Respir J.* 2016;47(2):520–530.
94. Rovner AJ, Stallings VA, Schall JI, Leonard MB, Zemel BS. Vitamin D insufficiency in children, adolescents, and young adults with cystic fibrosis despite routine oral supplementation. *Am J Clin Nutr.* 2007;86(6):1694–1699.
95. Barnes PJ. Inflammatory mechanisms in patients with chronic obstructive pulmonary disease. *J Allergy Clin Immunol.* 2016;138(1):16–27.

96. Oh JY, Sin DD. Lung inflammation in COPD: why does it matter? *F1000 Med Rep*. 2012;4:23.
97. Greenberg SB, Allen M, Wilson J, Atmar RL. Respiratory viral infections in adults with and without chronic obstructive pulmonary disease. *Am J Respir Crit Care Med*. 2000;162(1):167–173.
98. George SN, Garcha DS, Mackay AJ, et al. Human rhinovirus infection during naturally occurring COPD exacerbations. *Eur Respir J*. 2014;44(1):87–96.
99. Schneider D, Ganesan S, Comstock AT, et al. Increased cytokine response of rhinovirus-infected airway epithelial cells in chronic obstructive pulmonary disease. *Am J Respir Crit Care Med*. 2010;182(3):332–340.
100. Mallia P, Message SD, Gielen V, et al. Experimental rhinovirus infection as a human model of chronic obstructive pulmonary disease exacerbation. *Am J Respir Crit Care Med*. 2011;183(6):734–742.
101. Mallia P, Message SD, Contoli M, et al. Neutrophil adhesion molecules in experimental rhinovirus infection in COPD. *Respir Res*. 2013;14:72.
102. Mallia P, Message SD, Contoli M, et al. Lymphocyte subsets in experimental rhinovirus infection in chronic obstructive pulmonary disease. *Respir Med*. 2014;108 (1):78–85.
103. Footitt J, Mallia P, Durham AL, et al. Oxidative and nitrosative stress and histone deacetylase-2 activity in exacerbations of COPD. *Chest*. 2016;149(1):62–73.
104. Gagné F. Chapter 6—oxidative stress. In: Gagné F, ed. *Biochemical Ecotoxicology*. Oxford: Academic Press; 2014:103–115.
105. Dasgupta A, Klein K. Chapter 11— oxidative stress related to other diseases. In: Dasgupta A, Klein K, eds. *Antioxidants in Food, Vitamins and Supplements*. San Diego, CA: Elsevier; 2014:185–207.
106. Liu CW, Lee TL, Chen YC, et al. PM2.5-induced oxidative stress increases intercellular adhesion molecule-1 expression in lung epithelial cells through the IL-6/AKT/STAT3/NF-kappaB-dependent pathway. *Part Fibre Toxicol*. 2018;15(1):4.
107. Floreani AA, Wyatt TA, Stoner J, et al. Smoke and C5a induce airway epithelial intercellular adhesion molecule-1 and cell adhesion. *Am J Respir Cell Mol Biol*. 2003;29(4):472–482.
108. Ozguner F, Koyu A, Cesur G. Active smoking causes oxidative stress and decreases blood melatonin levels. *Toxicol Ind Health*. 2005;21(1–2):21–26.
109. Talukder MA, Johnson WM, Varadharaj S, et al. Chronic cigarette smoking causes hypertension, increased oxidative stress, impaired NO bioavailability, endothelial dysfunction, and cardiac remodeling in mice. *Am J Physiol Heart Circ Physiol*. 2011;300(1):H388–H396.
110. Shukla SD, Mahmood MQ, Weston S, et al. The main rhinovirus respiratory tract adhesion site (ICAM-1) is upregulated in smokers and patients with chronic airflow limitation (CAL). *Respir Res*. 2017;18(1):6.
111. Papi A, Johnston SL. Rhinovirus infection induces expression of its own receptor intercellular adhesion molecule 1 (ICAM-1) via increased NF-kappaB-mediated transcription. *J Biol Chem*. 1999;274(14):9707–9720.
112. Subauste MC, Jacoby DB, Richards SM, Proud D. Infection of a human respiratory epithelial cell line with rhinovirus. Induction of cytokine release and modulation of susceptibility to infection by cytokine exposure. *J Clin Invest*. 1995;96(1):549–557.
113. Berman R, Jiang D, Wu Q, Chu HW. Alpha1-antitrypsin reduces rhinovirus infection in primary human airway epithelial cells exposed to cigarette smoke. *Int J Chron Obstruct Pulmon Dis*. 2016;11:1279–1286.
114. Eddleston J, Lee RU, Doerner AM, Herschbach J, Zuraw BL. Cigarette smoke decreases innate responses of epithelial cells to rhinovirus infection. *Am J Respir Cell Mol Biol*. 2011;44(1):118–126.

115. Proud D, Hudy MH, Wiehler S, et al. Cigarette smoke modulates expression of human rhinovirus-induced airway epithelial host defense genes. *PLoS One*. 2012;7 (7):e40762.
116. Xu R, Li Q, Zhou XD, Perelman JM, Kolosov VP. Oxidative stress mediates the disruption of airway epithelial tight junctions through a TRPM2-PLCgamma1-PKCalpha signaling pathway. *Int J Mol Sci*. 2013;14(5):9475–9486.
117. Sajjan U, Wang Q, Zhao Y, Gruenert DC, Hershenson MB. Rhinovirus disrupts the barrier function of polarized airway epithelial cells. *Am J Respir Crit Care Med*. 2008;178(12):1271–1281.
118. Atkinson SK, Sadofsky LR, Morice AH. How does rhinovirus cause the common cold cough? *BMJ Open Respir Res*. 2016;3(1):e000118.
119. Looi K, Buckley AG, Rigby PJ, et al. Effects of human rhinovirus on epithelial barrier integrity and function in children with asthma. *Clin Exp Allergy*. 2018;48 (5):513–524.
120. Comstock AT, Ganesan S, Chattoraj A, et al. Rhinovirus-induced barrier dysfunction in polarized airway epithelial cells is mediated by NADPH oxidase 1. *J Virol*. 2011;85(13):6795–6808.
121. Banerjee A, Banerjee V, Czinn S, Blanchard T. Increased reactive oxygen species levels cause ER stress and cytotoxicity in andrographolide treated colon cancer cells. *Oncotarget*. 2017;8(16):26142–26153.
122. Hoffman SM, Tully JE, Nolin JD, et al. Endoplasmic reticulum stress mediates house dust mite-induced airway epithelial apoptosis and fibrosis. *Respir Res*. 2013;14 (1):141.
123. Jorgensen E, Stinson A, Shan L, Yang J, Gietl D, Albino AP. Cigarette smoke induces endoplasmic reticulum stress and the unfolded protein response in normal and malignant human lung cells. *BMC Cancer*. 2008;8:229.
124. Pathinayake PS, Hsu AC, Waters DW, Hansbro PM, Wood LG, Wark PAB. Understanding the unfolded protein response in the pathogenesis of asthma. *Front Immunol*. 2018;9:175.
125. Hanson PJ, Zhang HM, Hemida MG, Ye X, Qiu Y, Yang D. IRES-dependent translational control during virus-induced endoplasmic reticulum stress and apoptosis. *Front Microbiol*. 2012;3:92.
126. Lee KM, Chen CJ, Shih SR. Regulation mechanisms of viral IRES-driven translation. *Trends Microbiol*. 2017;25(7):546–561.
127. Smith JA. A new paradigm: innate immune sensing of viruses via the unfolded protein response. *Front Microbiol*. 2014;5:222.
128. Liu J, HuangFu WC, Kumar KG, et al. Virus-induced unfolded protein response attenuates antiviral defenses via phosphorylation-dependent degradation of the type I interferon receptor. *Cell Host Microbe*. 2009;5(1):72–83.
129. Smith JA, Turner MJ, DeLay ML, Klenk EI, Sowders DP, Colbert RA. Endoplasmic reticulum stress and the unfolded protein response are linked to synergistic IFN-beta induction via X-box binding protein 1. *Eur J Immunol*. 2008;38 (5):1194–1203.
130. Hu F, Yu X, Wang H, et al. ER stress and its regulator X-box-binding protein-1 enhance polyIC-induced innate immune response in dendritic cells. *Eur J Immunol*. 2011;41(4):1086–1097.
131. Ono JG, Worgall TS, Worgall S. 17q21 locus and ORMDL3: an increased risk for childhood asthma. *Pediatr Res*. 2013;75:165.
132. Cantero-Recasens G, Fandos C, Rubio-Moscardo F, Valverde MA, Vicente R. The asthma-associated ORMDL3 gene product regulates endoplasmic reticulum-mediated calcium signaling and cellular stress. *Hum Mol Genet*. 2010;19 (1):111–121.

133. Song DJ, Miller M, Beppu A, et al. Rhinovirus infection of ORMDL3 transgenic mice is associated with reduced rhinovirus viral load and airway inflammation. *J Immunol (Baltimore, MD: 1950)*. 2017;199(7):2215–2224.
134. Liu YP, Rajamanikham V, Baron M, et al. Association of ORMDL3 with rhinovirus-induced endoplasmic reticulum stress and type I Interferon responses in human leucocytes. *Clin Exp Allergy*. 2017;47(3):371–382.
135. Bhakta NR, Christenson SA, Nerella S, et al. IFN-stimulated gene expression, type 2 inflammation, and endoplasmic reticulum stress in asthma. *Am J Respir Crit Care Med*. 2018;197(3):313–324.
136. Zhu Z, Lee CG, Zheng T, et al. Airway inflammation and remodeling in asthma. Lessons from interleukin 11 and interleukin 13 transgenic mice. *Am J Respir Crit Care Med*. 2001;164(10 Pt 2):S67–S70.
137. Corren J. Role of interleukin-13 in asthma. *Curr Allergy Asthma Rep*. 2013;13 (5):415–420.
138. Lachowicz-Scroggins ME, Boushey HA, Finkbeiner WE, Widdicombe JH. Interleukin-13-induced mucous metaplasia increases susceptibility of human airway epithelium to rhinovirus infection. *Am J Respir Cell Mol Biol*. 2010;43(6):652–661.
139. Martino MB, Jones L, Brighton B, et al. The ER stress transducer IRE1beta is required for airway epithelial mucin production. *Mucosal Immunol*. 2013;6 (3):639–654.

CHAPTER 8

In vivo experimental models of infection and disease

Jason Girkin[1], Steven Maltby[1], Aran Singanayagam[2], Nathan Bartlett[1] and Patrick Mallia[2]
[1]Priority Research Center for Healthy Lungs, Faculty of Health and Medicine, University of Newcastle, Newcastle, Australia
[2]Faculty of Medicine, National Heart & Lung Institute, Imperial College London, London, United Kingdom

8.1 HUMAN MODELS OF RHINOVIRUS INFECTION

Experimental infection of human subjects with rhinovirus (RV) has long been used to study the pathogenesis of infection. Indeed, such studies were carried out even before RVs were identified as causative agents of the common cold. In 1914 almost 40 years prior to identification of RV, Kruse induced colds in volunteers by inoculating them with a filtrate of nasal washings.[1] Following the identification of RVs, viral challenge studies were used extensively in healthy volunteers to study numerous aspects of RV biology including viral infectivity, modes of transmission, role of environmental factors, host immune responses, and the effects of treatments. In the 1990s these studies were extended to subjects with asthma to study the pathogenesis of RV-induced asthma exacerbations, and over the last decade experimental RV infection has also been carried out in patients with chronic obstructive pulmonary disease (COPD). These studies have provided unique insights into the pathogenesis of RV infection that would have been difficult to obtain using studies of naturally acquired infections or in animal models.

8.2 RATIONALE FOR HUMAN INFECTION STUDIES

Viral respiratory tract infections are the commonest infectious syndrome in humans with adults experiencing two to four infections per annum. Given the frequency of viral infections it is pertinent to ask why studies

Rhinovirus Infections
DOI: https://doi.org/10.1016/B978-0-12-816417-4.00008-1

that deliberately expose humans to an infectious agent are required, when there are more than enough naturally acquired infections to yield sufficient subjects for studies. Despite the ubiquity of viral colds there are a number of factors that make studies of naturally acquired infections problematic. Although RVs are the commonest etiological agents of viral colds there are several other viral causes (and nonviral causes of upper airway symptoms), and the clinical syndromes caused by different virus types are indistinguishable.[2,3] Other factors contributing to variability in naturally acquired infections include different routes of inoculation (eye, nasopharynx, direct contact, aerosol, etc.), different inoculation doses, variability in the perception of symptoms leading to differences in time to presentation and host factors (immune status, smoking, age, etc.) that influence viral pathogenicity. Further, the heterogeneous nature of naturally occurring infections requires that large patient numbers are needed to identify statistically significant effects of treatment. Therefore human experimental infection studies are an attractive proposition as they allow for a known etiological agent to be administered at a standard dose, route of inoculation, and time point to a selected group of recipients with similar characteristics (e.g., age, smoking history, health/disease, and antibody status). Detailed follow up can be carried out in a controlled clinical setting with sample collection at defined time points in relation to the time of infection. As the clinical syndrome induced by RV challenge in young healthy volunteers is benign and self-limiting, experimental RV infection in this group is relatively uncontroversial. Perhaps the only risk to subjects was the possibility of additional infectious agents present in the inoculum and good manufacturing practice (GMP)-prepared stocks are now required by regulators for experimental infection studies in humans to contravene this risk.[4] Studies in healthy volunteers have been central to establishing the key aspects of the biology of RV infection including routes of acquisition of infection,[5–8] clinical symptoms,[8,9] inflammatory and immune responses,[10] involvement of the lower airway,[10,11] and correlates of immune protection of RV infection.[12]

Virus challenge studies have also been used in healthy volunteers to evaluate a vast array of potential treatments. However, none of these studies have led to the licensing of a single treatment for the common cold.[13–51] Licensing approval was sought for an antiviral drug, pleconaril, for the treatment of RV infection.[21] Approval was denied by the Food and Drug Administration as the adverse effects outweighed the benefits in healthy subjects with self-limiting colds. The lack of approval of any antiviral treatments casts doubt on whether continued investment in viral challenge studies is justified. However, the recognition that RV infection is associated with more severe clinical manifestations in people with chronic lung diseases such as asthma and COPD provided a new impetus to research and a new direction to human experimental infection studies.

8.3 RHINOVIRUS INFECTION AND EXACERBATIONS OF ASTHMA AND CHRONIC OBSTRUCTIVE PULMONARY DISEASE

Up until the early 1990s the prevailing view was that RV infection resulted in a self-limiting, mild upper respiratory tract syndrome only. There were occasional reports of RV infection associated with more severe clinical illness such as pneumonia[52] but both scientific and pharmacological research tended to be focused on other respiratory viruses such as influenza and respiratory syncytial virus, as these were considered to be more serious human pathogens. Colds had long been associated with asthma exacerbations but early studies investigating virus infection in asthma and COPD exacerbations reported low detection rates.[53–55] The consensus was that asthma exacerbations were predominantly triggered by allergen exposure and COPD exacerbations by acute bacterial infection. The development of highly sensitive and specific molecular diagnostic techniques using polymerase chain reaction (PCR) technology led to a revolution in viral diagnostics and a reevaluation of the role of respiratory viruses in a range of clinical syndromes. This was particularly pertinent to human RVs, which are either difficult or impossible to culture (e.g., RV-C strains) and due to the large number of serotypes diagnostic serology testing is not feasible. PCR-based diagnostic tests have a much greater sensitivity for the detection of RVs and studies using PCR revealed that the range of clinical illness associated with RV infection was much broader than previously recognized and included more severe disease syndromes such as pneumonia,[56] bronchiolitis,[57] acute rhinosinusitis,[58] and influenza-like illness.[59] In addition RVs could be detected in most asthma exacerbations,[60] and in a substantial proportion of COPD exacerbations.[61] Asthma is estimated to affect 360 million people worldwide and COPD affects 174.5 million people and was the cause of 3.2 million deaths in 2015.[62] Much of the enormous morbidity, mortality, and healthcare costs associated with asthma and COPD are related to acute exacerbations. Therefore the recognition that RVs are a major cause of asthma and COPD exacerbations stimulated new interest in their biology and treatment. As part of this research investigators considered whether experimental infection studies in humans could be extended from healthy volunteers to patients with asthma and COPD.

8.4 EXPERIMENTAL RHINOVIRUS INFECTION IN ASTHMA

Respiratory viruses can be detected in up to 80% of asthma exacerbations in children and 60%–80% of exacerbations in adults,[60,63–65] with RVs the commonest virus detected. The recognition of the role of RVs in asthma exacerbations stimulated research into their biology in an attempt to develop treatments for virus-induced exacerbations. This research included investigating whether experimental RV infection could be used in people with asthma in the same way it had been used in healthy individuals. The first experimental RV challenge of subjects with asthma was carried out in 1985 at Dalhousie University, Canada. Of the 21 volunteers inoculated, 19 became infected but only 4 had ≥10% decrease in forced expiratory volume in 1 second (FEV_1) and an increase in airway hyperreactivity (AHR). The authors felt that these findings suggested "that other viral pathogens may play a more important role in precipitating asthma attacks."[66] It is unclear why this study failed to induce features of asthma exacerbations but it would be almost another decade before further experimental RV infection studies in people with asthma were attempted. Experimental infection studies in allergic (nonasthmatic) subjects suggested that RV infection could induce changes in lower airway physiology similar to that seen in asthma.[67,68] In 1994 an experimental infection study from the University of Southampton, United Kingdom included a small group of people with allergic asthma and reported that upper respiratory symptoms were more severe in atopic subjects but did not report on lower airway symptoms or physiology.[69] Concurrently a study using PCR to detect viruses in naturally acquired asthma exacerbations strongly supported a role for RV.[60] Subsequent studies were carried out by research groups in the United Kingdom,[70] the Netherlands,[71–74] and the United States[75,76] in volunteers with mild, intermittent asthma. These studies demonstrated that RV infection induced airway obstruction,[77,78] increased AHR,[70–73,79] and airway inflammation[70,72,79–84] and RV could be detected in the lower airways,[75,85] thereby supporting a causative role for RVs in asthma exacerbations.

Having established that respiratory viruses are a trigger for asthma exacerbations, research focused on investigating why RVs cause a benign, self-limiting illness in healthy subjects but result in more severe manifestations in people with asthma. A study of naturally acquired infections in cohabiting couples discordant for asthma suggested that people with asthma were not more susceptible to virus infection but had more lower respiratory tract symptoms.[86] Airway inflammation during naturally acquired infections was greater in people with asthma compared with those without asthma but the number of subjects in this study was small and the viruses detected were different between the two groups.[87] Viral challenge studies are ideally suited to addressing this research question as people without asthma matched for characteristics such as age and gender can be infected simultaneously. Most of the earlier infection studies did not include a control group of healthy volunteers and therefore could not address the question as to whether host responses to infection differ in people with asthma. Studies that did include nonasthmatic controls produced somewhat inconsistent results with one study reporting no differences in lower airway inflammatory cells,[70] another reporting increased nasal inflammatory mediators in asthma[88] and discrepant results regarding virus-induced respiratory symptoms.[88,89] These divergent results were likely related to differences in sampling methods and timing, antibody status of subjects, and choice of healthy controls (e.g., atopic vs nonatopic).

The first study to show clear differences between subjects with and without asthma in their responses to RV infection was published in 2008.[90] In this study, RV challenge induced more respiratory symptoms, greater lung function impairment, increased bronchial hyperreactivity, and eosinophilic and neutrophilic lower airway inflammation in asthmatic compared with normal subjects with direct correlations between loss of lung function and the degree of neutrophilic, eosinophilic inflammation, and nasal viral load.[90] In addition, the study provided insights into potential mechanisms of differential responses to viral infection in people with asthma. Despite being infected with the same dose of virus, postinoculation virus loads tended to be higher in the asthmatic subjects compared with the healthy controls, suggesting that antiviral immunity may be impaired in people with asthma with subsequent failure to control viral replication. Virologic and clinical outcomes were related to deficient interferon (IFN)-γ and interleukin (IL)-10 responses and to augmented T-helper type 2 (T_H2) responses (IL-4, IL-5, and IL-13), indicating that excessive T_H2 or impaired T_H1 (or IL-10) immunity may be important mechanisms. When infected with RV in vitro, alveolar macrophages and bronchial epithelial cells from subjects with asthma demonstrated deficient production of IFNβ and IFNλ, and this was related to the severity of virus-induced asthma exacerbations.[91] Other reports subsequently confirmed that IFN production is deficient in asthma,[92,93] but this finding has not been replicated in all studies.[94–96] It may be that this phenomenon only occurs in a subset of people with asthma, or that it is seen in more severe or poorly controlled asthma. Such patients were not initially included in challenge studies as these were limited to mild, well-controlled asthma not requiring inhaled corticosteroids. In 2014 RV challenge was shown to be safe in a small group of people with well-controlled asthma requiring long-term use of inhaled corticosteroids.[97] A larger study confirmed this and reported significantly more upper and lower respiratory symptoms, greater reduction in peak expiratory flow and FEV_1, increased viral loads, increased bronchoalveolar lavage (BAL) eosinophils, and increased nasal IL-4, IL-5, and IL-13 in subjects with moderate asthma using inhaled corticosteroids.[98] This study also identified novel mediators of virus-induced asthma exacerbations including IL-33,[98] IL-25,[99] and IL-18.[100] Poor asthma control was associated with more severe virus-induced exacerbations, greater T_H2 inflammation and higher virus load.[101] Therefore responses to virus infection may differ depending on asthma severity and control, which may account for some of the discrepant results in earlier experimental infection studies. These successful viral challenge studies should pave the way for further studies in subjects with moderate asthma that should reveal new insights into the pathogenesis of exacerbations that may not have been obtained from studies in mild asthma.

The evidence that emerged from research, including experimental infection studies, that asthma is associated with deficient IFN responses led to the development of inhaled IFNβ as a treatment for asthma exacerbations. A clinical trial of inhaled IFNβ reported that treatment can reduce the severity of virus-induced exacerbations in a subgroup of patients assessed with more severe asthma.[102] The development of inhaled IFN as a novel asthma treatment is a clear demonstration of the potential of experimental RV infection studies to contribute to the discovery of new treatments for virus-induced asthma exacerbations.

8.5 EXPERIMENTAL RHINOVIRUS INFECTION IN CHRONIC OBSTRUCTIVE PULMONARY DISEASE

The Global Initiative for Obstructive Lung Disease defines COPD as "a common preventable and treatable disease, characterized by persistent airflow limitation that is usually progressive and associated with an enhanced chronic inflammatory response in the airways and the lung to noxious particles or gases. Exacerbations and comorbidities contribute to the overall severity in individual patients."[103] Acute exacerbations of COPD are the major drivers of morbidity, mortality, and healthcare costs in COPD and prevention of exacerbations a major unmet need.[104] Acute bacterial infection was believed to be the main cause of COPD exacerbations and this is reflected in the widespread use of antibiotics in COPD exacerbations.[105,106] Although COPD exacerbations are preceded by upper respiratory symptoms in up to two-thirds of cases, virus detection rates in the pre-PCR era were low.[53,107] As with asthma, the role of viruses in COPD exacerbations was reexamined using PCR-based detection methods. Although detection rates of respiratory viruses in COPD exacerbations are more variable than in asthma, respiratory viruses can be detected in 50%–64% of COPD exacerbations, with RVs the predominant virus type detected.[108–111] Despite this emerging evidence implicating respiratory viruses in a significant proportion of COPD exacerbations, both scientific and clinical research continued to focus on bacterial infection. As evidence of this, it was almost two decades after the first experimental RV infection study was carried out in asthma that a similar study in COPD was attempted. Despite the excellent safety record of experimental infection studies in asthma, caution was warranted in repeating these studies in COPD as there are major differences between these two populations. COPD patients are older, current or ex-smokers, and have impaired lung function with irreversible airflow obstruction. All these factors have the potential to result in a more severe response to experimental RV challenge, compared with the younger, nonsmoking patients with relatively normal baseline lung function recruited to the asthma infection studies.

The first experimental infection study in COPD was a small pilot study published in 2006 that established the safety of RV infection in four patients with moderate airflow obstruction (FEV_1 50%–80% predicted) and not using regular inhaled therapy.[112] The subjects developed symptoms consistent with a COPD exacerbation following RV inoculation, together with objective markers of exacerbation with falls in lung function and increases in upper airway inflammatory markers. All the subjects recovered completely without treatment and no adverse events were reported. Subsequently the same research group carried out two larger studies of experimental RV infection in subjects with COPD and non-COPD control subjects.[113,114] These studies replicated the findings of the pilot study, wherein RV infection manifested in cold symptoms, lower respiratory symptoms consistent with exacerbations of COPD, including airway inflammation and worsened airflow obstruction.[113,114] These studies provided important causal evidence linking virus infection to COPD exacerbations. Studies from naturally acquired infection were supportive of this link but do not provide definitive evidence as PCR evaluation detects viral nucleic acid and therefore does not prove the presence of live virus and samples are only collected after exacerbation onset. Respiratory virus nucleic acid can also be detected in COPD patients with stable disease, although is usually elevated during COPD exacerbations.[115] In experimental infection models, RV was present in airway samples prior to exacerbation onset, virus load increased in parallel with the increase in symptoms, airflow obstruction and inflammation, and clearance of virus was associated with exacerbation resolution.[113,114] Strong correlations were observed between virus load and airway neutrophil numbers, neutrophil elastase, IL-8, IL-6, and tumor necrosis factor-alpha (TNFα), granulocyte macrophage colony stimulating factor, most of which of also correlated with levels of epigenetic regulator, histone deacetylase 2 and these inflammatory responses were greater in patients with COPD. In these studies, RV infection is the sole experimental agent responsible for increased inflammatory markers in patients with COPD, providing strong evidence that RV infection directly causes exacerbations in COPD patients.

Another advantage of virus challenge studies over naturally acquired infections is the ability to carry out detailed and repeated lower airway sampling during the course of the exacerbations, including the use of bronchoscopy. This has provided a wealth of mechanistic data regarding the pathogenesis of virus-induced exacerbations including the presence of inflammatory mediators,[113,114] inflammatory cells,[113,116,117] oxidative and nitrosative stress,[114] and impaired antiviral IFN responses.[113] A novel observation that emerged from these studies was that secondary bacterial infections occurred in 60% of experimental virus-induced exacerbations,[118] whereas coinfection was rarely reported in naturally acquired exacerbations.[119] Analysis of the respiratory microbiome following experimental RV infection suggest that secondary infection occurs due to an outgrowth of previously present airway bacteria.[120] Potential mechanisms of secondary bacterial infection include reduced antimicrobial peptides[118] and increased glucose in the airways.[121] A subsequent study of naturally acquired exacerbations sampling at multiple time points during exacerbations confirmed the validity of this observation.[122] Therefore although the numbers of COPD subjects recruited to virus challenge studies to date is small, RV infection appears to be safe in this population and replicates the features of naturally acquired infection. Further studies including larger numbers of patients are needed to validate these findings and further investigate the mechanisms of virus-induced exacerbations in COPD.

8.6 FUTURE DIRECTIONS FOR HUMAN INFECTION MODELS

Since the first studies in healthy volunteers, experimental RV infection has been extended to patients with asthma and COPD and has contributed enormously to expanding our understanding of the biology of RV infection and how it affects patients with chronic airway diseases. A summary of the key findings from human experimental infection studies in people with asthma and COPD is provided in Table 8.1. These studies have tended to have a narrow focus on RV infection and host immune responses. It is clear from both in vivo and in vitro studies that there are interactions between respiratory virus infections and other factors that exacerbate asthma and COPD such as bacteria,[127] allergens,[128] and air pollution.[129] These factors have been somewhat neglected in viral challenge studies and are a promising field of future research that is starting to be addressed.[124,130]

As mentioned previously, the use of the viral challenge model in asthma is much further advanced compared with COPD. One study identified IFN-deficiency in COPD but this has not been replicated. With the development of inhaled IFN as a therapy option for asthma, the role of IFN in COPD requires urgent further investigation. Other areas of future research include the effects of virus infection on novel pathways such as lipidomics[131] and metabolomics[121] in asthma and COPD.

Respiratory viruses have also been identified as triggers of exacerbations in other airway diseases such as cystic fibrosis[132,133] and bronchiectasis[134,135] and there is evidence of impaired antiviral immunity in these diseases.[136] Experimental infection studies may help to define the role of virus infection in these patient populations.

Table 8.1 Human experimental infection studies in asthma and chronic obstructive pulmonary disease (COPD)

Study	Patient population	Controls	Main outcomes
Halperin et al. (1985)[66]	21 AA, 12 using SABA, 1 using ICS		No change in FEV_1 or AHR
Bardin et al. (1994)[69]	6 AA, mild asthma, SABA only	11 NANA, 6 ANA	More severe colds in atopic subjects
Fraenkel et al. (1995)[70]	6 AA, SABA only	11 NANA	No change in in FEV_1 or AHR, increase in bronchial mucosal lymphocytes in all groups but no difference between groups, increased eosinophils at convalescence in AA
Cheung et al. (1995)[71]	14 AA, SABA only. 7 infected, 7 sham infected		No change in FEV_1 or AHR, increased cold and asthma symptoms, increased blood neutrophils and reduced blood lymphocytes in the infected group
Grunberg et al. (1997, 1999)[77,80]	27 AA, SABA only. 19 infected, 8 sham infected		No change in laboratory FEV_1, fall in home FEV_1, increase in AHR, increased nasal IL-8, sputum ECP, IL-8 and IL-6, blood neutrophils and reduced blood lymphocytes in the infected group
Gern et al. (1997)[75]	5 AA, SABA only	3 ANA	Detection of RV in the lower airway
de Gouw et al. (1998)[79]	14 AA, SABA only. 7 infected, 7 sham infected		No change in FEV_1 or AHR, increase in FeNO in the infected group
Parry et al. (2000)[76]	17 AA	5 ANA	No significant differences in symptom scores, viral shedding, or cytokine responses between groups
Jarjour et al. (2000)[81]	8 AA, SABA only		Increased blood neutrophils and reduced blood lymphocytes, increased nasal IL-8 and G-CSF; no change in bronchial lavage IL-8, TNFα, IL-5, IL-1β, IFN-γ, LTB_4, or EDN; increase in BAL neutrophils and MPO

Gern et al. (2000)[82]	15 AA, SABA only	7 ANA. All 22 subjects analyzed together	No change in FEV_1, sputum inflammatory cells, eosinophils, lymphocytes, IL-8, IFN-γ and IL-5 protein; increased sputum neutrophils, G-CSF protein, IFN-γ and IL-5 mRNA, increased nasal neutrophils, IL-8 and G-CSF
Bardin et al. (2000)[78]	6 AA	11 NANA, 5 ANA	
Grunberg et al. (2000, 2001), de Kluijver et al. (2003)[74,88,123]	25 AA, SABA only. 12 received budesonide prior to inoculation and 13 placebo	12 NANA	No change in FEV_1 or AHR, increased bronchial biopsy T cells, increased biopsy expression of ICAM-1, nasal IL-8 and IL-1β in AA; increased nasal IL-1ra in NANA; no effect of budesonide
de Kluijver 2003[124]	11 AA exposed to allergen + placebo, 10 AA exposed to RV, 9 AA exposed to RV and allergen		Increase in cold scores, sputum neutrophils, neutrophil elastase, IL-8, nasal lavage neutrophils and IL-8; drop in FEV_1; no change in AHR or FeNO; no differences between RV only and RV + allergen groups
Zambrano et al. (2003)[89]	16 AA SABA only; 6 high IgE, 10 low IgE	9 NANA	No change in FEV_1 or AHR; lower respiratory symptoms and FeNO greater in AA with high IgE
Mosser et al. (2005)[85]	13 AA	6 NANA	No difference in upper or lower airway viral load between groups
Christiansen et al. (2008)[83]	4 AA, mild or intermittent asthma	4 ANA	Increased human tissue kallikrein activation activity in BAL in AA

(*Continued*)

Table 8.1 (Continued)

Study	Patient population	Controls	Main outcomes
Message et al. (2008)[90]	10 AA, SABA only	15 NANA	Increased chest symptom scores, sputum and BAL eosinophils, AHR in AA; significant falls in FEV_1 and PEF in AA; reduced blood CD4 + , CD8 + , and B cells in AA
Adura (2014)[97]	11 asthmatics using ICS		Increased cold and asthma scores, no adverse events
DeMore et al. (2009)[96]	15 AA, SABA only	18 ANA	No change in PEF from baseline; between groups no differences in cold scores, PEF, virus load, sputum or nasal lavage neutrophil, monocyte and lymphocytes, nasal lavage IL-6, IL-10, CXCL8, CCL2, and CCL5, serum CXCL10; increased sputum eosinophils in AA
Kloepfer et al. (2011)[125]	19 AA SABA only. 8 received montelukast, 11 placebo		No effect of montelukast on symptoms, PEF, viral load, sputum eosinophils, or neutrophils
Majoor et al. (2014)[84]	13 AA	11 NANA	Coagulant TF-exposing microparticles in BAL fluid reduced in AA
Jackson (2014)[98]	28 AA, 15 using ICS	11 NANA	Significantly greater upper and lower respiratory symptoms, greater reduction in PEF and FEV_1, increased viral loads, increased BAL eosinophils, increased nasal IL-4, IL-5, and IL-13 in AA
Silkoff et al. (2018)[126]	63 mild-to-moderate asthmatics, 63.5% using ICS. 32 randomized to CNTO3157		CNTO3157 had no effect on FEV1, PEF, symptom scores, viral load, or FeNO; more moderate and severe asthma exacerbations reported in subjects receiving CNTO3157

Mallia et al. (2006)[112]	4 COPD, FEV_1 50%–80% predicted		Increased upper and lower respiratory symptoms, falls in PEF and FEV_1, increased nasal IL-8
Mallia et al. (2011)[113]	11 COPD, FEV_1 50%–80% predicted	12 smokers with normal lung function	Upper and lower respiratory symptoms, falls in PEF, sputum neutrophils and NE, BAL lymphocytes, nasal lavage virus load higher in COPD
Footitt et al. (2016)[114]	9 COPD, FEV_1 50%–80% predicted	10 smokers, 11 nonsmokers with normal lung function	Sputum inflammatory cells, neutrophils, NE, IL-1β, GM-CSF, IL-8, TNFα, MMP-9, 8-OHdG, 3-NT, nitrite and 8-isoprostane higher in COPD; sputum HDAC2 activity reduced in COPD

Key findings from studies of experimental RV infection in human subjects with asthma and COPD, including information on the patient and control populations assessed and key experimental findings. *3-NT*, 3-Nitrotyrosine; *AA*, people with atopic asthma; *AHR*, airway hyperreactivity; *ANA*, atopic nonasthmatic; *BAL*, bronchoalveolar lavage; *ECP*, eosinophil cationic protein; *EDN*, eosinophil derived neurotoxin; *FeNO*, fraction of exhaled nitric oxide; *FEV_1*, forced expiratory volume in 1 second; *G-CSF*, granulocyte colony stimulating factor; *GM-CSF*, granulocyte macrophage colony stimulating factor; *HDAC*, histone deacetylase; *ICAM-1*, intercellular adhesion molecule 1; *ICS*, inhaled corticosteroid; *IFN*, interferon; *IgE*, immunoglobulin E; *IL*, interleukin; *MMP*, matrix metalloprotease; *MPO*, myeloperoxidase; *NANA*, nonatopic nonasthmatic; *NE*, neutrophil elastase; *PEF*, peak expiratory flow; *RV*, rhinovirus; *SABA*, short-acting β_2-agonist; *TNFα*, tumor necrosis factor-alpha.

Perhaps the most promising use of the virus challenge model will be to accelerate the process of drug development.[137,138] Virus challenge studies have been used to evaluate the effects of existing asthma therapies on virus-induced exacerbations.[123,125] Recently the first study using viral challenge to evaluate a novel, unlicensed drug in asthma was also published.[126] Although these studies had negative results they demonstrate the potential of the viral challenge model in drug development.

The key to successful drug development is the identification of clinically relevant mechanisms of RV infection or immunopathology that can be experimentally manipulated for therapeutic benefit. This is where human experimental RV infection is complemented by work in animal models. There are a number of options for modeling human RV infection in animal models and they provide the ability to investigate specific disease components and mechanisms that would be otherwise impossible in humans. Human experimental infection models have the advantage of identifying disease correlates, but the degree of experimental manipulation possible is extremely limited and the safety and effectiveness of interventions must first be evaluated in animals.

8.6.1 Animal models of rhinovirus infection

Animal models have proven useful for mechanistic studies across a range of diseases. Models of RV infection have been reported in several animal species, including mouse, cotton rat, and nonhuman primates. Each of these experimental systems provides its own advantages and disadvantages and a substantial contribution to the knowledge base of the biology of RV infection.

Animal models provide a range of benefits to complement human experimental approaches.[139] Experimental animals can be readily manipulated to induce consistent disease outcomes, such as the induction of allergic airway disease (AAD) to model asthma or cigarette smoke–induced COPD. Animal models have less variability than human populations providing more consistent experimental outcomes and statistical power in intervention studies. Experimental environment, exposures (e.g., previous infection history), endpoints, and interindividual variability can be controlled. Further, a broad array of tools are available to characterize disease outcomes, including reagents, genetically modified animal strains, experimental protocols, and assessment techniques (e.g., lung function testing).

Sample tissues can be easily isolated at experimental endpoints, which are difficult or impossible to sample in humans (e.g., lung tissues, draining lymph nodes, bone marrow). Further, animal models serve as a valuable preclinical system for the assessment of novel interventions, to provide proof-of-principle safety and efficacy findings prior to exposure of healthy human volunteers.

8.7 MOUSE MODELS

Mice in particular have been extensively used to model disease, including virus infection and exacerbations of asthma and COPD.[138–140] As a result, a wealth of tools and techniques are available to study immune mechanisms and pathophysiology in mice. Well-characterized protocols and reagents support the induction of disease states and allow detailed characterization of immune responses (e.g., fluorescently tagged monoclonal antibodies to quantify immune cell subsets). A range of transgenic and knockout mouse strains are available that allow for the careful dissection of relevant disease mechanisms. Further, reagents are available to assess the effects of novel interventions on disease outcomes (e.g., blocking antibodies and various forms of innate immunity activators; see Chapter 9: Emerging therapeutic approaches).

The expansion of mouse RV infection models has paralleled our understanding of RV biology in humans.[141] Initial approaches focused on understanding the effects of RV infection in isolation, identifying the mechanisms and cell types mediating lung pathology. Increasingly complex experimental models are now being used to characterize long-term effects of infection on airway function and the effects of RV infection on preexisting airway disease (e.g., asthma exacerbations).[138–140]

One of the biggest developments in mouse RV infection models was the protocol for isolation of highly purified, concentrated (high titers) of RV from Henrietta Lacks (HeLa) immortalized human epithelial cell lines and later, the intracellular adhesion molecule 1 (ICAM-1) transfected rhabdomyosarcoma cell lines.[142,143] Prior to this, clarified infected HeLa cell lysates were used for in vitro experiments.[144] Some investigators also used infected HeLa cell lysates in mouse models.[145] Efforts to improve the quality and validity of the model made use of a partial purification protocol to generate viral stocks.[146] For mouse models of RV infection or exacerbations, the refined, high-titer, RV purification protocol is the gold standard.

8.8 ORIGINS OF RHINOVIRUS MOUSE MODELS

A major barrier that hindered the early development of mouse models was the species specificity of RV infection. "Major-group" RV strains, which make up approximately 90% of all RV strains, enter the cell through binding of ICAM-1.[147,148] RV binding to ICAM-1 is limited to human and chimpanzee and does not occur in other species, including mouse.[149] As a result, major-group RV strains cannot infect mouse cells and fails to replicate or induce pathology in mouse models. Early attempts to develop RV mouse models failed to detect sufficient viral replication to induce disease.[150]

A major advance enabling the development of mouse RV infection models was the initial recognition that the minor-group RV-A1, which use the host cell receptor low-density lipoprotein receptor, can infect the mouse epithelial cell line LA-4.[151] This recognition suggested that minor-group RV viruses (e.g., RV-A1) may be useful to model infections in mice in vivo. Indeed, inoculation of wild-type BALB/c mice with RV-A1 induced lung pathology, mucus production, and inflammatory cytokine production.[152] Notably, RV infection was also sufficient to induce exacerbations of preexisting asthma in sensitized and challenged mice (as discussed in more detail below).[152]

An alternate approach was also sought to allow modeling of major-group RV infection in mice. The same study by Tuthill et al. demonstrated that transfection of LA-4 cells with a chimeric ICAM-1 receptor containing the human extracellular receptor domains allowed infection and replication by the major-group virus RV-A16.[151] This finding provided the basis for developing a transgenic mouse strain expressing a chimeric mouse–human ICAM-1 receptor. Chimeric receptor expression in hu-ICAM^{Tg} mice is sufficient to support in vivo infection with RV-A16, resulting in airway inflammation, mucus production, viral replication, and inflammatory cytokine production.[152] Of note, the hu-ICAM^{Tg} mouse was generated by random insertion of a chimeric transgene and little is known about the transgene insertion site within the genome. Use of this transgenic strain requires additional experimental considerations (e.g., genotyping and use of heterozygous animals in experiments), which has limited its broad utility.

Additional variations have also been reported in the literature, which aim to broaden the available mouse models. Genetic RV-A1 variants have been generated by serial passage through mouse embryonic fibroblasts in vitro and lung epithelial cells in vivo, which exhibit increased growth in mouse cells.[153] Inoculation of BALB/c mice with the RV-A1/M2M7 variant [8 × 10^6 plaque forming units (PFU)] allowed for recovery of virus from mouse lung after 24 hours, when mice were also pretreated with intranasal hydrochlorous acid to increase epithelial permeability.[153]

Successful use of mouse models also depended on the development of streamlined RV isolation protocols that have allowed consistent and rapid isolation of virus stocks, to limit variability between experiments. The current gold-standard protocols for RV isolation, RV-A1 infection of wild-type BALB/c mice, and the infection of transgenic mice expressing chimeric mouse–human ICAM-1 receptor with RV-A16 are published and readily available.[141]

8.9 TECHNICAL DETAILS AND MAIN FINDINGS FROM MOUSE RHINOVIRUS INFECTION MODELS

A range of studies have assessed the effects of primary RV infection in mice, contributing to our understanding of the mechanisms underlying disease pathogenesis. Studies have used similar protocols, typically performing intranasal inoculation of ~10^6–10^8 $TCID_{50}$ (tissue culture infective dose in 50% of culture, a titration of infection units of pathogens that do not form plaques in culture) RV-A1 and assessing responses over 1 week following infection in BALB/c or C57Bl.6 mice.

In the initial publication by Bartlett et al., intranasal inoculation of wild-type BALB/c mice with 5 × 10^6 $TCID_{50}$ RV-A1 induced a range of disease features similar to human disease.[152] RV infection induced airway inflammation characterized by increased neutrophil numbers at 24 and 48 hours postinfection and increased lymphocyte numbers persisting for 1-week postinfection.[152] Tissue pathology was characterized by perivascular and peribronchial inflammation, increased mucus production, and elevated inflammatory cytokine production (including MIP-2, KC, MIP-3α, IP-10, RANTES, ITAC, IL-6, IL-1β, IFNα/β/λ/γ).[152] Furthermore, RV infection resulted in the development of an RV-specific adaptive antibody response by 7 days postinfection.[152]

Further studies have provided insights into the effects of RV infection alone in mice. Following inoculation of wild-type C57Bl/6 mice with 1 × 10^8 $TCID_{50}$ RV-A1, detectable RV positive- and negative-strand RNA were recovered from the lung, indicative of active viral replication and viral RNA was detectable up to 7 days postinfection.[145] The study also noticed a small increase in airway neutrophils and lymphocytes in the presence of UV-inactivated RV-A1, although UV-inactivated RV-A1 (and major-group virus RV-39) failed to induce inflammatory cytokine production.[145] RV-A1 infection also increased airway responsiveness to methacholine challenge at days 1 and 4 postinfection.[145] Inoculation with RV-A1 or UV-inactivated RV-A1 induced PI3K activation in airway epithelial cells and pretreatment with the PI3K inhibitor LY294002 in vivo dampened neutrophilic inflammation and inflammatory cytokine production (KC, MIP-2, MIP-1α, IFNγ).[145] Inoculation of C57Bl/6 mice with RV-A1 (5×10^7 $TCID_{50}$) leads to discontinuous expression of zonula occludens-1, suggesting that infection disrupts airway epithelial barrier function.[154] RV infection also stimulated IL-15 production and release into the airways, which is dependent on type I IFN production and stimulates activation of natural killer (NK) and $CD8^+$ T cell responses.[155] Treatment with an IL-15–IL-15Ra complex increased expression of IL-15, IL-15Rα, IFNγ, and CXCL9 and stimulated increased NK, $CD8^+$, and $CD4^+$ T cell recruitment and activation.[155] CCL7 and IRF-7 are the most upregulated lung transcripts following RV-A1 infection.[156] Blocking CCL7 or IRF-7 function reduced lung neutrophil and macrophage accumulation and IFN responses and blocking CCL7 also reduced AHR.[156] This publication also delineated AHR from inflammatory infiltrates showing instead a relationship between classical proinflammatory transcription factors (NFκB) and AHR.

As alluded to previously, the use of knockout mice in particular has provided insights into a number of mechanisms regulating RV-induced pathology. Key roles for neutrophils and the neutrophil chemokine receptor CXCR2 in mediating RV-induced pathology were identified. Inoculation of $CXCR2^{-/-}$ mice with RV-A1 (4.5×10^6 $TCID_{50}$) resulted in reduced airway neutrophil numbers, reduced inflammatory cytokine production (TNFα, MIP-2, KC), decreased mucus production, and decreased cholinergic responsiveness, with no alteration in viral load, compared with wild-type control animals.[157] Further, antibody depletion of neutrophils and infection of $TNFR^{-/-}$ mice also reduced AHR, compared with control animals.[157] These findings provide evidence for a role of neutrophilic inflammation, potentially via TNFα production, on downstream pathology following RV infection. Roles for pattern recognition molecules have been demonstrated for MDA5, Toll-like receptor 3 (TLR3) and TLR7. Infection of $MDA5^{-/-}$ mice resulted in delayed type I IFN (IFNα/β) and suppressed type III IFN expression, with a slight early increase in viral load in the lung.[158] In contrast, inoculation of

$TLR3^{-/-}$ mice resulted in normal IFN responses and no difference in viral yield.[158] Both $MDA5^{-/-}$ and $TLR3^{-/-}$ mice had reduced neutrophil numbers, inflammatory cytokine production (CXCL1, CXCL2, CCL2, CXCL10) and airways responsiveness, compared with wild-type controls.[158] Differing roles for NFκB signaling pathways in RV-induced inflammation and type I INF responses in antiviral immunity have been demonstrated. Disruption of NFκB signaling in $p65^{+/-}$ mice resulted in reduced neutrophil numbers and inflammatory cytokine production (CXCL1, CXCL5, CXCL2), while IFN production and viral loads are unaltered.[159] In contrast, $IFNAR1^{-/-}$ mice have unaltered neutrophilic inflammation, a persistent increase in lymphocyte numbers and cytokines CCL5, CXCL10, and CXCL11, with reduced IFNα production and increased viral load.[159] A pathogenic role for the proinflammatory molecule MUC18 has also been demonstrated, with increased expression of antiviral genes (Mx1, IP-10), reduced neutrophil inflammation and viral load in $MUC18^{-/-}$ mice following RV-A1 inoculation (1×10^7 PFU).[160] Studies using $Tbet^{-/-}$ mice (a key regulator of T_H1 cell differentiation) have also demonstrated the key role for T_H1-polarized T cells in the response to RV infection. $Tbet^{-/-}$ mice developed a T_H2/T_H17-polarized immune response to RV infection (5×10^6 $TCID_{50}$) with increased IL-13 and IL-17A production, deficient NK cell responses, and decreased neutralizing antibody development.[161] $CD4^+$ T cells contributed to increased airway eosinophil numbers and mucus production following RV infection in $Tbet^{-/-}$ mice.[161] Studies using TSLP receptor-deficient mice ($TSLPR^{-/-}$) demonstrated that RV-A1 infection interferes with tolerance to an inhaled allergen, via a mechanism requiring TSLP, IL-33, and activation of OX40L on lung dendritic cells.[162]

After observing increased levels of the TNF super family member protein, Tnfsf10 (TRAIL or CD253) production over a time course of RV-A1 infection in mice, Girkin et al. compared RV-A1 infection in $Tnfsf10^{-/-}$ mice to wild-type BALB/c mice and observed an almost complete ablation of inflammatory responses to RV-A1.[163] Following RV infection, peribronchiolar inflammation and lung histopathology were reduced in $Tnfsf10^{-/-}$ mice; neutrophil and lymphocytes in BAL remained at baseline; and $CD4^+$ T cells, $CD8^+$ T cells, NKs, plasmacytoid dendritic cells (pDCs), and myeloid dendritic cells were all reduced in flow cytometry of total lung cells.[163] $Tnfsf10^{-/-}$ mice were protected from RV-induced AHR, and failed to develop RV-induced exacerbations of allergic airways disease.[163] An interesting proviral effect of TRAIL was also identified whereby Tnfsf10$^{-/-}$ mice had reduced viral load and anti-TRAIL antibodies reduced viral load (whereas recombinant TRAIL administration increased viral load) in BEAS2B cells infected with RV-A1 in vitro.[163] This effect on viral load was independent of IFN responses and may be associated with an unidentified role of apoptosis in RV replication, which remains to be explored.

Some studies have assessed the effect of primary RV infection on clinically relevant sequelae, including secondary bacterial infection and the effects of premature birth on infection. Exposure of epithelial cells in culture to RV-A1 resulted in increased bacterial attachment and translocation through an epithelial monolayer (nontypeable *Haemophilus influenzae* (NTHi), *Pseudomonas aeruginosa*, *Staphylococcus aureus*),[154] suggesting a potential mechanism underlying secondary bacterial infections following viral infection. A subsequent study demonstrated that primary inoculation with RV-A1 (5×10^6 $TCID_{50}$) delayed the clearance of NTHi in vivo, associated with suppressed chemokine production (KC, MIP-2) and neutrophilic inflammation through a TLR2-mediated mechanism.[164] The model has also been used to assess immune alterations relevant to premature birth and bronchopulmonary dysplasia, risk factors for viral-induced exacerbations. Exposure of neonatal mice to hyperoxia (75% oxygen) in early life increased inflammatory cytokine expression (IL-12, IFNγ, TNFα, CCL2, CCL3, CCL4) and suppressed early IFN responses following RV-A1 infection (9×10^6 PFU) at 14 days of age.[165] One study has also assessed the effect of RV infection timing on subsequent development of AAD. Inoculation of 7-day-old mice with RV-A1 and subsequent induction of house dust mite (HDM)-induced allergic airways disease had additive effects with increased neutrophilia and AHR in female mice, although RV inoculation had no additional impact in male mice.[166] These studies extend the use of RV infection in mice to new areas, including mechanisms of early life infection susceptibility, to mechanisms of secondary bacterial infection/compromised antimicrobial immunity and experimental exploration of clinical risk factors associated with increased likelihood to develop virus-induced exacerbations of respiratory diseases.

8.10 PRECLINICAL TESTING IN MOUSE MODELS OF RHINOVIRUS INFECTION

Mouse models are valuable tools for the preclinical testing of novel treatments. Several studies have used the mouse RV infection model to assess intervention strategies, including vaccine development and drug treatment. Primary inoculation of BALB/c with RV-A1 (1×10^6 $TCID_{50}$) rapidly induced circulating RV-specific IgG antibody production within 4 days, which binds capsid protein VP1 and those antibodies were cross-reactive to another minor strain RV (RV-29).[167] Repeated RV infections were necessary to induce RV-specific IgA responses and neutralizing antibodies, but administration of CpG or subcutaneous immunization with Freund's adjuvant promoted neutralizing antibody development and may inform potential vaccine strategies.[167] Pretreatment with the plant flavanol quercetin before and during RV-A1 infection effectively reduced viral replication, inflammatory cytokine production (KC, MIP-2, TNFα, CCL2, IFNα, and IFNλ2), and AHR.[168] Treatment with the cancer therapeutic gemcitabine (20,20-difluorodeoxycytidine) reduced RV load, inflammatory cytokine levels (TNFα, IL-1β), and reduced lung lymphocyte numbers.[169] Treatment with corticosteroid therapy (fluticasone proprionate) suppressed IFN responses to RV and reduced airway inflammation, leading to increased mucus production and reduced antimicrobial responses.[170] Effects on viral load, mucin production, and antibacterial response could be reversed by administration of recombinant IFN-β.[170] Despite promising findings in mouse models, quercetin has not entered clinical trials for the treatment of RV infection, likely due to a previous randomized community clinical trial in 2010 that showed little benefit of quercetin supplementation on upper respiratory infections.[171] These findings may highlight the limitation of mouse models, which (while valuable) do not always fully recapitulate human disease mechanisms.

8.11 RHINOVIRUS-INDUCED DISEASE EXACERBATION MODELS IN MICE

Animal models to study RV-mediated exacerbations of airway disease have also been developed. These models combine experimental RV infection with models of airways disease, including asthma, COPD, and chronic rhinosinusitis. Models of asthma typically consist of administration of a sensitizing agent [e.g., ovalbumin (OVA) or HDM] and subsequent challenge in the airways to induce an eosinophilic, allergic airways disease. COPD is typically induced by prolonged and repeated exposure of mice to cigarette smoke or treatment with elastase. After airways disease is established, mice are then inoculated with RV to induce disease exacerbations. These models are explained and expanded upon in the following sections.

Researchers have also used double-stranded RNA (dsRNA) administration as a surrogate for virus infection to exacerbate preexisting disease (reviewed in Refs. [139,140]). However, these approaches fail to model the complexity of virus infection and are beyond the scope of the current book chapter.

8.11.1 Mouse asthma exacerbation models

Many studies have characterized the effects of RV infection on preexisting asthma and have provided insights into the immune cell types involved, key molecules, and responses to potential therapies. In the initial report of an RV (RV-A1) exacerbation model, OVA-sensitized and challenged BALB/c mice were inoculated with RV-A1 during the allergen challenge phase.[152] The combination of virus and allergen challenge increased airway neutrophil, eosinophil, and lymphocyte numbers; increased cytokine production (IL-4, IL-13, and IFNγ), increased AHR; and increased mucus gene expression.[152]

Subsequent studies have identified key functional roles for macrophages, gamma-delta (γδ) T cells, dendritic cell subsets, and neutrophils in RV-induced immunopathology. In a similar model, RV-A1 inoculation into OVA-sensitized/challenged mice increased macrophage lung infiltration and eotaxin-1/CCL11 expression.[172] Eotaxin was expressed by pulmonary macrophages in the lung after combined virus infection and allergen challenge.[172] Further, macrophage depletion or antieotaxin treatment reduced RV-induced airway eosinophilia and AHR.[172] Macrophage activation state also modulates the response to RV infection in allergen sensitized/challenged mice and shapes the resulting pattern of inflammation. RV infection in asthma exacerbation models induced an IL-13-expressing macrophage population, with M2 polarized phenotype.[173] Depletion of IL-13-producing cells in CD11b-DTR mice or $CCR2^{-/-}$ mice reduced airway inflammation and AHR.[173] Interestingly, while RV infection of OVA-treated wild-type mice contributes to mixed neutrophilic and eosinophilic airway inflammation and M2 macrophage phenotype, $IL\text{-}4R^{-/-}$ mice exhibit neutrophil inflammation alone and increased M2 polarization of pulmonary macrophages but still have exacerbated airway responses.[174] γδT cells dampen exacerbation responses. γδT cells are increased in RV-induced asthma exacerbation models and blocking responses with anti-γδTCR antibody worsened exacerbations with increased AHR, and increased numbers of T_H2 cells and eosinophils, with no effect on virus load.[175] pDCs were recruited to the lung during RV-induced inflammation and subsequently promoted T_H2 responses in the lung draining lymph nodes, in a process mediated by IL-25.[176] Depletion of pDCs with an antibody or treatment with anti-IL-25 reduced eosinophil numbers, decreased lung pathology, reduced cytokine production (IL-5, IL-13), and reduced AHR.[176] Functional roles for neutrophils, and neutrophil extracellular traps (NETs), have also been provided in RV-induced asthma exacerbation models. Chronic low-dose HDM exposure and RV infection have additive effects on neutrophilia and induce AHR.[177] A more recent study demonstrated that RV infection in an HDM-mediated asthma model results in double-stranded DNA release into the airways and administration of genomic DNA alone was sufficient to mimic characteristic components of RV-induced exacerbations.[178] Further, blocking neutrophil elastase or degrading NETs by applying DNase into the airways reduced eosinophil and lymphocyte numbers, tissue pathology, and cytokine production (IL-5, IL-13).[178]

A number of studies have highlighted the functional roles of specific molecules in RV-induced mouse exacerbation models, as potential therapeutic targets to validate in patient populations. In addition to roles during RV infection alone highlighted above, MDA5 and TLR3 are also involved in RV-induced exacerbations. $MDA5^{-/-}$ and $TLR3^{-/-}$ mice have decreased inflammatory responses and AHR, while $MDA5^{-/-}$ also had decreased IFN responses (IFNβ/λ2/λ3).[158] Midline 1 (a E3 ubiquitin ligase) is upregulated in an HDM-induced model, and short interfering RNA–mediated inhibition prior to RV inoculation reduced neutrophil numbers and mucus production production.[179] The monocyte chemotactic protein CCL2 is produced by epithelial cells and macrophages following RV-induced exacerbation and administration of an anti-CCL2 antibody reduced eosinophil numbers and AHR.[180] Foxa3-overexpressing transgenic mice produce excess mucus in their airways and RV infection increased Foxa3 expression.[181] In Foxa3-deficient mice ($Foxa3^{-/-}$), RV clearance is increased, with increased IFNβ activation.[181] IL-25 expression is also increased in RV-induced exacerbations and blocking the IL-25 receptor reduced type 2 cytokine production (IL-4, IL-5, IL-13, IL-25, IL-33, TSLP), mucus production, and numbers of eosinophils, neutrophils, T cells, and innate lymphoid type 2 cells.[99] Combining dsRNA administration with RV-A1 inoculation worsened preexisting allergic airways disease. Repeated dsRNA administration after OVA sensitization/challenge resulted in neutrophilic lung inflammation and tissue pathology and combined dsRNA and RV-A1 inoculation increased expression of TSLP, TNFα, and IFNλ in the lung.[182] Key roles for pattern recognition receptors have also been demonstrated in RV asthma exacerbation models. HDM-allergic TLR7$^{-/-}$ mice had a decreased antiviral response, with reduced IFN release (IFNα/β/λ1/λ2/λ3) and increased virus replication, associated with increased eosinophil and lymphocyte numbers, increased IL-5 and CCL11, and AHR.[183] Administration of IFN or transfer of wild-type TLR7-competent pDCs could restore antiviral responses and reduce disease exacerbation.[183] OVA-allergic TLR2$^{-/-}$ mice also had reduced macrophage, neutrophil, and eosinophil numbers and suppressed AHR after RV inoculation.[146] Bone marrow transfer experiments demonstrated that TLR2$^{-/-}$ bone marrow could protect from exacerbations, while transfer of wild-type bone marrow restored responses in TLR2$^{-/-}$ mice.[146] Transfer of wild-type macrophages into TLR2$^{-/-}$ mice could also restore exacerbations.[146] As previously mentioned, a role for TRAIL has also been demonstrated, with HDM-allergic TRAIL-deficient mice (*Tnfsf10*$^{-/-}$) protected from RV-induced AHR and induction of airway inflammation.[163]

RV-induced asthma exacerbation models have also been used to assess the responses to existing therapies and as preclinical models for novel therapies. Treatment of HDM-allergic mice with the long acting beta-2 agonist salmeterol reduced AHR and eosinophil numbers during RV exacerbation, and limited chemokine levels (CCL11, CCL20, CXCL2) through modulation of PP2A.[184] The findings of this study were focused on PP2A as a novel therapeutic target rather than promoting salmeterol monotherapy (which was associated with adverse events and tolerance to β2-agonists with chronic salmeterol use[185]). An approach to block major-group RV virus infection was assessed through administration of antihuman ICAM-1 antibody, which prevented entry of RV-A16 and RV-14 and reduced neutrophil and lymphocyte numbers, cytokine production (IL-4, IL-5, IL-6, CCL1, CCL11), mucus production, and virus load in human transgenic mice in an OVA-allergic model.[186] Treatment with a nontoxic anthraquinone derivate reduced RV-induced AHR, neutrophil, and eosinophil airway inflammation; inflammatory cytokine production; and mucus hypersecretion while also boosting type 1 IFN response and reducing viral yields, with associated decreased AKT, HIF-1α, and VEGF production.[187] Treatment with an antiinflammatory VAP-1/SSAO inhibitor, PXS-4728A, or the macrolide antibiotic azithromycin also reduced neutrophil numbers and PXS-4728A reduced AHR following RV-A1 inoculation in HDM-allergic mice.[188]

8.11.2 Mouse chronic obstructive pulmonary disease exacerbation models

RV also plays an important role in virus-induced exacerbations of COPD. Several studies have assessed the effects of RV inoculation in animal models of COPD. Exposure to elastase and lipopolysaccharide (LPS) once per week for 4 weeks induces features of COPD, including airway inflammation, goblet cell metaplasia, and altered lung function.[189] Addition of RV-A1 led to persistence of viral RNA ($>$14 days postinfection), deficient IFN responses (IFNα/β/γ) and increased AHR, lung volume, cytokine production (TNFα, IL-5, IL-13), and mucus production, compared with elastase/LPS administration alone.[189] A subsequent study attempting to replicate the elastase/LPS model of COPD found that a single elastase treatment followed by RV-A1 inoculation was enough to increase airway neutrophil and lymphocyte numbers, increased inflammatory cytokine production (TNFα, CXCL10, CCL5), mucus hypersecretion, and AHR.[190] In the same elastase-induced model, fluticasone proprionate treatment reduced IFN responses, increased viral load, suppressed airway immune cell numbers (lymphocytes and neutrophils), suppressed inflammatory cytokines (IL-6, TNFα), and increased mucus production, following RV-A1 exacerbation.[170] The differences in the experimental approaches required to elicit an RV-induced exacerbation in these different studies is likely due to the quality of virus inoculum used by the different investigators, which as explained previously, is influenced by purification approach. The first study used only crude virus-infected cell lysates, whereas the later study employed a highly purified virus inoculum.

Several studies have also reported on RV infection in a cigarette smoke–induced COPD model. In an initial study, 8 weeks of cigarette smoke exposure resulted in increased viral persistence, neutrophilia, and increased mucus production following RV infection.[191] Subsequent studies demonstrated that goblet cell gene expression was reduced following treatment with a gamma-secretase inhibitor (GSK L685,458) to limit NOTCH activation.[192] Further, supplementation of feed with quercetin reduced RV-induced lung inflammation (including neutrophilia), goblet cell metaplasia, and AHR.[193]

8.11.3 Mouse chronic sinusitis exacerbation models

To our knowledge, only one study has assessed the effect of RV infection in a chronic sinusitis model. A mouse model of chronic allergic rhinosinusitis was induced by 5 weeks of repetitive nasal OVA challenges.[194] Increased RV-A1 yields were reported in the nasal tissue of mice with rhinosinusitis, although inflammatory cytokine production and histopathology were unaffected.[194] This study served to illustrate the range of additional diseases where RV infection has been shown to be relevant in human populations where animal models are available for future research (e.g., cystic fibrosis).

8.12 OTHER RHINOVIRUS ANIMAL MODELS

A limited number of studies reported the use of RV infection in other animals, namely cotton rats and nonhuman primates. We note that historical studies assessing "RV" infection in other animal species are referring to genetically distinct viral genera and should not be confused with human RV (e.g., equine RV and bovine RV). For example, while human RV and equine RV were both originally assigned to the *Rhinovirus* genus, they have been reclassified into *Enterovirus* and *Apthovirus*, respectively. Equine RV has subsequently been renamed "equine rhinitis virus."

8.12.1 Cotton rat

The cotton rat (*Sigmodon hispidus*) is a recognized model for human respiratory infection, particularly for respiratory syncytial virus, as well as adenoviruses, parainfluenza virus, measles, and human metapneumovirus (reviewed in Ref. [195]). To date, two studies have reported on RV infection in cotton rats, providing evidence that cotton rats are partially permissive to RV major-group infection. Intranasal inoculation of RV-A16 (10^7 PFU) into cotton rats induced lower respiratory histopathology, increased mucus production, and induction of INF-activated genes.[196] Immunization with live RV-A16-induced high levels of circulating antibodies and protected from subsequent infection, while prophylactic transfer of anti-RV-A16 serum also protected from disease.[196] Further, this protection was transferred effectively from mother to newborn, limiting viral yields in subsequently infected progeny.[196] In a later study, the same group provided evidence that infection with RV-B14 (10^6 PFU) induced similar disease pathology. Furthermore, immunization with RV-B14 provided protection from subsequent infection with either RV-B14 (an RV-B group virus) or RV-A16 (an RV-A group virus), demonstrating some degree of cross-reactivity to very different major-group viruses.[197]

8.13 NONHUMAN PRIMATES

Chimpanzees and gibbons are the only nonhuman primates that support RV infection, although RV infection has also been reported in the vervet monkey cells, with consistent infection requiring high dose exposure.[198] Initial RV infection studies in chimpanzees were reported in 1968, using RV-B14 and RV-A43[199] and in gibbons in 1969.[200] Subsequent studies in chimpanzees and gibbons assessed the antiviral effects of drug treatments on RV infection, using bis-benzimidazole and triazinoindole, respectively.[201,202] Administration of soluble truncated form of human ICAM-1 can prevent subsequent infection in chimpanzees.[203] However, it has been noted that neither chimpanzees nor gibbons develop "cold" symptoms following RV infection and the high costs and logistics of these studies has limited further progress.

Chimpanzees are an endangered species and require considerable resources and facilities for research. Current chimpanzee research is limited to the United States and Gabon. However, the National Institutes of Health in the United States have indicated that they are seeking to eliminate the use of chimpanzees in research. All but one species of gibbon are endangered. Thus clinical research using nonhuman primates in the future to characterize RV infection are likely to be limited or nonexistent.

8.14 ANIMAL MODELS USING OTHER VIRUSES

As RV does not normally infect rodents, an attenuated mengovirus infection model has been proposed as an alternative option to model RV infection. Mengovirus also belongs to the *Picornaviridae* family and normally causes systemic infection in rodents. Using an attenuated mengovirus, intranasal inoculation of 10^7 PFU into rats increased airway neutrophil and lymphocyte numbers, induced lung tissue pathology, and increased expression of CXCL1 and CCL2.[204] A subsequent report using a genetically attenuated mengovirus vMC(0) in mice also induced lower respiratory tract infection with increased lung neutrophil and lymphocyte numbers, expression of CXCL1, CXCL2, CXCL5, IL-17A, INFs, and chemokines CXCL10 and CCL2.[205]

Other respiratory virus infections are associated with acute exacerbations of asthma and COPD, including respiratory syncytial virus, influenza, human coronavirus, human parainfluenza virus, human metapneumoviruses, and adenoviruses.[206] Animal models for these infections have largely been limited by the specificity of viruses to humans. It is unclear to what extent the mechanisms causing pathology differ between different viruses (or between strains of the same virus). A detailed discussion of the disease processes induced by each of these different virus infections is beyond the scope of this chapter. A detailed analysis of the relevant disease mechanisms in each infection setting is necessary to inform our understanding of disease exacerbations and ideally to identify common mechanisms between viruses that can be targeted for therapy.

8.14.1 Considerations, cautions, and limitations of animal infection models

No animal model can completely recapitulate naturally occurring human RV infection. While animal models provide important insights into disease mechanisms, it is important to also recognize their limitations.

There are recognized limitations of mice as models of human respiratory disease.[207] These include differences in response/symptoms between other species and humans. There are differences in respiratory tract architecture in human, nonhuman primate, and mouse airways. They range from dichotomas (each airway splits into two), trichotomas (airways split into threes), or monopodial branching (central airway continues while subordinate airways branch out) with differences in airflow inhomogeneities covered in detail by Miller et al.[208] There are also differences in mucus production processes in mouse compared with human airways. The short lifespans of laboratory animals do not capture the life-course of human disease, mice do not naturally develop asthma or COPD, and most current models of asthma represent eosinophilic, allergic patterns of disease. It remains unclear to what extent the current models and pathophysiology truly reflect human disease (particularly considering recognized heterogeneity of the human population).

RV has evolved for efficient replication in the human respiratory tract. Due to the decreased efficiency of RV entry into nonhuman epithelial cells (and likely differences in the nuances of cellular machinery required for replication), a high amount of viral load is required to elicit a biological response to RV in laboratory animals (e.g., 10^6 $TCID_{50}$ in mouse vs 5–10 $TCID_{50}$ in experimental human infection models). Human RV strains also demonstrate limited viral replication and different replication kinetic between mouse and man. These differences highlight the importance of confirming findings in relevant patient cohorts/samples and the utility of using human experimental models in parallel with animal models. This point is not purely for academic consideration. More so, it is important to take into consideration clinical trial design and outcomes. For example, mouse models highlighted the key relevance of IL-5 in asthma pathology through use of knockout mice[209] and antibody blockade.[210] However, the initial randomized control trial assessing anti-IL-5 therapy (mepolizumab) in a broad asthma population failed to demonstrate any clinical effect.[211] It was not until subsequent trials limited recruitment to patients with demonstrable eosinophilic asthma (a patient subset that is more closely modeled by the experimental mouse system) that clinical improvements were observed.[212,213]

8.14.2 Future directions for animal models

In a similar way to experimental human infection models, there has been a narrow focus in animal models. In mice, focus has largely been on RV infection alone with a growing body of literature assessing asthma exacerbations. While difficult to model in mice, RV-induced COPD exacerbations models are emerging through use of elastase administration and cigarette smoke–induced COPD. A summary of key findings from mouse models of RV-induced exacerbations of airway disease is presented in Table 8.2. Limited studies have reported on RV effects or potential interventions in these models. As with human experimental infections, animal infection models may also be relevant to an expanding array of diseases in the future (e.g., CF, bronchiectasis).

To date, there has been limited assessment across different RV strains in both animal and human studies. The primary focus of RV models has been on RV-A1 in mice, or RV-A16 and RV-A39 in human, possibly due to the availability of these strains and the ease of growing these strains in cell culture. In particular (due to its relatively recent discovery), RV-C infection has yet to be assessed in animal models. There has so far been difficulty in generating sufficient quantities of RV-C for research purposes (particularly at infectious titers required for mouse models). With the recent establishment of a suitable cell line (E8 HeLa cells) that supports RV-C replication this gap in the literature will likely be rectified.[214]

Table 8.2 Key experimental infection studies in mouse models of exacerbations

Mouse asthma exacerbations

Authors	Model	Interventions	Main findings
Toussaint et al. (2017)[178]	HDM + RV-A1 infection	DNAse and NETosis inhibition	Treatment suppressed type 2 immunopathology
Girkin et al. (2017)[163]	HDM + RV-A1	Deletion of TNF-related apoptosis inducing ligand ($Tnfsf10^{-/-}$)	Deletion suppressed cellular infiltration and AHR Reduced apoptotic cell death and reduced IFN- λ2/3
Han et al. (2016)[146]	OVA + RV-A1	Gene targeted deletion of TLR2 ($TLR2^{-/-}$)	Deletion reduced neutrophilic and eosinophilic inflammation and AHR
Hatchwell et al. (2015)[183]	HDM + RV-A1	Targeted deletion of TLR7 Exogenous IFN Adoptive transfer of TLR7-competent pDCs	Treatment attenuated eosinophilic inflammation and AHR
Phan et al. (2014)[177]	HDM + RV-A1	Nil	HDM and RV had additive increases in neutrophilia and tissue elastance
Hatchwell (2014)[184]	HDM + RV-A1	Salmeterol	Treatment reduced inflammation via increased PPA2 activity
Chen et al. (2014)[181]	HDM + RV-A1	Deletion of Foxa3 ($Foxa3^{-/-}$)	Deletion inhibited RV clearance
Beale et al. (2014)[99]	OVA + RV-A1	IL-25 receptor blockade	Treatment attenuated type 2 cytokine expression, mucus production and inflammatory cell recruitment
Hong et al. (2014)[174]	OVA + RV-A1	Gene targeted deletion of IL-4 receptor (IL-4R)	Deletion shifted from type 2 to type 1 response and increased neutrophilic inflammation
de Souza Alves (2013)[187]	HDM + RV-A1	Anthraquinone (PI3K-mediated AKT phosphorylation inhibitor)	Treatment reduced AHR, viral replication, neutrophilic and eosinophilic inflammation
Traub et al. (2013)[186]	OVA + RV-A16	Anti-ICAM-1	Treatment suppressed T_H2 cytokine/chemokine production
Glanville et al. (2013)[175]	OVA + RV-A1	Anti-gamma-delta-T-cell receptor antibody	Treatment increased T_H2 inflammation and AHR
Collison et al. (2013)[179]	HDM + RV-A1	Inhibition of the E3 ubiquitin ligase MID1	Treatment suppressed allergic airway inflammation and AHR
Schneider et al. (2013)[180]	OVA + RV-A1	CCL2 neutralizing antibody	Treatment reduced airway inflammation and AHR
Nagarkar et al. (2010)[172]	OVA + RV-A1	Antieotaxin-1	Treatment reduced airway eosinophilia and AHR
Bartlett et al. (2008)[152]	OVA + RV-A1	Nil	RV infection exacerbated allergic airway inflammation and AHR

Mouse COPD exacerbations

Jing et al. (2018)[192]	Cigarette smoke + RV-A1	Gamma-secretase inhibitor (Notch inhibitor)	Attenuated mucin expression
Farazuddin et al. (2018)[193]	Cigarette smoke + RV-A1	Quercetin	Reduced inflammation, goblet cell metaplasia, and AHR
Singanayagam et al. (2018)[170]	Elastase + RV-A1	Fluticasone propionate	Suppressed antiviral immunity Impaired virus clearance Mucus hypersecretion Increased bacterial loads
Singanayagam et al. (2015)[190]	Elastase + RV-A1	Nil	Enhanced airway inflammation Increased mucus production Exaggerated AHR
Ganesan, et al. (2014)[191]	Cigarette smoke + RV-A1	Nil	Viral persistence Increased neutrophilia Increased mucus production
Sajjan et al. (2009)[189]	Elastase/LPS + RV-A1	Nil	Deficient antiviral immunity Increased inflammation Exaggerated AHR Increased mucin expression

Key findings from studies of RV infection in mouse models with underlying allergic airway disease (asthma) and COPD, including information on the approach used, interventions where applicable, and key experimental findings. *AHR*, airway hyperresponsiveness; *HDM*, house dust mite; *ICAM-1*, intracellular adhesion molecule 1; *IL*, interleukin; *IFN*, interferon; *LPS*, lipopolysaccharide; *MID1*, midline 1; *OVA*, ovalbumin; *pDCs*, plasmacytoid dendritic cells; *RV*, rhinovirus; T_H2, T-helper type 2; *TLR*, Toll-like receptor.

As is the case for the majority of human virus–mouse infection models, the mouse is semipermissive to RV infection and as such a high-titer inoculum is required to induce prolonged replication and robust, reproducible host immune responses. A mouse-adapted RV strain (RV-1BM) has been generated by serial passage in mouse epithelial cells (LA-4 cells) though this mouse-adapted virus has only been characterized in vitro with primary mouse tracheal epithelial cells[215] and has not yet been tested in vivo.

The clinical translation of novel therapies identified in animal models for the treatment of RV infection in humans is yet to come to fruition. However, there are multiple molecules currently in the drug development pipeline, ranging from virus-targeting molecules, drugs targeting host factors of the viral replication cycle, and biologics such as innate immune stimulators and cytokine blocking monoclonal antibodies, all of which are elaborated on in Chapter 9, Emerging therapeutic approaches.

8.15 CONCLUSION

There are significant opportunities for further research in both human and nonhuman models, including assessment of infections in various unexplored disease backgrounds that are exacerbated by RV infection (e.g., cystic fibrosis) and expansion of studies using newly identified RV strains (e.g., RV-C strains). Human and mouse RV experimental infection models effectively complement each other and have contributed immensely to our understanding the mechanisms shaping RV-induced pathology. Human experimental RV challenge studies have shed light on the biology of RV infection and the mechanisms associated with RV-induced exacerbations of chronic respiratory diseases. Mouse models of RV infection in particular are readily manipulatable to identify cause and effect between specific molecules and disease outcomes for preclinical testing. An excellent example of how human and mouse models complement each other is the growing understanding of the disease mechanisms during RV-induced asthma exacerbations. Human experimental infection revealed a potential role for induced type 2 immunity following RV infection in individuals with asthma.[90,98] Subsequent mouse model studies have demonstrated a causal role for RV-induced type 2 immune effector molecules in exacerbations,[99,152,176,179] allowing preclinical assessment of the efficacy and safety of novel therapies. Findings from these studies have not yet resulted in the development of approved therapies for RV infections, cold symptoms, or exacerbations of respiratory diseases associated with RV infection. However, the wealth of knowledge derived from experimental RV infections has broadened our understanding and identified many potential therapeutic approaches.

REFERENCES

1. Kruse. *Münchener medizinische Wochenschrift*. 1914;61:1547.
2. Ma X, Conrad T, Alchikh M, Reiche J, Schweiger B, Rath B. Can we distinguish respiratory viral infections based on clinical features? A prospective pediatric cohort compared to systematic literature review. *Rev Med Virol*. 2018;28:e1997.
3. Thiberville SD, Ninove L, Vu Hai V, et al. The viral etiology of an influenza-like illness during the 2009 pandemic. *J Med Virol*. 2012;84(7):1071–1079.
4. Fullen DJ, Murray B, Mori J, et al. A tool for investigating asthma and COPD exacerbations: a newly manufactured and well characterised GMP wild-type human rhinovirus for use in the human viral challenge model. *PLoS One*. 2016;11(12): e0166113.
5. Couch RB, Cate TR, Douglas Jr. RG, Gerone PJ, Knight V. Effect of route of inoculation on experimental respiratory viral disease in volunteers and evidence for airborne transmission. *Bacteriol Rev*. 1966;30(3):517–529.
6. Dick EC, Jennings LC, Mink KA, Wartgow CD, Inhorn SL. Aerosol transmission of rhinovirus colds. *J Infect Dis*. 1987;156(3):442–448.
7. Gwaltney Jr. JM, Hendley JO. Transmission of experimental rhinovirus infection by contaminated surfaces. *Am J Epidemiol*. 1982;116(5):828–833.
8. Douglas Jr RG. Pathogenesis of rhinovirus common colds in human volunteers. *Ann Otol Rhinol Laryngol*. 1970;79(3):563–571.
9. Halperin SA, Eggleston PA, Hendley JO, Suratt PM, Groschel DH, Gwaltney Jr. JM. Pathogenesis of lower respiratory tract symptoms in experimental rhinovirus infection. *Am Rev Respir Dis*. 1983;128(5):806–810.
10. Bush RK, Busse W, Flaherty D, Warshauer D, Dick EC, Reed CE. Effects of experimental rhinovirus 16 infection on airways and leukocyte function in normal subjects. *J Allergy Clin Immunol*. 1978;61(2):80–87.
11. Cate TR, Couch RB, Fleet WF, Griffith WR, Gerone PJ, Knight V. Production of tracheobronchitis in volunteers with rhinovirus in a small-particle aerosol. *Am J Epidemiol*. 1965;81:95–105.
12. Hendley JO, Edmondson Jr WP, Gwaltney Jr JM. Relation between naturally acquired immunity and infectivity of two rhinoviruses in volunteers. *J Infect Dis*. 1972;125(3):243–248.
13. Winther B, Buchert D, Turner RB, Hendley JO, Tschaikin M. Decreased rhinovirus shedding after intranasal oxymetazoline application in adults with induced colds compared with intranasal saline. *Am J Rhinol Allergy*. 2010;24(5):374–377.
14. Turner RB, Bauer R, Woelkart K, Hulsey TC, Gangemi JD. An evaluation of *Echinacea angustifolia* in experimental rhinovirus infections. *N Engl J Med*. 2005;353 (4):341–348.
15. Turner RB, Riker DK, Gangemi JD. Ineffectiveness of *Echinacea* for prevention of experimental rhinovirus colds. *Antimicrob Agents Chemother*. 2000;44(6):1708–1709.
16. Sperber SJ, Hendley JO, Hayden FG, Riker DK, Sorrentino JV, Gwaltney Jr JM. Effects of naproxen on experimental rhinovirus colds. A randomized, double-blind, controlled trial. *Ann Intern Med*. 1992;117(1):37–41.

17. Sperber SJ, Hunger SB, Schwartz B, Pestka S. Anti-rhinoviral activity of recombinant and hybrid species of interferon alpha. *Antiviral Res.* 1993;22(2–3):121–129.
18. Sperber SJ, Shah LP, Gilbert RD, Ritchey TW, Monto AS. *Echinacea purpurea* for prevention of experimental rhinovirus colds. *Clin Infect Dis.* 2004;38(10):1367–1371.
19. Sperber SJ, Sorrentino JV, Riker DK, Hayden FG. Evaluation of an alpha agonist alone and in combination with a nonsteroidal antiinflammatory agent in the treatment of experimental rhinovirus colds. *Bull NY Acad Med.* 1989;65(1):145–160.
20. Turner RB, Sperber SJ, Sorrentino JV, et al. Effectiveness of clemastine fumarate for treatment of rhinorrhea and sneezing associated with the common cold. *Clin Infect Dis.* 1997;25(4):824–830.
21. Hayden FG, Turner RB, Gwaltney JM, et al. Phase II, randomized, double-blind, placebo-controlled studies of ruprintrivir nasal spray 2-percent suspension for prevention and treatment of experimentally induced rhinovirus colds in healthy volunteers. *Antimicrob Agents Chemother.* 2003;47(12):3907–3916.
22. Turner RB, Cetnarowski WE. Effect of treatment with zinc gluconate or zinc acetate on experimental and natural colds. *Clin Infect Dis.* 2000;31(5):1202–1208.
23. Abisheganaden JA, Avila PC, Kishiyama JL, et al. Effect of clarithromycin on experimental rhinovirus-16 colds: a randomized, double-blind, controlled trial. *Am J Med.* 2000;108(6):453–459.
24. Turner RB, Wecker MT, Pohl G, et al. Efficacy of tremacamra, a soluble intercellular adhesion molecule 1, for experimental rhinovirus infection: a randomized clinical trial. *JAMA.* 1999;281(19):1797–1804.
25. Pitkaranta A, Wecker MT, Korts DC, Hayden FG. Combined intranasal ipratropium bromide and oxymetazoline in experimental rhinovirus infection. *Am J Rhinol.* 1998;12(2):125–129.
26. Gwaltney Jr JM, Druce HM. Efficacy of brompheniramine maleate for the treatment of rhinovirus colds. *Clin Infect Dis.* 1997;25(5):1188–1194.
27. Gwaltney Jr JM, Park J, Paul RA, Edelman DA, O'Connor RR, Turner RB. Randomized controlled trial of clemastine fumarate for treatment of experimental rhinovirus colds. *Clin Infect Dis.* 1996;22(4):656–662.
28. Gustafson LM, Proud D, Hendley JO, Hayden FG, Gwaltney Jr JM. Oral prednisone therapy in experimental rhinovirus infections. *J Allergy Clin Immunol.* 1996;97(4):1009–1014.
29. Doyle WJ, Riker DK, McBride TP, et al. Therapeutic effects of an anticholinergic-sympathomimetic combination in induced rhinovirus colds. *Ann Otol Rhinol Laryngol.* 1993;102(7):521–527.
30. Turner RB, Dutko FJ, Goldstein NH, Lockwood G, Hayden FG. Efficacy of oral WIN 54954 for prophylaxis of experimental rhinovirus infection. *Antimicrob Agents Chemother.* 1993;37(2):297–300.
31. al-Nakib W, Higgins PG, Barrow GI, et al. Suppression of colds in human volunteers challenged with rhinovirus by a new synthetic drug (R61837). *Antimicrob Agents Chemother.* 1989;33(4):522–525.
32. Hayden FG, Andries K, Janssen PA. Safety and efficacy of intranasal pirodavir (R77975) in experimental rhinovirus infection. *Antimicrob Agents Chemother.* 1992;36(4):727–732.
33. Hayden FG, Hipskind GJ, Woerner DH, et al. Intranasal pirodavir (R77,975) treatment of rhinovirus colds. *Antimicrob Agents Chemother.* 1995;39(2):290–294.
34. Higgins PG, al-Nakib W, Barrow GI, Tyrrell DA. Recombinant human interferon-gamma as prophylaxis against rhinovirus colds in volunteers. *J Interferon Res.* 1988;8(5):591–596.

35. Higgins PG, Al-Nakib W, Willman J, Tyrrell DA. Interferon-beta ser as prophylaxis against experimental rhinovirus infection in volunteers. *J Interferon Res.* 1986;6 (2):153–159.
36. Higgins PG, Barrow GI, al-Nakib W, Tyrrell DA, DeLong DC, Lenox-Smith I. Failure to demonstrate synergy between interferon-alpha and a synthetic antiviral, enviroxime, in rhinovirus infections in volunteers. *Antiviral Res.* 1988;10 (1–3):141–149.
37. Higgins PG, Barrow GI, Tyrrell DA. A study of the efficacy of the bradykinin antagonist, NPC 567, in rhinovirus infections in human volunteers. *Antiviral Res.* 1990;14 (6):339–344.
38. Higgins PG, Barrow GI, Tyrrell DA, Isaacs D, Gauci CL. The efficacy of intranasal interferon alpha-2a in respiratory syncytial virus infection in volunteers. *Antiviral Res.* 1990;14(1):3–10.
39. Hsia J, Simon GL, Higgins N, Goldstein AL, Hayden FG. Immune modulation by aspirin during experimental rhinovirus colds. *Bull NY Acad Med.* 1989;65(1):45–56.
40. Gaffey MJ, Gwaltney Jr JM, Dressler WE, Sorrentino JV, Hayden FG. Intranasally administered atropine methonitrate treatment of experimental rhinovirus colds. *Am Rev Respir Dis.* 1987;135(1):241–244.
41. Gaffey MJ, Gwaltney Jr JM, Sastre A, Dressler WE, Sorrentino JV, Hayden FG. Intranasally and orally administered antihistamine treatment of experimental rhinovirus colds. *Am Rev Respir Dis.* 1987;136(3):556–560.
42. Gaffey MJ, Hayden FG, Boyd JC, Gwaltney Jr JM. Ipratropium bromide treatment of experimental rhinovirus infection. *Antimicrob Agents Chemother.* 1988;32 (11):1644–1647.
43. Hayden FG, Gwaltney Jr JM, Colonno RJ. Modification of experimental rhinovirus colds by receptor blockade. *Antiviral Res.* 1988;9(4):233–247.
44. Al-Nakib W, Higgins PG, Barrow I, Tyrrell DA, Lenox-Smith I, Ishitsuka H. Intranasal chalcone, Ro 09-0410, as prophylaxis against rhinovirus infection in human volunteers. *J Antimicrob Chemother.* 1987;20(6):887–892.
45. Al-Nakib W, Willman J, Higgins PG, Tyrrell DA, Shepherd WM, Freestone DS. Failure of intranasally administered 4', 6-dichloroflavan to protect against rhinovirus infection in man. *Arch Virol.* 1987;92(3–4):255–260.
46. Phillpotts RJ, Wallace J, Tyrrell DA, Freestone DS, Shepherd WM. Failure of oral 4′,6-dichloroflavan to protect against rhinovirus infection in man. *Arch Virol.* 1983;75 (1–2):115–121.
47. Phillpotts RJ, Wallace J, Tyrrell DA, Tagart VB. Therapeutic activity of enviroxime against rhinovirus infection in volunteers. *Antimicrob Agents Chemother.* 1983;23 (5):671–675.
48. Pachuta DM, Togo Y, Hornick RB, Schwartz AR, Tominaga S. Evaluation of isoprinosine in experimental human rhinovirus infection. *Antimicrob Agents Chemother.* 1974;5(4):403–408.
49. Togo Y, Durr FE, Laurenzana DA. Clinical evaluation of prophylactic intranasal 1-phenyl-3-(4-phenyl-2-thiazolyl) guanidine (CL 88,277) medication against rhinovirus 44 challenge. *Med Microbiol Immunol.* 1977;163(1):37–44.
50. Togo Y, Schwartz AR, Hornick RB. Antiviral effect of 3, 4-dihydro-1-isoquinolineacetamide hydrochloride in experimental human rhinovirus infection. *Antimicrob Agents Chemother.* 1973;4(6):612–616.
51. Togo Y, Schwartz AR, Hornick RB. Failure of a 3-substituted triazinoindole in the prevention of experimental human rhinovirus infection. *Chemotherapy.* 1973;18 (1):17–26.
52. George RB, Mogabgab WJ. Atypical pneumonia in young men with rhinovirus infections. *Ann Intern Med.* 1969;71(6):1073–1078.

53. Buscho RO, Saxtan D, Shultz PS, Finch E, Mufson MA. Infections with viruses and Mycoplasma pneumoniae during exacerbations of chronic bronchitis. *J Infect Dis.* 1978;137(4):377–383.
54. Mitchell I, Inglis JM, Simpson H. Viral infection as a precipitant of wheeze in children. Combined home and hospital study. *Arch Dis Child.* 1978;53(2):106–111.
55. Beasley R, Coleman ED, Hermon Y, Holst PE, O'Donnell TV, Tobias M. Viral respiratory tract infection and exacerbations of asthma in adult patients. *Thorax.* 1988;43(9):679–683.
56. Jain S, Self WH, Wunderink RG, et al. Community-acquired pneumonia requiring hospitalization among U.S. adults. *N Engl J Med.* 2015;373(5):415–427.
57. Mansbach JM, Piedra PA, Teach SJ, et al. Prospective multicenter study of viral etiology and hospital length of stay in children with severe bronchiolitis. *Arch Pediatr Adolesc Med.* 2012;166(8):700–706.
58. Pitkaranta A, Arruda E, Malmberg H, Hayden FG. Detection of rhinovirus in sinus brushings of patients with acute community-acquired sinusitis by reverse transcription-PCR. *J Clin Microbiol.* 1997;35(7):1791–1793.
59. Chang ML, Jordan-Villegas A, Evans A, et al. Respiratory viruses identified in an urban children's hospital emergency department during the 2009 influenza A(H1N1) pandemic. *Pediatr Emerg Care.* 2012;28(10):990–997.
60. Johnston SL, Pattemore PK, Sanderson G, et al. Community study of role of viral infections in exacerbations of asthma in 9-11 year old children. *BMJ.* 1995;310 (6989):1225–1229.
61. Zwaans WA, Mallia P, van Winden ME, Rohde GG. The relevance of respiratory viral infections in the exacerbations of chronic obstructive pulmonary disease—a systematic review. *J Clin Virol.* 2014;61(2):181–188.
62. GBD 2015 Chronic Respiratory Disease Collaborators. Global, regional, and national deaths, prevalence, disability-adjusted life years, and years lived with disability for chronic obstructive pulmonary disease and asthma, 1990-2015: a systematic analysis for the Global Burden of Disease Study 2015. *Lancet Respir Med.* 2017;5 (9):691–706.
63. Wark PA, Johnston SL, Moric I, Simpson JL, Hensley MJ, Gibson PG. Neutrophil degranulation and cell lysis is associated with clinical severity in virus-induced asthma. *Eur Respir J.* 2002;19(1):68–75.
64. Grissell TV, Powell H, Shafren DR, et al. Interleukin-10 gene expression in acute virus-induced asthma. *Am J Respir Crit Care Med.* 2005;172(4):433–439.
65. Arden KE, Chang AB, Lambert SB, Nissen MD, Sloots TP, Mackay IM. Newly identified respiratory viruses in children with asthma exacerbation not requiring admission to hospital. *J Med Virol.* 2010;82(8):1458–1461.
66. Halperin SA, Eggleston PA, Beasley P, et al. Exacerbations of asthma in adults during experimental rhinovirus infection. *Am Rev Respir Dis.* 1985;132(5):976–980.
67. Lemanske Jr RF, Dick EC, Swenson CA, Vrtis RF, Busse WW. Rhinovirus upper respiratory infection increases airway hyperreactivity and late asthmatic reactions. *J Clin Invest.* 1989;83(1):1–10.
68. Calhoun WJ, Swenson CA, Dick EC, Schwartz LB, Lemanske Jr RF, Busse WW. Experimental rhinovirus 16 infection potentiates histamine release after antigen bronchoprovocation in allergic subjects. *Am Rev Respir Dis.* 1991;144(6):1267–1273.
69. Bardin PG, Fraenkel DJ, Sanderson G, et al. Amplified rhinovirus colds in atopic subjects. *Clin Exp Allergy.* 1994;24(5):457–464.
70. Fraenkel DJ, Bardin PG, Sanderson G, Lampe F, Johnston SL, Holgate ST. Lower airways inflammation during rhinovirus colds in normal and in asthmatic subjects. *Am J Respir Crit Care Med.* 1995;151(3 Pt 1):879–886.

71. Cheung D, Dick EC, Timmers MC, de Klerk EP, Spaan WJ, Sterk PJ. Rhinovirus inhalation causes long-lasting excessive airway narrowing in response to methacholine in asthmatic subjects in vivo. *Am J Respir Crit Care Med.* 1995;152(5 Pt 1):1490–1496.
72. Grunberg K, Timmers MC, Smits HH, et al. Effect of experimental rhinovirus 16 colds on airway hyperresponsiveness to histamine and interleukin-8 in nasal lavage in asthmatic subjects in vivo. *Clin Exp Allergy.* 1997;27(1):36–45.
73. Grunberg K, Kuijpers EA, de Klerk EP, et al. Effects of experimental rhinovirus 16 infection on airway hyperresponsiveness to bradykinin in asthmatic subjects in vivo. *Am J Respir Crit Care Med.* 1997;155(3):833–838.
74. Grunberg K, Sharon RF, Hiltermann TJ, et al. Experimental rhinovirus 16 infection increases intercellular adhesion molecule-1 expression in bronchial epithelium of asthmatics regardless of inhaled steroid treatment. *Clin Exp Allergy.* 2000;30 (7):1015–1023.
75. Gern JE, Galagan DM, Jarjour NN, Dick EC, Busse WW. Detection of rhinovirus RNA in lower airway cells during experimentally induced infection. *Am J Respir Crit Care Med.* 1997;155(3):1159–1161.
76. Parry DE, Busse WW, Sukow KA, Dick CR, Swenson C, Gern JE. Rhinovirus-induced PBMC responses and outcome of experimental infection in allergic subjects. *J Allergy Clin Immunol.* 2000;105(4):692–698.
77. Grunberg K, Timmers MC, de Klerk EP, Dick EC, Sterk PJ. Experimental rhinovirus 16 infection causes variable airway obstruction in subjects with atopic asthma. *Am J Respir Crit Care Med.* 1999;160(4):1375–1380.
78. Bardin PG, Fraenkel DJ, Sanderson G, van Schalkwyk EM, Holgate ST, Johnston SL. Peak expiratory flow changes during experimental rhinovirus infection. *Eur Respir J.* 2000;16(5):980–985.
79. de Gouw HW, Grunberg K, Schot R, Kroes AC, Dick EC, Sterk PJ. Relationship between exhaled nitric oxide and airway hyperresponsiveness following experimental rhinovirus infection in asthmatic subjects. *Eur Respir J.* 1998;11(1):126–132.
80. Grunberg K, Smits HH, Timmers MC, et al. Experimental rhinovirus 16 infection. Effects on cell differentials and soluble markers in sputum in asthmatic subjects. *Am J Respir Crit Care Med.* 1997;156(2 Pt 1):609–616.
81. Jarjour NN, Gern JE, Kelly EA, Swenson CA, Dick CR, Busse WW. The effect of an experimental rhinovirus 16 infection on bronchial lavage neutrophils. *J Allergy Clin Immunol.* 2000;105(6 Pt 1):1169–1177.
82. Gern JE, Vrtis R, Grindle KA, Swenson C, Busse WW. Relationship of upper and lower airway cytokines to outcome of experimental rhinovirus infection. *Am J Respir Crit Care Med.* 2000;162(6):2226–2231.
83. Christiansen SC, Eddleston J, Bengtson SH, et al. Experimental rhinovirus infection increases human tissue kallikrein activation in allergic subjects. *Int Arch Allergy Immunol.* 2008;147(4):299–304.
84. Majoor CJ, van de Pol MA, Kamphuisen PW, et al. Evaluation of coagulation activation after rhinovirus infection in patients with asthma and healthy control subjects: an observational study. *Respir Res.* 2014;15:14.
85. Mosser AG, Vrtis R, Burchell L, et al. Quantitative and qualitative analysis of rhinovirus infection in bronchial tissues. *Am J Respir Crit Care Med.* 2005;171(6):645–651.
86. Corne JM, Marshall C, Smith S, et al. Frequency, severity, and duration of rhinovirus infections in asthmatic and non-asthmatic individuals: a longitudinal cohort study. *Lancet.* 2002;359(9309):831–834.
87. Pizzichini MM, Pizzichini E, Efthimiadis A, et al. Asthma and natural colds. Inflammatory indices in induced sputum: a feasibility study. *Am J Respir Crit Care Med.* 1998;158(4):1178–1184.

88. de Kluijver J, Grunberg K, Pons D, et al. Interleukin-1beta and interleukin-1ra levels in nasal lavages during experimental rhinovirus infection in asthmatic and non-asthmatic subjects. *Clin Exp Allergy*. 2003;33(10):1415–1418.
89. Zambrano JC, Carper HT, Rakes GP, et al. Experimental rhinovirus challenges in adults with mild asthma: response to infection in relation to IgE. *J Allergy Clin Immunol*. 2003;111(5):1008–1016.
90. Message SD, Laza-Stanca V, Mallia P, et al. Rhinovirus-induced lower respiratory illness is increased in asthma and related to virus load and Th1/2 cytokine and IL-10 production. *Proc Natl Acad Sci USA*. 2008;105(36):13562–13567.
91. Contoli M, Message SD, Laza-Stanca V, et al. Role of deficient type III interferon-lambda production in asthma exacerbations. *Nat Med*. 2006;12(9):1023–1026.
92. Sykes A, Edwards MR, Macintyre J, et al. Rhinovirus 16-induced IFN-alpha and IFN-beta are deficient in bronchoalveolar lavage cells in asthmatic patients. *J Allergy Clin Immunol*. 2012;129(6):1506–1514.e6.
93. Parsons KS, Hsu AC, Wark PA. TLR3 and MDA5 signalling, although not expression, is impaired in asthmatic epithelial cells in response to rhinovirus infection. *Clin Exp Allergy*. 2014;44(1):91–101.
94. Sykes A, Edwards MR, Macintyre J, et al. TLR3, TLR4 and TLRs7-9 induced interferons are not impaired in airway and blood cells in well controlled asthma. *PLoS One*. 2013;8(6):e65921.
95. Sykes A, Macintyre J, Edwards MR, et al. Rhinovirus-induced interferon production is not deficient in well controlled asthma. *Thorax*. 2014;69(3):240–246.
96. DeMore JP, Weisshaar EH, Vrtis RF, et al. Similar colds in subjects with allergic asthma and nonatopic subjects after inoculation with rhinovirus-16. *J Allergy Clin Immunol*. 2009;124(2):245–252. 252 e241–e243.
97. Adura PT, Reed E, Macintyre J, et al. Experimental rhinovirus 16 infection in moderate asthmatics on inhaled corticosteroids. *Eur Respir J*. 2014;43(4):1186–1189.
98. Jackson DJ, Makrinioti H, Rana BM, et al. IL-33-dependent type 2 inflammation during rhinovirus-induced asthma exacerbations in vivo. *Am J Respir Crit Care Med*. 2014;190(12):1373–1382.
99. Beale J, Jayaraman A, Jackson DJ, et al. Rhinovirus-induced IL-25 in asthma exacerbation drives type 2 immunity and allergic pulmonary inflammation. *Sci Transl Med*. 2014;6(256):256ra134.
100. Jackson DJ, Glanville N, Trujillo-Torralbo MB, et al. Interleukin-18 is associated with protection against rhinovirus-induced colds and asthma exacerbations. *Clin Infect Dis*. 2015;60(10):1528–1531.
101. Jackson DJ, Trujillo-Torralbo MB, Del-Rosario J, et al. The influence of asthma control on the severity of virus-induced asthma exacerbations. *J Allergy Clin Immunol*. 2015;136(2):497–500.e3.
102. Djukanovic R, Harrison T, Johnston SL, et al. The effect of inhaled IFN-beta on worsening of asthma symptoms caused by viral infections. A randomized trial. *Am J Respir Crit Care Med*. 2014;190(2):145–154.
103. Vogelmeier CF, Criner GJ, Martinez FJ, et al. Global strategy for the diagnosis, management, and prevention of chronic obstructive lung disease 2017 report: GOLD executive summary. *Am J Respir Crit Care Med*. 2017;195(5):557–582.
104. Halpin DM, Miravitlles M, Metzdorf N, Celli B. Impact and prevention of severe exacerbations of COPD: a review of the evidence. *Int J Chron Obstruct Pulmon Dis*. 2017;12:2891–2908.
105. Beasley V, Joshi PV, Singanayagam A, Molyneaux PL, Johnston SL, Mallia P. Lung microbiology and exacerbations in COPD. *Int J Chron Obstruct Pulmon Dis*. 2012;7:555–569.

106. Amin AN, Bollu V, Stensland MD, Netzer L, Ganapathy V. Treatment patterns for patients hospitalized with chronic obstructive pulmonary disease. *Am J Health Syst Pharm*. 2018;75(6):359–366.
107. Stenhouse AC. Rhinovirus infection in acute exacerbations of chronic bronchitis: a controlled prospective study. *Br Med J*. 1967;3(5563):461–463.
108. Wilkinson TMA, Aris E, Bourne S, et al. A prospective, observational cohort study of the seasonal dynamics of airway pathogens in the aetiology of exacerbations in COPD. *Thorax*. 2017;72(10):919–927.
109. Biancardi E, Fennell M, Rawlinson W, Thomas PS. Viruses are frequently present as the infecting agent in acute exacerbations of chronic obstructive pulmonary disease in patients presenting to hospital. *Intern Med J*. 2016;46(10):1160–1165.
110. Hosseini SS, Ghasemian E, Jamaati H, Tabaraie B, Amini Z, Cox K. Association between respiratory viruses and exacerbation of COPD: a case-control study. *Infect Dis (Lond)*. 2015;47(8):523–529.
111. Seemungal T, Harper-Owen R, Bhowmik A, et al. Respiratory viruses, symptoms, and inflammatory markers in acute exacerbations and stable chronic obstructive pulmonary disease. *Am J Respir Crit Care Med*. 2001;164(9):1618–1623.
112. Mallia P, Message SD, Kebadze T, Parker HL, Kon OM, Johnston SL. An experimental model of rhinovirus induced chronic obstructive pulmonary disease exacerbations: a pilot study. *Respir Res*. 2006;7:116.
113. Mallia P, Message SD, Gielen V, et al. Experimental rhinovirus infection as a human model of chronic obstructive pulmonary disease exacerbation. *Am J Respir Crit Care Med*. 2011;183(6):734–742.
114. Footitt J, Mallia P, Durham AL, et al. Oxidative and nitrosative stress and histone deacetylase-2 activity in exacerbations of COPD. *Chest*. 2016;149(1):62–73.
115. Rohde G, Wiethege A, Borg I, et al. Respiratory viruses in exacerbations of chronic obstructive pulmonary disease requiring hospitalisation: a case-control study. *Thorax*. 2003;58(1):37–42.
116. Mallia P, Message SD, Contoli M, et al. Lymphocyte subsets in experimental rhinovirus infection in chronic obstructive pulmonary disease. *Respir Med*. 2014;108(1):78–85.
117. Mallia P, Message SD, Contoli M, et al. Neutrophil adhesion molecules in experimental rhinovirus infection in COPD. *Respir Res*. 2013;14(1):72.
118. Mallia P, Footitt J, Sotero R, et al. Rhinovirus infection induces degradation of antimicrobial peptides and secondary bacterial infection in chronic obstructive pulmonary disease. *Am J Respir Crit Care Med*. 2012;186(11):1117–1124.
119. Dimopoulos G, Lerikou M, Tsiodras S, et al. Viral epidemiology of acute exacerbations of chronic obstructive pulmonary disease. *Pulm Pharmacol Ther*. 2012;25(1):12–18.
120. Molyneaux PL, Mallia P, Cox MJ, et al. Outgrowth of the bacterial airway microbiome after rhinovirus exacerbation of chronic obstructive pulmonary disease. *Am J Respir Crit Care Med*. 2013;188(10):1224–1231.
121. Mallia P, Webber J, Gill SK, et al. Role of airway glucose in bacterial infections in patients with chronic obstructive pulmonary disease. *J Allergy Clin Immunol*. 2018;142(3):815–823.e6.
122. George S, Garcha DS, Patel A, et al. Human rhinovirus infection and secondary bacterial infection in COPD exacerbations. *Am J Respir Crit Care Med*. 2013;187:A2164.
123. Grunberg K, Sharon RF, Sont JK, et al. Rhinovirus-induced airway inflammation in asthma: effect of treatment with inhaled corticosteroids before and during experimental infection. *Am J Respir Crit Care Med*. 2001;164(10 Pt 1):1816–1822.

124. de Kluijver J, Evertse CE, Sont JK, et al. Are rhinovirus-induced airway responses in asthma aggravated by chronic allergen exposure? *Am J Respir Crit Care Med.* 2003;168(10):1174–1180.
125. Kloepfer KM, DeMore JP, Vrtis RF, et al. Effects of montelukast on patients with asthma after experimental inoculation with human rhinovirus 16. *Ann Allergy Asthma Immunol.* 2011;106(3):252–257.
126. Silkoff PE, Flavin S, Gordon R, et al. Toll-like receptor 3 blockade in rhinovirus-induced experimental asthma exacerbations: a randomized controlled study. *J Allergy Clin Immunol.* 2018;141(4):1220–1230.
127. Kloepfer KM, Lee WM, Pappas TE, et al. Detection of pathogenic bacteria during rhinovirus infection is associated with increased respiratory symptoms and asthma exacerbations. *J Allergy Clin Immunol.* 2014;133(5):1301–1307. e1301-1303.
128. Akbarshahi H, Menzel M, Ramu S, Mahmutovic Persson I, Bjermer L, Uller L. House dust mite impairs antiviral response in asthma exacerbation models through its effects on TLR3. *Allergy.* 2018;73(5):1053–1063.
129. Capistrano SJ, Zakarya R, Chen H, Oliver BG. Biomass smoke exposure enhances rhinovirus-induced inflammation in primary lung fibroblasts. *Int J Mol Sci.* 2016;17 (9):1–12.
130. Lehtinen MJ, Hibberd AA, Mannikko S, et al. Nasal microbiota clusters associate with inflammatory response, viral load, and symptom severity in experimental rhinovirus challenge. *Sci Rep.* 2018;8(1):11411.
131. Nguyen A, Guedan A, Mousnier A, et al. Host lipidome analysis during rhinovirus replication in human bronchial epithelial cells identifies potential therapeutic targets. *J Lipid Res.* 2018;59(9):1671–1684.
132. Stelzer-Braid S, Liu N, Doumit M, et al. Association of rhinovirus with exacerbations in young children affected by cystic fibrosis: preliminary data. *J Med Virol.* 2017;89(8):1494–1497.
133. Kieninger E, Singer F, Tapparel C, et al. High rhinovirus burden in lower airways of children with cystic fibrosis. *Chest.* 2013;143(3):782–790.
134. Gao YH, Guan WJ, Xu G, et al. The role of viral infection in pulmonary exacerbations of bronchiectasis in adults: a prospective study. *Chest.* 2015;147(6):1635–1643.
135. Kapur N, Mackay IM, Sloots TP, Masters IB, Chang AB. Respiratory viruses in exacerbations of non-cystic fibrosis bronchiectasis in children. *Arch Dis Child.* 2014;99(8):749–753.
136. Schogler A, Stokes AB, Casaulta C, et al. Interferon response of the cystic fibrosis bronchial epithelium to major and minor group rhinovirus infection. *J Cyst Fibros.* 2016;15(3):332–339.
137. Lambkin-Williams R, Noulin N, Mann A, Catchpole A, Gilbert AS. The human viral challenge model: accelerating the evaluation of respiratory antivirals, vaccines and novel diagnostics. *Respir Res.* 2018;19(1):123.
138. Del Vecchio AM, Branigan PJ, Barnathan ES, Flavin SK, Silkoff PE, Turner RB. Utility of animal and in vivo experimental infection of humans with rhinoviruses in the development of therapeutic agents for viral exacerbations of asthma and chronic obstructive pulmonary disease. *Pulm Pharmacol Ther.* 2015;30:32–43.
139. Maltby S, Tay HL, Yang M, Foster PS. Mouse models of severe asthma: understanding the mechanisms of steroid resistance, tissue remodelling and disease exacerbation. *Respirology.* 2017;22(5):874–885.
140. Kumar RK, Foster PS, Rosenberg HF. Respiratory viral infection, epithelial cytokines, and innate lymphoid cells in asthma exacerbations. *J Leukoc Biol.* 2014;96(3):391–396.
141. Bartlett NW, Singanayagam A, Johnston SL. Mouse models of rhinovirus infection and airways disease. *Methods Mol Biol.* 2015;1221:181–188.

142. Lee W-M, Chen Y, Wang W, Mosser A. Growth of human rhinovirus in H1-HeLa cell suspension culture and purifiation of virions. Methods in Molecular Biology In: Jans DA, Ghildyal R, eds. *Rhinoviruses: Methods and Protocols*. vol. 1221. New York: Springer Science + Business Media; 2015:49–61.
143. Bartlett NW, Singanayagam A, Johnston SL. Mouse models of rhinovirus infection and airways disease. In: Jans DA, Ghildyal R, eds. *Rhinoviruses: Methods and Protocols*. New York: Springer New York; 2015:181–188.
144. Papi A, Johnston SL. Rhinovirus infection induces expression of its own receptor intercellular adhesion molecule 1 (ICAM-1) via increased NF-kappaB-mediated transcription. *J Biol Chem*. 1999;274(14):9707–9720.
145. Newcomb DC, Sajjan US, Nagarkar DR, et al. Human rhinovirus 1B exposure induces phosphatidylinositol 3-kinase-dependent airway inflammation in mice. *Am J Respir Crit Care Med*. 2008;177(10):1111–1121.
146. Han M, Chung Y, Young Hong J, et al. Toll-like receptor 2-expressing macrophages are required and sufficient for rhinovirus-induced airway inflammation. *J Allergy Clin Immunol*. 2016;138(6):1619–1630.
147. Greve JM, Davis G, Meyer AM, et al. The major human rhinovirus receptor is ICAM-1. *Cell*. 1989;56(5):839–847.
148. Tomassini JE, Graham D, DeWitt CM, Lineberger DW, Rodkey JA, Colonno RJ. cDNA cloning reveals that the major group rhinovirus receptor on HeLa cells is intercellular adhesion molecule 1. *Proc Natl Acad Sci USA*. 1989;86(13):4907–4911.
149. Register RB, Uncapher CR, Naylor AM, Lineberger DW, Colonno RJ. Human-murine chimeras of ICAM-1 identify amino acid residues critical for rhinovirus and antibody binding. *J Virol*. 1991;65(12):6589–6596.
150. Yin FH, Lomax NB. Establishment of a mouse model for human rhinovirus infection. *J Gen Virol*. 1986;67(Pt 11):2335–2340.
151. Tuthill TJ, Papadopoulos NG, Jourdan P, et al. Mouse respiratory epithelial cells support efficient replication of human rhinovirus. *J Gen Virol*. 2003;84(Pt 10):2829–2836.
152. Bartlett NW, Walton RP, Edwards MR, et al. Mouse models of rhinovirus-induced disease and exacerbation of allergic airway inflammation. *Nat Med*. 2008;14(2):199–204.
153. Rasmussen AL, Racaniello VR. Selection of rhinovirus 1A variants adapted for growth in mouse lung epithelial cells. *Virology*. 2011;420(2):82–88.
154. Sajjan U, Wang Q, Zhao Y, Gruenert DC, Hershenson MB. Rhinovirus disrupts the barrier function of polarized airway epithelial cells. *Am J Respir Crit Care Med*. 2008;178(12):1271–1281.
155. Jayaraman A, Jackson DJ, Message SD, et al. IL-15 complexes induce NK- and T-cell responses independent of type I IFN signaling during rhinovirus infection. *Mucosal Immunol*. 2014;7(5):1151–1164.
156. Girkin J, Hatchwell L, Foster P, et al. CCL7 and IRF-7 mediate hallmark inflammatory and IFN responses following rhinovirus 1B infection. *J Immunol*. 2015;194(10):4924–4930.
157. Nagarkar DR, Wang Q, Shim J, et al. CXCR2 is required for neutrophilic airway inflammation and hyperresponsiveness in a mouse model of human rhinovirus infection. *J Immunol*. 2009;183(10):6698–6707.
158. Wang Q, Miller DJ, Bowman ER, et al. MDA5 and TLR3 initiate pro-inflammatory signaling pathways leading to rhinovirus-induced airways inflammation and hyperresponsiveness. *PLoS Pathog*. 2011;7(5):e1002070.
159. Bartlett NW, Slater L, Glanville N, et al. Defining critical roles for NF-kappaB p65 and type I interferon in innate immunity to rhinovirus. *EMBO Mol Med*. 2012;4(12):1244–1260.

160. Berman R, Jiang D, Wu Q, Stevenson CR, Schaefer NR, Chu HW. MUC18 regulates lung rhinovirus infection and inflammation. *PLoS One*. 2016;11(10):e0163927.
161. Glanville N, Peel TJ, Schroder A, et al. Tbet deficiency causes T helper cell dependent airways eosinophilia and mucus hypersecretion in response to rhinovirus infection. *PLoS Pathog*. 2016;12(9):e1005913.
162. Mehta AK, Duan W, Doerner AM, et al. Rhinovirus infection interferes with induction of tolerance to aeroantigens through OX40 ligand, thymic stromal lymphopoietin, and IL-33. *J Allergy Clin Immunol*. 2016;137(1):278–288.e6.
163. Girkin JL, Hatchwell LM, Collison AM, et al. TRAIL signaling is proinflammatory and proviral in a murine model of rhinovirus 1B infection. *Am J Physiol Lung Cell Mol Physiol*. 2017;312(1):L89–L99.
164. Unger BL, Faris AN, Ganesan S, Comstock AT, Hershenson MB, Sajjan US. Rhinovirus attenuates non-typeable *Haemophilus influenzae*-stimulated IL-8 responses via TLR2-dependent degradation of IRAK-1. *PLoS Pathog*. 2012;8(10):e1002969.
165. Cui TX, Maheshwer B, Hong JY, Goldsmith AM, Bentley JK, Popova AP. Hyperoxic exposure of immature mice increases the inflammatory response to subsequent rhinovirus infection: association with danger signals. *J Immunol*. 2016;196(11):4692–4705.
166. Phan JA, Kicic A, Berry LJ, Sly PD, Larcombe AN. Early life rhinovirus infection exacerbates house-dust-mite induced lung disease more severely in female mice. *Exp Lung Res*. 2016;42(1):24–36.
167. McLean GR, Walton RP, Shetty S, et al. Rhinovirus infections and immunisation induce cross-serotype reactive antibodies to VP1. *Antiviral Res*. 2012;95(3):193–201.
168. Ganesan S, Faris AN, Comstock AT, et al. Quercetin inhibits rhinovirus replication in vitro and in vivo. *Antiviral Res*. 2012;94(3):258–271.
169. Song JH, Kim SR, Heo EY, et al. Antiviral activity of gemcitabine against human rhinovirus in vitro and in vivo. *Antiviral Res*. 2017;145:6–13.
170. Singanayagam A, Glanville N, Girkin JL, et al. Corticosteroid suppression of antiviral immunity increases bacterial loads and mucus production in COPD exacerbations. *Nat Commun*. 2018;9(1):2229.
171. Heinz SA, Henson DA, Austin MD, Jin F, Nieman DC. Quercetin supplementation and upper respiratory tract infection: a randomized community clinical trial. *Pharmacol Res*. 2010;62(3):237–242.
172. Nagarkar DR, Bowman ER, Schneider D, et al. Rhinovirus infection of allergen-sensitized and -challenged mice induces eotaxin release from functionally polarized macrophages. *J Immunol*. 2010;185(4):2525–2535.
173. Chung Y, Hong JY, Lei J, Chen Q, Bentley JK, Hershenson MB. Rhinovirus infection induces interleukin-13 production from CD11b-positive, M2-polarized exudative macrophages. *Am J Respir Cell Mol Biol*. 2015;52(2):205–216.
174. Hong JY, Chung Y, Steenrod J, et al. Macrophage activation state determines the response to rhinovirus infection in a mouse model of allergic asthma. *Respir Res*. 2014;15:63.
175. Glanville N, Message SD, Walton RP, et al. γδT cells suppress inflammation and disease during rhinovirus-induced asthma exacerbations. *Mucosal Immunol*. 2013;6(6):1091–1100.
176. Chairakaki AD, Saridaki MI, Pyrillou K, et al. Plasmacytoid dendritic cells drive acute asthma exacerbations. *J Allergy Clin Immunol*. 2018;142(2):542–566.e12.
177. Phan JA, Kicic A, Berry LJ, et al. Rhinovirus exacerbates house-dust-mite induced lung disease in adult mice. *PLoS One*. 2014;9(3):e92163.
178. Toussaint M, Jackson DJ, Swieboda D, et al. Host DNA released by NETosis promotes rhinovirus-induced type-2 allergic asthma exacerbation. *Nat Med*. 2017;23(6):681–691.

179. Collison A, Hatchwell L, Verrills N, et al. The E3 ubiquitin ligase midline 1 promotes allergen and rhinovirus-induced asthma by inhibiting protein phosphatase 2A activity. *Nat Med.* 2013;19(2):232–237.
180. Schneider D, Hong JY, Bowman ER, et al. Macrophage/epithelial cell CCL2 contributes to rhinovirus-induced hyperresponsiveness and inflammation in a mouse model of allergic airways disease. *Am J Physiol Lung Cell Mol Physiol.* 2013;304(3): L162–L169.
181. Chen G, Korfhagen TR, Karp CL, et al. Foxa3 induces goblet cell metaplasia and inhibits innate antiviral immunity. *Am J Respir Crit Care Med.* 2014;189(3):301–313.
182. Mahmutovic-Persson I, Akbarshahi H, Bartlett NW, et al. Inhaled dsRNA and rhinovirus evoke neutrophilic exacerbation and lung expression of thymic stromal lymphopoietin in allergic mice with established experimental asthma. *Allergy.* 2014;69 (3):348–358.
183. Hatchwell L, Collison A, Girkin J, et al. Toll-like receptor 7 governs interferon and inflammatory responses to rhinovirus and is suppressed by IL-5-induced lung eosinophilia. *Thorax.* 2015;70(9):854–861.
184. Hatchwell L, Girkin J, Dun MD, et al. Salmeterol attenuates chemotactic responses in rhinovirus-induced exacerbation of allergic airways disease by modulating protein phosphatase 2A. *J Allergy Clin Immunol.* 2014;133(6):1720–1727.
185. Morales DR. LABA monotherapy in asthma: an avoidable problem. *Br J Gen Pract.* 2013;63(617):627–628.
186. Traub S, Nikonova A, Carruthers A, et al. An anti-human icam-1 antibody inhibits rhinovirus-induced exacerbations of lung inflammation. *PLoS Pathog.* 2013;9(8): e1003520.
187. de Souza Alves CC, Collison A, Hatchwell L, et al. Inhibiting AKT phosphorylation employing non-cytotoxic anthraquinones ameliorates TH2 mediated allergic airways disease and rhinovirus exacerbation. *PLoS One.* 2013;8(11):e79565.
188. Schilter HC, Collison A, Russo RC, et al. Effects of an anti-inflammatory VAP-1/SSAO inhibitor, PXS-4728A, on pulmonary neutrophil migration. *Respir Res.* 2015;16:42.
189. Sajjan U, Ganesan S, Comstock AT, et al. Elastase- and LPS-exposed mice display altered responses to rhinovirus infection. *Am J Physiol Lung Cell Mol Physiol.* 2009;297(5):L931–L944.
190. Singanayagam A, Glanville N, Walton RP, et al. A short-term mouse model that reproduces the immunopathological features of rhinovirus induced exacerbation of COPD. *Clin Sci (Lond).* 2015;129(3):245–258.
191. Ganesan S, Comstock AT, Kinker B, Mancuso P, Beck JM, Sajjan US. Combined exposure to cigarette smoke and nontypeable *Haemophilus influenzae* drives development of a COPD phenotype in mice. *Respir Res.* 2014;15:11.
192. Jing Y, Gimenes JA, Mishra R, et al. NOTCH3 contributes to rhinovirus-induced goblet cell hyperplasia in COPD airway epithelial cells. *Thorax.* 2018;:18–32.
193. Farazuddin M, Mishra R, Jing Y, Srivastava V, Comstock AT, Sajjan US. Quercetin prevents rhinovirus-induced progression of lung disease in mice with COPD phenotype. *PLoS One.* 2018;13(7):e0199612.
194. Lee SB, Song JA, Choi GE, Kim HS, Jang YJ. Rhinovirus infection in murine chronic allergic rhinosinusitis model. *Int Forum Allergy Rhinol.* 2016;6 (11):1131–1138.
195. Blanco JC, Boukhvalova MS, Perez DR, Vogel SN, Kajon A. Modeling human respiratory viral infections in the cotton rat (*Sigmodon hispidus*). *J Antivir Antiretrovir.* 2014;6:40–42.
196. Blanco JC, Core S, Pletneva LM, March TH, Boukhvalova MS, Kajon AE. Prophylactic antibody treatment and intramuscular immunization reduce infectious

human rhinovirus 16 load in the lower respiratory tract of challenged cotton rats. *Trials Vaccinol.* 2014;3:52–60.

197. Patel MC, Pletneva LM, Boukhvalova MS, Vogel SN, Kajon AE, Blanco JCG. Immunization with live human rhinovirus (HRV) 16 induces protection in cotton rats against HRV14 infection. *Front Microbiol.* 2017;8:1646.
198. Martin GV, Heath RB. Rhinovirus infection of vervet monkeys. A model of human rhinovirus disease. *Br J Exp Pathol.* 1969;50(5):516–519.
199. Dick EC. Experimental infections of chimpanzees with human rhinovirus types 14 and 43. *Proc Soc Exp Biol Med.* 1968;127(4):1079–1081.
200. Pinto CA, Haff RF. Experimental infection of gibbons with rhinovirus. *Nature.* 1969;224(226):1310–1311.
201. Shipkowitz NL, Bower RR, Schleicher JB, Aquino F, Appell RN, Roderick WR. Antiviral activity of a bis-benzimidazole against experimental rhinovirus infections in chimpanzees. *Appl Microbiol.* 1972;23(1):117–122.
202. Pinto CA, Bahnsen HP, Ravin LJ, Haff RF, Pagano JF. The antiviral effect of a triazinoindole (SK&F 40491) in rhinovirus infected gibbons. *Proc Soc Exp Biol Med.* 1972;141(2):467–474.
203. Huguenel ED, Cohn D, Dockum DP, et al. Prevention of rhinovirus infection in chimpanzees by soluble intercellular adhesion molecule-1. *Am J Respir Crit Care Med.* 1997;155(4):1206–1210.
204. Rosenthal LA, Amineva SP, Szakaly RJ, Lemanske Jr RF, Gern JE, Sorkness RL. A rat model of picornavirus-induced airway infection and inflammation. *Virol J.* 2009;6:122.
205. Rosenthal LA, Szakaly RJ, Amineva SP, et al. Lower respiratory tract infection induced by a genetically modified picornavirus in its natural murine host. *PLoS One.* 2012;7(2):e32061.
206. Jackson DJ, Johnston SL. The role of viruses in acute exacerbations of asthma. *J Allergy Clin Immunol.* 2010;125(6):1178–1187. quiz 1188-1179.
207. Persson CG. Con: mice are not a good model of human airway disease. *Am J Respir Crit Care Med.* 2002;166(1):6–7. discussion 8.
208. Miller FJ, Mercer RR, Crapo JD. Lower respiratory tract structure of laboratory animals and humans: dosimetry implications. *Aerosol Sci Technol.* 1993;18(3):257–271.
209. Foster PS, Hogan SP, Ramsay AJ, Matthaei KI, Young IG. Interleukin 5 deficiency abolishes eosinophilia, airways hyperreactivity, and lung damage in a mouse asthma model. *J Exp Med.* 1996;183(1):195–201.
210. Hogan SP, Koskinen A, Foster PS. Interleukin-5 and eosinophils induce airway damage and bronchial hyperreactivity during allergic airway inflammation in BALB/c mice. *Immunol Cell Biol.* 1997;75(3):284–288.
211. Leckie MJ, ten Brinke A, Khan J, et al. Effects of an interleukin-5 blocking monoclonal antibody on eosinophils, airway hyper-responsiveness, and the late asthmatic response. *Lancet.* 2000;356(9248):2144–2148.
212. Nair P, Pizzichini MM, Kjarsgaard M, et al. Mepolizumab for prednisone-dependent asthma with sputum eosinophilia. *N Engl J Med.* 2009;360(10):985–993.
213. Pavord ID, Korn S, Howarth P, et al. Mepolizumab for severe eosinophilic asthma (DREAM): a multicentre, double-blind, placebo-controlled trial. *Lancet.* 2012;380 (9842):651–659.
214. Bochkov YA, Watters K, Basnet S, et al. Mutations in VP1 and 3A proteins improve binding and replication of rhinovirus C15 in HeLa-E8 cells. *Virology.* 2016;499:350–360.
215. Foxman EF, Storer JA, Fitzgerald ME, et al. Temperature-dependent innate defense against the common cold virus limits viral replication at warm temperature in mouse airway cells. *Proc Natl Acad Sci USA.* 2015;112(3):827–832.

CHAPTER 9

Emerging therapeutic approaches

Gary McLean[1,2], Jason Girkin[3] and Roberto Solari[2]
[1]Molecular Immunology, Cellular Molecular and Immunology Research Centre, London Metropolitan University, London, United Kingdom
[2]National Heart and Lung Institute, Imperial College London, London, United Kingdom
[3]University of Newcastle, Callaghan, NSW, Australia

There are currently no approved drugs for the treatment of rhinovirus (RV) or indeed of any picornavirus infection, despite decades of drug discovery efforts. Likewise there are no licensed vaccines for RV even though trials began in the late 1960s. However, a large number of experimental approaches that target RV itself, the RV-induced inflammatory response, or promote broad RV-specific immunity have been through preclinical development and clinical trials with mixed success. This chapter will amalgamate these studies and highlight the most promising and applicable therapeutic approaches.

9.1 CHEMOTHERAPEUTICS

Picornaviruses are a family of nonenveloped single positive-strand RNA viruses with a genome of about 7.5 kb that encodes a single polyprotein that is posttranslationally cleaved into four capsid proteins (VP1–VP4) and seven nonstructural proteins. A number of these viral proteins have been the subject of historical drug discovery efforts including the capsid, the 3C protease, and the RNA dependent RNA polymerase.[1–4] These viral targets have the potential advantage of avoiding cellular toxicity but face the challenge of viral resistance emerging due to mutation. The earliest efforts at antipicornaviral drug discovery targeted these viral proteins and more recently efforts have turned to host targets involved in viral replication.

Rhinovirus Infections
DOI: https://doi.org/10.1016/B978-0-12-816417-4.00009-3

9.1.1 Viral targets

9.1.1.1 Entry inhibitors

The first documented types of antipicornaviral drugs were discovered fortuitously as compounds that bound the viral capsid and appeared to block cell entry and viral uncoating. The viral capsid is an icosahedral structure made up of 60 copies of the four capsid protein subunits. Viral entry into cells requires capsid binding to cell surface receptors followed by endocytosis and capsid uncoating, which releases the viral genome into the cell and the three RV species, RV-A, B, and C, utilize different cellular receptors for entry. The majority of RV-A and all RV-B use the intercellular adhesion molecule 1 (ICAM-1), some RV-A types use the low-density lipoprotein receptor, and RV-C types use cadherin-related family member 3 (CDHR3). These cell surface receptors bind close to a so-called canyon, which is a depression in the surface of the capsid and it is into this canyon that the capsid binding drugs fit, by stabilizing the capsid block viral entry and uncoating. The first generation of capsid binders were the WIN compounds,[5] which tended to have relatively narrow serotype specificity, but a derivative of these compounds, Picovir (Pleconaril) did progress into clinical trials. After a New Drug Application the drug was rejected by the U.S. Food and Drug Administration (FDA) in 2003 for safety reasons but an intranasal form was licensed to Schering-Plough in 2003 and went into trials for RV exacerbations in asthma. Janssen in collaboration with Biota also have developed a capsid binder called Pirodavir for the same indication. However, Pirodavir was not efficacious at blocking natural infections when administered intranasally.[6] The only capsid binder that has shown positive clinical data was Vapendavir (BTA798) (ClinicalTrials.gov Identifier: NCT01175226).[7] The trial showed improvement in its primary efficacy endpoint, the Wisconsin Upper Respiratory Symptom Survey-21, showing significant improvement in symptom score from days 2–4 postinfection.[7] However, there is no information in the public domain about further developments of this agent. One potential issue with capsid binding drugs that is frequently cited is the rapid emergence of resistance mutants.

9.1.1.2 Protease inhibitors

The RV genome encodes two proteases: $2A^{pro}$ and $3C^{pro}$. Structurally, both of these proteases resemble trypsin-like serine proteases although they both contain a cysteine as the active site nucleophile. The $2A^{pro}$

protease may provide the first cleavage of the polyprotein, separating the capsid precursor from the remaining nonstructural proteins, but appears to have a limited activity in cleaving the remaining viral polyprotein and is thought to mainly target host proteins, in particular eukaryotic translation initiation factor 4 G (eIF4G), which may cause shut off of host protein translation whilst allowing viral protein translation. Most drug discovery effort however has focused on the $3C^{pro}$, which has been considered an excellent target for antiviral intervention. The most advanced molecule is Rupintrivir, developed by Agouron/Pfizer, who progressed it as a nasal spray to a phase II viral challenge study in healthy volunteers.[8] Although the drug showed efficacy in numerous symptom scores it did not decrease the frequency of colds and appears to have been discontinued.

9.1.1.3 Polymerase inhibitors

Replication of the RV RNA genome is catalyzed by the RNA dependent RNA polymerase, or $3D^{pol}$. Although nucleoside analogues have proven excellent inhibitors of viral polymerases and represent a major antiviral therapeutic strategy for multiple pathogens, there has been little or no progress in targeting the picornavirus $3D^{pol}$.

9.1.2 Host targets

Picornaviruses, like all viruses, hijack host cellular components to complete their replication cycle, and the hope is that inhibition of these host targets may provide novel opportunities with a greater barrier to the emergence of viral resistance. The counterargument is that host targets represent a greater risk of toxicity, and clearly this balance of efficacy versus toxicity is critical to the development of drugs against host targets.

9.1.2.1 Entry inhibitors

Boehringer Ingelheim developed soluble ICAM (Tremacamra) as an inhibitor of viral entry as the major group human RV uses ICAM as a cellular receptor to enter cells. In an experimental viral challenge study Tremacamra clearly inhibited the severity of RV induced cold symptoms,[9] however this too appears to have been discontinued.

9.1.2.2 The replication complex

Picornaviruses use the cytoplasmic face of the endoplasmic reticulum (ER) and Golgi membranes as a platform for genome replication and the morphology and composition of these membranes is greatly remodeled by the virus. Studies on the picornavirus nonstructural proteins 2B, 2C, and 3A have shown that they associate with the ER and Golgi membranes and suggested that these proteins play an important role in membrane protein and lipid remodeling to generate replication complexes. A greater understanding of the biology of 2B, 2C, and 3A and the host proteins with which they interact has led to the identification of novel potential host antiviral targets. One of the first antiviral compounds found to influence these membrane remodeling effects was Enviroxime[10] and enviroxime-like compounds, which originally came from black-box screens looking for inhibitors of viral replication. Enviroxime has been in clinical development but failed due to poor exposure, toxicity, and lack of efficacy when administered both orally and intranasally.[11,12] Clinical trials of orally administered enviroxime caused nausea, vomiting, and stomach pain in 60% of subjects whereas intranasal enviroxime, although relatively well tolerated, showed little efficacy.[12,13]

Phosphatidylinositol-4-kinase IIIβ (PI4KIIIβ) is well known to play a role in membrane traffic in the Golgi, and Picornaviruses appear to recruit PI4KIIIβ via the 3A protein, although the exact mechanism is not fully clear.[14–18] The viral RNA polymerase ($3D^{pol}$) binds and assembles on phosphatidylinositol-4-phosphate (PtdIns4P) enriched membranes and inhibitors of PI4KIIIβ such as PIK93 very effectively block viral replication.[19] Consequently, this enzyme has been actively targeted by a number of companies although there is currently no clinical development reported in the public domain. Black-box antiviral screens from Galapagos,[20] Boehringer Ingelheim,[21] and academic groups[22–24] have also identified PI4KIIIβ inhibitors as having antipicornaviral activity and it has since been discovered that this enzyme is also the target of the Enviroxime compounds and the antipicornaviral compounds GW5074 and T-00127-HEV1.[25,26] Novartis have also published on PI4KIIIβ inhibitors for the treatment of HCV infection[27] although this compound appears to have been terminated due to toxicity.[28] Animals dosed with recently discovered PI4KIIIβ inhibitors also show clear signs of toxicity raising doubts that this is a sufficiently well-tolerated target[21] and resistance also appears to emerge to PI4KIIIβ inhibitors with mutations in the 3A protein appearing to be the main effector.[29]

The picornavirus 3A protein is also known to interact with the Arf guanosine-5'-triphosphate (GTP) exchange protein GBF1.[30–33] The small GTP binding protein Arf1 is well known for its involvement on ER/Golgi membrane trafficking and normally recruits PI4KIIIβ. It is also implicated in RV replication since the GBF1 inhibitory chemical BrefeldinA (BFA) completely blocks viral replication.[34] Further components of the replication complex have also been identified. Oxysterol binding proteins (OSBP) bind PtdIns4P and Arf1 and are sterol regulated controllers of cholesterol homeostasis in the ER. OSBP has been identified as the target of some antipicornaviral drugs such as the minor-enviroxime compound, 25-OH cholesterol.[35] Recently it has been shown that the antifungal drug Itraconazol inhibits enterovirus replication by targeting OSBP.[36] In addition, the natural product OSW-1, which has been mainly studied for its anticancer properties, has now been shown to potently block replication of a range of picornaviruses through inhibition of OSBP.[37] Interactome studies using 3A proteins from multiple picornaviruses as affinity ligands identified a number of additional interacting membrane proteins including VAP-A (vesicle associated membrane protein (VAMP)-associated protein-A), an ER protein that binds OSBP and is essential for the stimulation of sphingomyelin (SM) synthesis by 25-OH cholesterol.[15,38,39] Further strong evidence for the role of OSBP and VAP-A comes from the observation that the antiviral effector protein interferon (IFN)-inducible transmembrane protein 3 interacts with VAP A and prevents its association with OSBP, thereby disrupting intracellular cholesterol homeostasis and inhibiting viral entry.[40] This large body of evidence has started to give us a picture of the host components of the viral replication complex and suggests host targets involved in membrane remodeling that may represent a novel antipicornaviral strategy.

9.1.2.3 *Novel membrane targets*

This accumulating evidence from both chemical inhibitors and interactome studies implicated Arf/GBF1, PI4KIIIβ, and OSBP as potential antiviral targets acting at the level of the formation of replication complexes on membranes of the secretory pathway. Looking for known druggable targets likely to be present in these membrane complexes identified protein kinase D (PKD) as a possible candidate. PKD is present along with Arf, PI4KIIIβ, 14-3-3γ, and C-terminal-binding protein 1/brefeldin A ADP-ribosylated substrate (CtBP1/BARS) proteins in a complex that mediates vesicle budding and fission from the Golgi.[41] Since CtBP1/BARS is adenosine diphosphate (ADP)-ribosylated by BFA, there was strong circumstantial evidence that this complex may be involved in the RV replication complex and based on the published literature one could build a hypothetical model for the key components at the Golgi/ER interface involved in the formation of the replication complex (Fig. 9.1). The RV protein 3A interacts with the host proteins ACBD3 and PI4KIIIβ as well as the Arf GTP exchange protein, GBF1. BFA, a drug that inhibits Arf GTP exchange and blocks the function of CtBP1/BARS, is potently antiviral, as are PI4KIIIβ inhibitors and OSBP inhibitors and PKD is known to be an upstream regulator of both PI4KIIIβ and OSBP and is recruited to the Golgi by Arf.

PKD also appears to play an important role in the architecture and function of the Golgi apparatus. At the trans-Golgi network, PKD activates PI4KIIIβ to generate PI4P, which mediates the Golgi localization of ceramide transfer protein (CERT) and OSBP proteins via binding to their pleckstrin homology domains. PKD-mediated phosphorylation of CERT at Ser132 and OSBP at Ser240 impairs their Golgi localization and inhibits their functions in integrating the cholesterol and SM metabolism. Since the PI4KIIIβ and OSBP pathways are clearly important for viral replication and PKD regulates both these pathways,[42,43] it was clearly of interest as a potential antiviral target and recent evidence has confirmed that the PKD inhibitor CTR0066101 is capable of inhibiting replication of RV, PV, and foot-and-mouth disease virus (FMDV).[44]

Picornavirus replication not only modifies the morphology of the secretory pathway in host cells, it also greatly influences the lipid composition of the cell. A recent comprehensive lipidomics study has revealed the extent to which the host lipidome is modified over the time course of a single RV replication cycle and identified a number of lipid modifying enzymes as potential antiviral targets.[45] This study revealed that inhibitors of fatty acid synthase (FAS), ceramidase, and SM synthase were highly

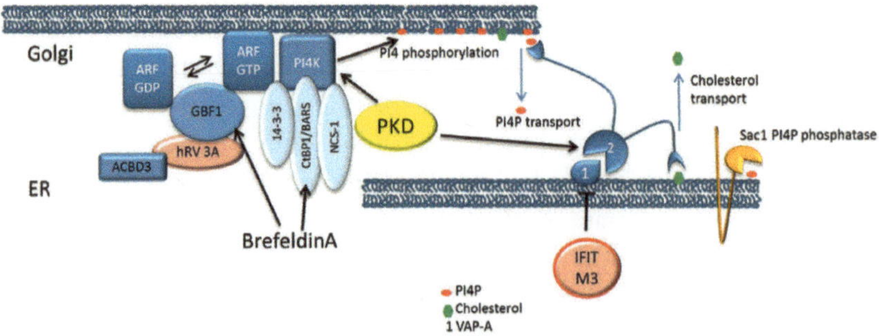

Figure 9.1 Model of the possible components of the HRV replication complex. *HRV*, Human RV.

effective at inhibiting RV replication. This confirmed previous findings that described FAS as an antiviral target for coxsackievirus B3 (CVB3);[46] indeed amentoflavone, a FAS inhibitor, reduced CVB3 replication.[47]

9.1.2.4 Assembly inhibitors

The four capsid proteins are expressed at the N-terminus of the viral polyprotein and the first proteolytic cleavage separates the capsid precursor from the nonstructural proteins. The capsid precursor is then further processed into three capsid proteins, VP0, VP3, and VP1, which trigger a cascade of protein self-assembly that ultimately leads to the formation of infectious virions. This cascade starts with the formation of a VP0/VP1/VP3 complex termed a protomer; five protomers assemble into pentamers, and 12 pentamers and the viral RNA genome then assemble to form an icosahedral virion. The final maturation step involves cleavage of VP0 into VP4 and VP2 in the intact viral capsid. VP0 is encoded at the N terminus of the viral polyprotein and in many picornaviruses it is *N*-myristoylated by the host cell *N*-myristoyltransferase (NMT). A recent study has shown that a highly selective and potent inhibitor of human NMT is an effective inhibitor of RV, PV, and FMDV replication.[48] The inhibitor, IMP-1088, by blocking the myristoylation of VP0 inhibits the cleavage of VP0 into VP2 and VP4 and so inhibits the assembly of intact virions.

Folding and assembly of the picornavirus capsid also requires the activity of the host chaperone Hsp90 and pharmacological inhibition of this target impaired the replication of poliovirus, RV, and coxsackievirus without the emergence of drug resistance, suggesting this was an attractive antiviral strategy.[49]

9.2 BIOTHERAPEUTICS AND IMMUNOTHERAPIES

9.2.1 Innate immune stimulators

An alternative strategy to delivering a direct antiviral is to boost the host antiviral immunity and a number of direct and indirect strategies have been attempted. One approach has been to administer IFN-β by inhalation. A phase II trial has been completed in a natural RV infection study design in which 147 people with asthma on inhaled corticosteroids, with a history of virus-associated exacerbations, were randomized to 14-day treatment with inhaled IFN-β or placebo within 24 hours of developing cold symptoms.[50] Although the study did not meet its primary endpoint there was clear evidence of benefit from the IFN-β treatment and proof of concept that a supplement to the host antiviral response could be delivered in a timely fashion to be efficacious.

An indirect approach is to deliver an activator of the innate immune response such as a Toll-like receptor (TLR) agonist to stimulate a broad antiviral response. There are numerous synthetic ligands to TLRs at various stages of development and their applications include add-ons to chemotherapy, vaccine adjuvants, rheumatoid arthritis treatments, antiviral therapies, and more. TLR-agonists are too numerous to cover in full detail and many fall outside of the scope of this chapter but every effort has been made to include evidence for the use of TLR-agonists that could enhance the antiviral response to RV. One such example is a clinical trial of the intranasal delivery of polyriboinosinic and polyribocytidylic acid (Poly I:C named PrEP-001), a TLR3 agonist. TLR3 recognizes double stranded RNA present during viral replication.[51,52] Healthy subjects randomly received two doses of drug (PrEP-001) or placebo, 48 and 24 hours prechallenge with HRV-A16 and the result was a cellular innate immune response in the nasal epithelium and a clear reduction in symptom score.[53,54]

TLR7 and 8 are located on endosomal membranes and recognize viral RNA[55] and have quickly emerged as targets for the development of antiviral therapies, ranging from topically applied interventions for warts, genital warts, and to vaccine adjuvants or add-on therapies.[56–66] Due to the ability of TLR7 or TLR7/8 agonists to promote type I IFN and Th1 responses and the inhibitory effect this has on Th2-mediated allergic responses, an inhaled TLR7 agonist GSK-2245035 was developed for the treatment of allergic asthma and allergic rhinitis.[67,68] Due to the established antiviral capacity of TLR7/8 agonists and the nasal (upper respiratory) route of administration of GSK-2245035 (the primary site of initial infection of RV and other respiratory viruses) it would follow that GSK-2245035 may protect against RV infection.

Receptors for antimicrobial immunity can also be exploited to promote resistance to viral infection. PUL-042 is a combination of TLR2/6 and TLR9 agonists that has completed phase I trials and was developed by Pulmotech Inc.[69] for broad spectrum protection against pneumonia in susceptible individuals.[70–72] TLR2/6 heterodimers recognize Gram-positive bacterial cell wall components and TLR9 recognizes intracellular viral nucleic acid (by detecting unmethylated CpG) but activation of these receptors (alone or in combination with oseltamivir, a neuraminidase inhibitor) elicits antiviral protection in proof of principle animal models of influenza infection.[73] If delivered intranasally as with PrEP-001, it would be reasonable to assume PUL-042 could also protect against RV infection, due to its established broad-spectrum profile of protection against bacterial and viral pneumonia in preclinical studies. Another TLR2 agonist, Pam2Cys is in preclinical developed as a standalone prophylactic antiviral, which protected mice from lethal influenza infection without compromising adaptive immune responses to influenza.[74,75] Similar versions of Pam2cys are now in preclinical development for the treatment of RV infection.[76]

9.2.2 Targeting virus induced pathogenic responses

The innate immune response to RV is pivotal in the control of infection but a number of mechanisms involved in the innate immune response can precipitate exacerbations of respiratory disease. Attempts to prevent exacerbations have culminated in the development of monoclonal antibody therapies (mAbs) against Th2-associated inflammatory mediators produced by the respiratory epithelium in response to RV infection. RV-induced lower respiratory illness (particularly in people with asthma) is directly related to the extent of production of type-2 cytokines, IL-4, IL-5, and IL-13.[77] There are a large number of mAbs in development for the treatment of respiratory disease and coverage of most of these interventions falls outside of the scope of this book. There are however, some key examples relevant to the prevention of RV-induced exacerbations. The most advanced of all of the Th2 cytokine targeting therapies is the anti-IL-5 mAb, Mepolizumab, which is now an approved asthma therapy after successful completion of multiple clinical trials, with positive results including decreased rates of exacerbation, better asthma control and, reduced blood eosinophils numbers.[78–93] Mepolizumab was listed in phase III clinical trials for the treatment of RV-induced exacerbations of asthma[94] though the clinical trial records show that the study is past its completion date and there has not been an update for some time aside from evidence for reduced eosinophil numbers following RV16 challenge revealed at the American Thoracic Society conference in 2018.[95]

Other mAbs relevant to the treatment of RV-induced exacerbation include mAbs against epithelial-derived cytokines thymic stromal lymphopoietin (TSLP), IL-33, and IL-25, known to promote Th2-associated inflammation by activation of pivotal type 2 lymphocytes and subsequent production of the effector molecules IL-4, IL-5, and -13. Most of the following examples have not been specifically designed to prevent RV-induced immunopathogenesis, but the mediators they target are involved in RV-induced exacerbations and could therefore be adapted for prevention or treatment of exacerbations or chronic respiratory diseases caused by RV.

TSLP is released by the epithelium in response to RV infection in children.[96] TSLP-activated dendritic cells promote naïve T helper cells to produce the proallergenic Th2 cytokines IL-4, IL-5, and IL-13.[97,98] Levels of TSLP are increased in severe asthma sufferers.[99] Tezepelumab is a mAb against TSLP. Tezepelumab delivered subcutaneously every 2–4 weeks (depending on dosing) reduced exacerbation rates by 61%–66% overall, and in asthmatic subgroups with high eosinophil counts, exacerbation rates were reduced even further (62%–73%) in phase 2 clinical trials.[100] Anti-TSLP mAb therapies are promising interventions for inhibition of pathogenic inflammatory responses induced by RV infection in severe asthma sufferers.

The popularity of targeting epithelial cytokines has resulted in development of four different mAbs to IL-33 in various stages of clinical testing. IL-33 is an epithelial derived alarmin known to promote Th2 cytokine production from T cells and type II innate lymphoid cells (ILC2s) following RV infection.[101] After successful phase 2a testing in atopic dermatitis and peanut allergy patients with ANB020 (antiIL33 mAB) Anaptysbio is in the process of phase 2a testing of ANB020 in severe adult eosinophilic asthma.[102] MEDI3506 is an antiIL-33 mAB in phase I for chronic obstructive pulmonary disease (COPD) treatment.[103,104] REGN3500 (another antiIL-33 mAb) is in phase I development as a monotherapy, an add-on to steroid therapy (fluticasone propionate), or in combination with Dupilumab (an antiIL-4Ra mAb which blocks IL-4 and IL-13 signaling).[105–107] ST2, the receptor for IL-33, is also a therapeutic target with two mAbs in clinical testing for the treatment of asthma.[87,108] A positive outcome from these clinical trials could have promising potential to prevent/alleviate the exacerbation of lung inflammation caused by RV infections in people with chronic respiratory disease.

Other opportunities remain to target epithelial derived cytokines. IL-25 is produced by bronchial epithelial cells in response to RV infection and initiates type 2 immune responses that cause exacerbations.[109] Blocking IL-25 signaling ameliorated airways hyperresponsiveness[110] and allergic pulmonary inflammation by reducing the production of IL-4, IL-5, and IL-13 from T helper and non−T type 2 lymphocytes (likely ILC2s).[109]

9.2.3 The use of heat against the cold

One of the many home remedies against the common cold (and therefore against RV infections), among chicken soup and vitamin C, is the use of steam and heat. There is a fairly sound scientific basis to this practice. Foxman et al. published a temperature sensitive mechanism of innate immunity against RV in mouse nasal epithelial cells that were warmed from a 33°C (comparable to nasal temperatures) to 37°C (comparable to lung temperatures).[111] RV replicated more efficiently at 33°C due to diminished antiviral responses of nasal epithelial cells at 33°C. Warming the nasal epithelial cells promoted more efficient antiviral responses and viral clearance. Unfortunately, this translated poorly to clinical testing. When reviewing six trials across five publications in a periodically updated systematic review, Singh[112] found that heated, humidified air delivered into the upper respiratory tract through the RhinoTherm device had no effect on viral shedding, and there was no significant difference in resolution of symptoms. One Israeli study published a positive effect on nasal airflow with the use of the RhinoTherm device but a study in the United States published a reduction in flow.[112] The authors concluded that there is still a need for more double-blind, randomized trials that include standardized treatment modalities.

Despite the lack of efficacy demonstrated by the RhinoTherm, the inhalation of heated, humidified air remains a common home remedy and is even recommended on some occasions by general practitioner (GPs) and is referred to as steam inhalation therapy (SIT). Unfortunately, the risks of scalds and burns are high (depending on the particular approach to generating the steam and the techniques used for inhalation). The risk is particularly high in children (cases ranging from just 9 months old to 15 years of age) who often receive burns directly to the face (from contact with hot water or steam) or to large areas of the body from spillages of containers of hot water, with burns ranging from 0.25% to 22% of total body area, sometimes requiring surgical intervention (skin grafts).[113−117] In 2016, 150 regional GPs were surveyed in Wales about their use and knowledge of SIT. Twenty-one responded (14%), 17 had recommended

SIT, 8 had recommended it to children <5 years old, and only 5 responders reported the potential risk of burns associated with SIT.[113] SIT is no longer recommended for treating conditions of the upper respiratory tract due to the high risk of burns and lack of efficacy.

9.3 THE POTENTIAL FOR A SUCCESSFUL HUMAN RHINOVIRUS VACCINE

A vaccine, described by the World Health Organization as "a biological preparation that improves immunity to a disease," must fulfill several key criteria before use in humans (see Box 9.1). Overall they must be safe, they must be efficacious by inducing protective immunity, they should be simple to administer, and they should be cost-effective enough for incorporation into a widespread immunization program. Furthermore, because these are biological preparations, issues can surround vaccine stability via production pipelines, the chemical formulation, and transport and storage situations—particularly when application to large numbers and geographically distinct individuals is required. Arguably, the greatest achievement in vaccination has been the eradication of smallpox in 1979—an eventuality attributed to pioneering work by Edward Jenner in 1796 with the first demonstrations of the efficacy of vaccination using cowpox. Since then vaccines have further developed with numerous effective vaccines now available and they continue to save millions of lives following the introduction of mandatory vaccination programs for children.[118]

For RV, the major challenge with vaccine development is to cover points 5 and 6 due to the large number (approximately 160) antigenically and serologically distinct strains or serotypes.[119,120] Additionally, knowledge of what constitutes a protective immune response or immunological correlate of protection for RV remains incompletely understood;

BOX 9.1 The main criteria for a successful vaccine

1. Safe and easy to administer
2. None or very limited adverse effects
3. Cost effectiveness for large-scale application
4. Preparations are stable and storable
5. Induce a protective immune response with immunological memory
6. Generate broad protection and longevity of immune response

however, antibodies are known to protect against reinfection with the same serotype[121] and serotype-specific neutralizing epitopes have been described.[122] Knowledge about T cell responses to RV is even more limited but they are thought to be CD4-dominated and more cross-serotype reactive than antibody responses.[123] Knowledge of T cell epitopes is restricted to RV structural proteins and studies have demonstrated cross-serotype reactivity but protective capabilities are unknown.[124,125] This section of the chapter will review past attempts and the current knowledge regarding the prospects for a protective RV vaccine.

9.3.1 The first attempts at rhinovirus vaccines: human clinical trials

The Common Cold Unit (CCU) in Salisbury, United Kingdom was established in 1946 and pioneered research into the common cold and vaccine development. RVs, as one of the causative agents of the common cold, were first discovered in 1956 and the first known clinical trials investigating a "common cold vaccine" were performed not long after at the CCU in the early 1960s. It was shown that intranasal immunization with live or formalin inactivated single RV serotype could induce virus specific antibodies.[126] However it was not known if these antibodies could protect from RV infection, or their duration of persistence. Other CCU attempts showed that protection to intranasal RV challenge was achieved with prior immunization but only to the homologous virus strain found in the vaccine formulation.[127] Trials in the late 1960s at the National Institute of Allergy and Infectious Disease in the United States showed similar results with a single serotype (RV-A13) immunization regimen and unsurprisingly found that intranasal immunization rather than intramuscular administration was required to generate nasal secretory antibodies, which were defined as the correlate of protection.[128] These protective nasal antibodies were subsequently found to persist for almost 1 year following immunization.[129] Further trials in the United States could not demonstrate strong protective effects when immunizing with a single serotype[130] and because of the continued discovery of serologically distinct RVs, a multivalent approach containing more serotypes was therefore initiated.

New studies testing formalin inactivated multivalent RV vaccines containing 10 serotypes were initiated in the mid-1970s.[131] The rationale for these vaccines was to address the issue of weak cross-serotype protection induced by monovalent vaccination, however perhaps due to the low titer of viral antigens within the formulation, they failed to induce significant cross protection amongst RV serotypes. It was concluded that selection of vaccine constituents containing serotypes based on potential cross-reaction profiles might be more suitable. Furthermore, even though virus inactivation is important for vaccine safety, it is also suspected that inactivation might be unfavorable for the generation of significant immune responses due to loss of protective epitopes. Formalin treatment was almost exclusively used as the method for RV inactivation in the studies mentioned above although alternative methods such as heat treatment (pasteurization), low pH, and ultra violet treatment are also effective and retain immunogenicity.[132] These methods, whilst largely safe for human application, which is a fundamental criterion for safe vaccine application, are likely to destroy critical epitopes required for optimal immune responses and therefore can impact negatively on vaccine efficacy by reducing preparation immunogenicity.[133] Secondly, another potential reason as to why these prior studies displayed limited success is that these inactivated formulations require an adjuvant to amplify the immune responses. The use of adjuvants would most likely have improved vaccine efficacy significantly;[134] however, at that time, Alum was the only approved adjuvant for use in humans whilst there are now several others available.[135] Third, concentration of virus to achieve higher titers of antigen in the formulation would be necessary when creating a multivalent vaccine. As noted previously in the Hamory study,[131] low titer virus preparations were used, which would create difficulties when including numerous RV serotypes in a single formulation, whilst at the same time, maintaining a limited administration volume intranasally.

9.3.2 Second attempts at rhinovirus vaccines: animal studies

After the minimal success of the human trials, experimental studies in immunized animals (rabbits and mice) were initiated. These studies began to determine properties of antibody cross-reactivity,[136–138] which encouraged renewed hope for a RV vaccine due to the convincing demonstration of cross-serotype neutralizing antibodies. Despite these positive steps, RV vaccine research studies in the scientific literature seemingly disappeared for more than 20 years. Studies then began to appear using animals immunized with recombinant RV capsid protein subunits and synthetic peptides, which again proposed possibilities for cross-serotype protective antibodies generation. Short conserved regions at the N-terminus of the capsid protein VP4 eliciting cross-serotype protective antibodies were identified[139] and the entire VP1 polypeptide displayed similar effects.[140] Despite these encouraging studies and the application of modern molecular analyses, the formal demonstration of protective vaccine responses to RVs in in vivo settings remained elusive largely because of the absence of a small animal in vivo model of RV infections. Thus, challenge studies to demonstrate protective capabilities still required immediate translation to human clinical trials, bypassing a critical step in the vaccine development pathway.

9.3.3 Recent approaches: application of mouse and rat models of human rhinovirus infection

A mouse model of human RV infection was developed 10 years ago[141] and has permitted new approaches for RV vaccine development where specific RV challenge following immunization strategies can be addressed. Before this model was available, infection of mouse cells with human RV's was ineffective due to significant sequence differences between the major group entry receptor human ICAM-1 (required for 90% of type A and B serotypes) and its mouse counterpart.[142] The demonstration that mouse cells could be infected with minor group RV through entry via the mouse low-density lipoprotein (LDL) receptor (required for 10% of type A and B serotypes) was mitigated by the fact that there was very limited intracellular viral replication.[143] For the improved mouse model of RV infection, mice transgenic for human ICAM-1 and better methods for generating high titer RV inoculum allowed for the intranasal infection of mice with RVs.[141] Additionally, more recently it has been shown that cotton rats are permissive for RV infection and display characteristics similar to the mouse RV infection model.[144] Thus in these animal models, intranasal RV infection causes acute lung inflammation, activates innate immune responses, initiates adaptive immune responses, and permits limited viral replication. Many of these features reproduce what is known to occur in humans following RV infection except that viral replication is significantly more vigorous in humans. Immunization and challenge strategies have subsequently been investigated in these model systems,[144–147] providing a basis for evaluating the immunological correlates of protection to RVs in vivo that have been induced by immunization strategies. In one such study, hyperimmunization of mice with inactivated RV, followed by homologous intranasal challenge, generated very strong cross-serotype humoral immune responses binding the capsid protein VP1.[146] However, neutralizing antibody was serotype-specific. Using the cotton rat model, prior immunization with inactivated RV via the intramuscular route but not by the intranasal route produced significant neutralizing antibody responses and reduced the viral load in the lungs upon homotypic challenge.[144] In further experiments both prophylactic antibody administration and maternal immunity transfer to neonates were protective[144] and heterotypic responses were noted in newer studies.[147] Collectively, these studies in immunized animals with live RV challenge generated similar results to those obtained in the earlier human trials. Thus, although antibodies were demonstrated as the protective correlate and that immunization could protect against homologous RV challenge, cross-serotype protection was again rather limited.

The role of T cell immunity in protective immune responses to RV was still rather poorly understood and animal studies attempting to induce broadly reactive T cell immunity had not been initiated. However, RV genome sequence information was intensifying, and the discovery plus sequencing of new type C RVs had recently occurred.[119] Thus, bioinformatic analyses of the known type A and B RVs to identify conserved regions of the RV polyprotein allowed the discovery of VP0 as a potential immunogen generating broadly reactive immunity. In subsequent studies, a bacterially produced recombinant version of the RV-A16 VP0 protein was used to immunize mice.[145] Here, recombinant VP0 was immunogenic in vivo, inducing immunogen and RV-specific antibodies as well as cross-serotypic systemic cellular immune responses. The formulation of VP0 within a Th1-promoting adjuvant induced cross-serotype cellular T lymphocytes producing the Th1 cytokine IFNγ and improved Th1-associated RV-specific antibody responses. Interestingly, VP0-immunized mice challenged with heterologous RV strains displayed enhanced cross-reactive cellular, increased memory CD4 T cell numbers and stronger humoral immune responses suggesting broad cross-serotype reactivity was obtained with this strategy. Unexpectedly, VP0 immunization followed by live RV challenge improved the generation of neutralizing antibodies to RV and caused more rapid virus clearance in vivo. This in vivo study of VP0 represents the first subunit vaccine approach for an RV vaccine demonstrating protective cross-serotype immunity. Further experimentation in this model system of RV infection and translation to humans awaits.

9.3.4 Revisiting the multivalent inactivated rhinovirus approach

The repeated limited success of single valency RV vaccination strategies in humans[126–130,148] and in animal models[136–140,144–147] has necessitated revisiting a multivalent approach pioneered by Hamory in 1975.[131] Experiments reinvestigating the Hamory decavalent vaccine revealed that the likely reason for its limited initial success in clinical trials was due to low immunogen levels in the preparation.[149,150] Thus, immunization of mice with an optimized decavalent formulation with higher titers of the RV serotypes included resulted in improved induction of broadly neutralizing antibodies in mice.[149] Such a multivalent vaccine is dependent upon maintaining immunogen levels at a high level without increasing the administration volume to an unacceptable size. Using this approach it was possible to incorporate 50 RV serotypes into a single vaccine formulation administered to rhesus macaques and increase those given to mice to 25 RV serotypes.[149] This remarkable achievement generated neutralizing antibody responses to all of the 25 or 50 serotypes found in the vaccine. However, there was no evidence for cross-serotype reactive antibodies to RV strains not included within the formulation and no in vivo protection studies were performed. Nevertheless, this is an approach that has generated the largest breadth of protection to date and could be the most likely to succeed in future human clinical trials. In fact, polyvalent vaccines in general are becoming more frequent with many now licensed for use in humans, and are perhaps the future when considering that many pathogens have numerous strains and serotypes known.[151]

REFERENCES

1. Barnard DL. Current status of anti-picornavirus therapies. *Curr Pharm Des*. 2006;12 (11):1379–1390.
2. Bauer L, Lyoo H, van der Schaar HM, Strating JR, van Kuppeveld FJ. Direct-acting antivirals and host-targeting strategies to combat enterovirus infections. *Curr Opin Virol*. 2017;24:1–8.
3. McKinlay MA, Collett MS, Hincks JR, et al. Progress in the development of poliovirus antiviral agents and their essential role in reducing risks that threaten eradication. *J Infect Dis*. 2014;210(Suppl 1):S447–S453.
4. Rollinger JM, Schmidtke M. The human rhinovirus: human-pathological impact, mechanisms of antirhinoviral agents, and strategies for their discovery. *Med Res Rev*. 2011;31(1):42–92.

5. Smith TJ, Kremer MJ, Luo M, et al. The site of attachment in human rhinovirus 14 for antiviral agents that inhibit uncoating. *Science.* 1986;233(4770):1286–1293.
6. Hayden FG, Hipskind GJ, Woerner DH, et al. Intranasal pirodavir (R77,975) treatment of rhinovirus colds. *Antimicrob Agents Chemother.* 1995;39(2):290–294.
7. Biota Scientific Management Pty L. *Vaxart. A Phase 2 Study of BTA798 in Asthmatic Adults With Symptomatic Human Rhinovirus Infection*; 2012.
8. Hayden FG, Turner RB, Gwaltney JM, et al. Phase II, randomized, double-blind, placebo-controlled studies of ruprintrivir nasal spray 2-percent suspension for prevention and treatment of experimentally induced rhinovirus colds in healthy volunteers. *Antimicrob Agents Chemother.* 2003;47(12):3907–3916.
9. Turner RB, Wecker MT, Pohl G, et al. Efficacy of tremacamra, a soluble intercellular adhesion molecule 1, for experimental rhinovirus infection: a randomized clinical trial. *JAMA.* 1999;281(19):1797–1804.
10. Wikel JH, Paget CJ, DeLong DC, et al. Synthesis of syn and anti isomers of 6-[[(hydroxyimino)phenyl]methyl]-1-[(1-methylethyl)sulfonyl]-1H-benzimidazol-2-amine. Inhibitors of rhinovirus multiplication. *J Med Chem.* 1980;23(4):368–372.
11. Hayden FG, Gwaltney Jr. JM. Prophylactic activity of intranasal enviroxime against experimentally induced rhinovirus type 39 infection. *Antimicrob Agents Chemother.* 1982;21(6):892–897.
12. Phillpotts RJ, Jones RW, Delong DC, Reed SE, Wallace J, Tyrrell DA. The activity of enviroxime against rhinovirus infection in man. *Lancet.* 1981;1(8234):1342–1344.
13. Phillpotts RJ, Wallace J, Tyrrell DA, Tagart VB. Therapeutic activity of enviroxime against rhinovirus infection in volunteers. *Antimicrob Agents Chemother.* 1983;23(5):671–675.
14. Dorobantu CM, Ford-Siltz LA, Sittig SP, et al. GBF1- and ACBD3-independent recruitment of PI4KIIIbeta to replication sites by rhinovirus 3A proteins. *J Virol.* 2015;89(3):1913–1918.
15. Greninger AL, Knudsen GM, Betegon M, Burlingame AL, Derisi JL. The 3A protein from multiple picornaviruses utilizes the Golgi adaptor protein ACBD3 to recruit PI4KIIIbeta. *J Virol.* 2012;86(7):3605–3616.
16. Lei X, Xiao X, Zhang Z, et al. The Golgi protein ACBD3 facilitates Enterovirus 71 replication by interacting with 3A. *Sci Rep.* 2017;7:44592.
17. Sasaki J, Ishikawa K, Arita M, Taniguchi K. ACBD3-mediated recruitment of PI4KB to picornavirus RNA replication sites. *EMBO J.* 2012;31(3):754–766.
18. Xiao X, Lei X, Zhang Z, Ma Y, Qi J, Wu C, et al. Enterovirus 3A facilitates viral replication by promoting phosphatidylinositol 4-kinase IIIβ-ACBD3 interaction. *J Virol.* 2017;91(19):e00791–17.
19. Hsu NY, Ilnytska O, Belov G, et al. Viral reorganization of the secretory pathway generates distinct organelles for RNA replication. *Cell.* 2010;141(5):799–811.
20. van der Schaar HM, Leyssen P, Thibaut HJ, et al. A novel, broad-spectrum inhibitor of enterovirus replication that targets host cell factor phosphatidylinositol 4-kinase IIIbeta. *Antimicrob Agents Chemother.* 2013;57(10):4971–4981.
21. Spickler C, Lippens J, Laberge MK, et al. Phosphatidylinositol 4-kinase III beta is essential for replication of human rhinovirus and its inhibition causes a lethal phenotype in vivo. *Antimicrob Agents Chemother.* 2013;57(7):3358–3368.
22. Arita M, Wakita T, Shimizu H. Characterization of pharmacologically active compounds that inhibit poliovirus and enterovirus 71 infectivity. *J Gen Virol.* 2008;89(Pt 10):2518–2530.
23. Arita M, Wakita T, Shimizu H. Cellular kinase inhibitors that suppress enterovirus replication have a conserved target in viral protein 3A similar to that of enviroxime. *J Gen Virol.* 2009;90(Pt 8):1869–1879.

24. Ford Siltz LA, Viktorova EG, Zhang B, et al. New small-molecule inhibitors effectively blocking picornavirus replication. *J Virol.* 2014;88(19):11091–11107.
25. Arita M, Kojima H, Nagano T, Okabe T, Wakita T, Shimizu H. Phosphatidylinositol 4-kinase III beta is a target of enviroxime-like compounds for antipoliovirus activity. *J Virol.* 2011;85(5):2364–2372.
26. van der Schaar HM, van der Linden L, Lanke KH, et al. Coxsackievirus mutants that can bypass host factor PI4KIIIbeta and the need for high levels of PI4P lipids for replication. *Cell Res.* 2012;22(11):1576–1592.
27. Keaney EP, Connolly M, Dobler M, et al. 2-Alkyloxazoles as potent and selective PI4KIIIbeta inhibitors demonstrating inhibition of HCV replication. *Bioorg Med Chem Lett.* 2014;24(16):3714–3718.
28. Lamarche MJ, Borawski J, Bose A, et al. Anti-hepatitis C virus activity and toxicity of type III phosphatidylinositol-4-kinase beta inhibitors. *Antimicrob Agents Chemother.* 2012;56(10):5149–5156.
29. Arita M. Mechanism of poliovirus resistance to host phosphatidylinositol-4 kinase III beta inhibitor. *ACS Infect Dis.* 2016;2(2):140–148.
30. Wessels E, Duijsings D, Niu TK, et al. A viral protein that blocks Arf1-mediated COP-I assembly by inhibiting the guanine nucleotide exchange factor GBF1. *Dev Cell.* 2006;11(2):191–201.
31. Wessels E, Duijsings D, Lanke KH, et al. Effects of picornavirus 3A proteins on protein transport and GBF1-dependent COP-I recruitment. *J Virol.* 2006;80(23):11852–11860.
32. Belov GA, Altan-Bonnet N, Kovtunovych G, Jackson CL, Lippincott-Schwartz J, Ehrenfeld E. Hijacking components of the cellular secretory pathway for replication of poliovirus RNA. *J Virol.* 2007;81(2):558–567.
33. Wessels E, Duijsings D, Lanke KH, Melchers WJ, Jackson CL, van Kuppeveld FJ. Molecular determinants of the interaction between coxsackievirus protein 3A and guanine nucleotide exchange factor GBF1. *J Virol.* 2007;81(10):5238–5245.
34. Arita M, Philipov S, Galabov AS. Phosphatidylinositol 4-kinase III beta is the target of oxoglaucine and pachypodol (Ro 09-0179) for their anti-poliovirus activities, and is located at upstream of the target step of brefeldin A. *Microbiol Immunol.* 2015;59(6):338–347.
35. Arita M, Kojima H, Nagano T, Okabe T, Wakita T, Shimizu H. Oxysterol-binding protein family I is the target of minor enviroxime-like compounds. *J Virol.* 2013;87(8):4252–4260.
36. Strating JR, van der Linden L, Albulescu L, et al. Itraconazole inhibits enterovirus replication by targeting the oxysterol-binding protein. *Cell Rep.* 2015;10(4):600–615.
37. Albulescu L, Strating JR, Thibaut HJ, et al. Broad-range inhibition of enterovirus replication by OSW-1, a natural compound targeting OSBP. *Antiviral Res.* 2015;117:110–114.
38. Furuita K, Jee J, Fukada H, Mishima M, Kojima C. Electrostatic interaction between oxysterol-binding protein and VAMP-associated protein A revealed by NMR and mutagenesis studies. *J Biol Chem.* 2010;285(17):12961–12970.
39. Wyles JP, Ridgway ND. VAMP-associated protein-A regulates partitioning of oxysterol-binding protein-related protein-9 between the endoplasmic reticulum and Golgi apparatus. *Exp Cell Res.* 2004;297(2):533–547.
40. Amini-Bavil-Olyaee S, Choi YJ, Lee JH, et al. The antiviral effector IFITM3 disrupts intracellular cholesterol homeostasis to block viral entry. *Cell Host Microbe.* 2013;13(4):452–464.
41. Valente C, Turacchio G, Mariggio S, et al. A 14-3-3gamma dimer-based scaffold bridges CtBP1-S/BARS to PI(4)KIIIbeta to regulate post-Golgi carrier formation. *Nat Cell Biol.* 2012;14(4):343–354.

42. Nhek S, Ngo M, Yang X, et al. Regulation of oxysterol-binding protein Golgi localization through protein kinase D-mediated phosphorylation. *Mol Biol Cell.* 2010;21(13):2327–2337.
43. Hausser A, Storz P, Martens S, Link G, Toker A, Pfizenmaier K. Protein kinase D regulates vesicular transport by phosphorylating and activating phosphatidylinositol-4 kinase IIIbeta at the Golgi complex. *Nat Cell Biol.* 2005;7(9):880–886.
44. Guedan A, Swieboda D, Charles M, et al. Investigation of the role of protein kinase D in human rhinovirus replication. *J Virol.* 2017;91(9).
45. Nguyen A, Guedan A, Mousnier A, et al. Host lipidome analysis during rhinovirus replication in HBECs identifies potential therapeutic targets. *J Lipid Res.* 2018;59(9):1671–1684.
46. Rassmann A, Henke A, Jarasch N, Lottspeich F, Saluz HP, Munder T. The human fatty acid synthase: a new therapeutic target for coxsackievirus B3-induced diseases?. *Antiviral Res.* 2007;76(2):150–158.
47. Wilsky S, Sobotta K, Wiesener N, et al. Inhibition of fatty acid synthase by amentoflavone reduces coxsackievirus B3 replication. *Arch Virol.* 2012;157(2):259–269.
48. Mousnier A, Bell AS, Swieboda DP, et al. Fragment-derived inhibitors of human N-myristoyltransferase block capsid assembly and replication of the common cold virus. *Nat Chem.* 2018;10(6):599–606.
49. Geller R, Vignuzzi M, Andino R, Frydman J. Evolutionary constraints on chaperone-mediated folding provide an antiviral approach refractory to development of drug resistance. *Genes Dev.* 2007;21(2):195–205.
50. Djukanovic R, Harrison T, Johnston SL, et al. The effect of inhaled IFN-beta on worsening of asthma symptoms caused by viral infections. A randomized trial. *Am J Respir Crit Care Med.* 2014;190(2):145–154.
51. Alexopoulou L, Holt AC, Medzhitov R, Flavell RA. Recognition of double-stranded RNA and activation of NF-kappaB by Toll-like receptor 3. *Nature.* 2001;413(6857):732–738.
52. Liu L, Botos I, Wang Y, et al. Structural basis of toll-like receptor 3 signaling with double-stranded RNA. *Science.* 2008;320(5874):379–381.
53. Malcolm BA, Aerts CA, Dubois KJ, et al. PrEP-001 prophylactic effect against rhinovirus and influenza virus—RESULTS of 2 randomized trials. *Antiviral Res.* 2018;153:70–77.
54. Prep Biopharm Limited, Hvivo. *A study to examine the duration of effect of PrEP-001 in healthy subjects challenged with HRV-16.* <https://ClinicalTrials.gov/show/NCT03338556>; 2016.
55. Heil F, Hemmi H, Hochrein H, et al. Species-specific recognition of single-stranded RNA via toll-like receptor 7 and 8. *Science.* 2004;303(5663):1526–1529.
56. Graceway Pharmaceuticals LLC. *Phase 3 study of imiquimod creams in the treatment of external genital warts,* <https://ClinicalTrials.gov/show/NCT00735462>; 2009.
57. Medical University of Vienna. *Immunevasion of human papillomavirus (HPV) in vulvar intraepithelial neoplasia 2/3 and anogenital warts and efficiency and mechanisms of imiquimod treatment.* <https://ClinicalTrials.gov/show/NCT00941811>; 2009.
58. National Institute of Allergy and Infectious Diseases. *Safety and immunogenicity of inactivated influenza A/H5N1 vaccine administered with or without topical Aldara.* <https://ClinicalTrials.gov/show/NCT03472976>; 2019.
59. The University of Hong Kong. *Intradermal influenza vaccine in the Young.* <https://ClinicalTrials.gov/show/NCT02103023>; 2014.
60. The University of Hong Kong. *ID HBV vaccination with imiquimod in OBI.* <https://ClinicalTrials.gov/show/NCT03307902>; 2018.
61. University of Lausanne Hospitals. *Imiquimod and influenza vaccine for immunocompromised patients.* <https://ClinicalTrials.gov/show/NCT02960815>; 2017.

62. Graceway Pharmaceuticals LLC. *Study with a topical gel to treat common warts in adults.* <https://ClinicalTrials.gov/show/NCT00117923>; 2007.
63. Graceway Pharmaceuticals LLC. *An up to twelve week safety and efficacy study with a topical gel to treat common warts in pediatric subjects.* <https://ClinicalTrials.gov/show/NCT00116662>.
64. Graceway Pharmaceuticals LLC. *Twelve week safety and efficacy study with a topical gel to treat common warts in adults.* <https://ClinicalTrials.gov/show/NCT00114920>.
65. University of British Columbia. *Optimizing hepatitis B vaccine response through the use of a topical immune modulator.* <https://ClinicalTrials.gov/show/NCT00175435>; 2007.
66. University of British Columbia. *Enhancing influenza vaccination in seniors with TLR (Toll like receptor) agonists.* <https://ClinicalTrials.gov/show/NCT01737580>; 2013.
67. GlaxoSmithKline. *Effect of the GSK2245035 on the allergen-induced asthmatic response.* <https://ClinicalTrials.gov/show/NCT02833974>; 2018.
68. Tsitoura D, Ambery C, Price M, et al. Early clinical evaluation of the intranasal TLR7 agonist GSK2245035: use of translational biomarkers to guide dosing and confirm target engagement. *Clin Pharmacol Ther.* 2015;98(4):369–380.
69. Pulmotect Inc. *Safety and tolerability of single ascending doses of PUL-042 inhalation solution in healthy subjects.* <https://ClinicalTrials.gov/show/NCT02124278>; 2014.
70. Duggan JM, You D, Cleaver JO, et al. Synergistic interactions of TLR2/6 and TLR9 induce a high level of resistance to lung infection in mice. *J Immunol.* 2011;186 (10):5916–5926.
71. Leiva-Juarez MM, Ware HH, Kulkarni VV, Zweidler-McKay PA, Tuvim MJ, Evans SE. Inducible epithelial resistance protects mice against leukemia-associated pneumonia. *Blood.* 2016;128(7):982–992.
72. Tuvim MJ, Gilbert BE, Dickey BF, Evans SE. Synergistic TLR2/6 and TLR9 activation protects mice against lethal influenza pneumonia. *PLoS One.* 2012;7(1):e30596.
73. Leiva-Juarez MM, Kirkpatrick CT, Gilbert BE, et al. Combined aerosolized Toll-like receptor ligands are an effective therapeutic agent against influenza pneumonia when co-administered with oseltamivir. *Eur J Pharmacol.* 2018;818:191–197.
74. Mifsud EJ, Tan AC, Brown LE, Chua BY, Jackson DC. Generation of adaptive immune responses following influenza virus challenge is not compromised by pretreatment with the TLR-2 Agonist Pam2Cys. *Front Immunol.* 2015;6:290.
75. Tan AC, Mifsud EJ, Zeng W, et al. Intranasal administration of the TLR2 agonist Pam2Cys provides rapid protection against influenza in mice. *Mol Pharm.* 2012;9 (9):2710–2718.
76. Bartlett NW, Girkin J, Wong CY, et al. Upper airway TLR2 immune modulators prime broad respiratory immunity against rhinovirus and influenza infection and inhibit subsequent lung inflammation. *D107. Host Pathog Interact.* 2018;197:A7803.
77. Message SD, Laza-Stanca V, Mallia P, et al. Rhinovirus-induced lower respiratory illness is increased in asthma and related to virus load and Th1/2 cytokine and IL-10 production. *Proc Natl Acad Sci USA.* 2008;105(36):13562–13567.
78. GlaxoSmithKline. *Study of mepolizumab autoinjector in asthmatics.* <https://ClinicalTrials.gov/show/NCT03099096>; 2017.
79. GlaxoSmithKline. *Efficacy and safety study of mepolizumab adjunctive therapy in participants with severe eosinophilic asthma on markers of asthma control.* <https://ClinicalTrials.gov/show/NCT02281318>; 2016.
80. GlaxoSmithKline. *Pharmacokinetics and pharmacodynamics of mepolizumab administered subcutaneously in children.* <https://ClinicalTrials.gov/show/NCT02377427>; 2016.
81. GlaxoSmithKline. *A study to determine long-term safety of mepolizumab in asthmatic subjects.* <https://ClinicalTrials.gov/show/NCT01842607>; 2015.
82. GlaxoSmithKline. *Efficacy and safety study of mepolizumab adjunctive therapy in subjects with severe uncontrolled refractory asthma.* <https://ClinicalTrials.gov/show/NCT01691521>; 2014.

83. GlaxoSmithKline. *Mepolizumab steroid-sparing study in subjects with severe refractory asthma.* <https://ClinicalTrials.gov/show/NCT01691508>; 2013.
84. GlaxoSmithKline. *Dose ranging efficacy and safety with mepolizumab in severe asthma.* <https://ClinicalTrials.gov/show/NCT01000506>; 2012.
85. GlaxoSmithKline. *Dose ranging pharmacokinetics and pharmacodynamics study with mepolizumab in asthma patients with elevated eosinophils.* <https://ClinicalTrials.gov/show/NCT01366521>; 2012.
86. GlaxoSmithKline. *Japanese phase 1 study of mepolizumab.* <https://ClinicalTrials.gov/show/NCT01471327>; 2012.
87. Amgen. *A first-in-human, double blind, single dose study in healthy subjects and subjects with mild atopic asthma.* <https://ClinicalTrials.gov/show/NCT01928368>; 2016.
88. St. Joseph's Healthcare Hamilton, GlaxoSmithKline. *The prednisone-sparing effect of antiIL-5 antibody (SB-240563).* <https://ClinicalTrials.gov/show/NCT00292877>; 2008.
89. Queen's University Belfast, Rasp, Medical Research Council, et al. *Exploring asthma exacerbations in mepolizumab treated patients.* <https://ClinicalTrials.gov/show/NCT03324230>; 2020.
90. McMaster University, Teva Pharmaceuticals, St. Joseph's Healthcare. *AntiInterleukin-5 (IL5) monoclonal antibody (MAb) in prednisone-dependent eosinophilic asthma.* <https://ClinicalTrials.gov/show/NCT02559791>; 2017.
91. GlaxoSmithKline. *MEA112997 open-label long term extension safety study of mepolizumab in asthmatic subjects.* <https://ClinicalTrials.gov/show/NCT01691859>; 2017.
92. GlaxoSmithKline. *A phase 3a, repeat dose, open-label, long-term safety study of mepolizumab in asthmatic subjects.* <https://ClinicalTrials.gov/show/NCT02135692>; 2017.
93. GlaxoSmithKline. *Study of mepolizumab safety syringe in asthmatics.* <https://ClinicalTrials.gov/show/NCT03021304>; 2017.
94. Academisch Medisch Centrum—Universiteit van Amsterdam, The Netherlands Asthma Foundation, GlaxoSmithKline. *Mepolizumab treatment for rhinovirus-induced asthma exacerbations.* <https://ClinicalTrials.gov/show/NCT01520051>; 2013.
95. Bal S, Pol MA Van De, Smids-Dierdorp B, et al. Anti-IL5 treatment alters eosinophil responses to a rhinovirus-16 challenge in mild asthma patients and also that of neutrophils, macrophages and B cells. *A28. Adv. COPD Asthma.* 2017;197:A1196.
96. Perez GF, Pancham K, Huseni S, et al. Rhinovirus infection in young children is associated with elevated airway TSLP levels. *Eur Respir J.* 2014;44(4):1075–1078.
97. Liu YJ. Thymic stromal lymphopoietin: master switch for allergic inflammation. *J Exp Med.* 2006;203(2):269–273.
98. Soumelis V, Reche PA, Kanzler H, et al. Human epithelial cells trigger dendritic cell mediated allergic inflammation by producing TSLP. *Nat Immunol.* 2002;3(7):673–680.
99. Shikotra A, Choy DF, Ohri CM, et al. Increased expression of immunoreactive thymic stromal lymphopoietin in patients with severe asthma. *J Allergy Clin Immunol.* 2012;129(1):104–111.e19.
100. Corren J, Parnes JR, Wang L, et al. Tezepelumab in adults with uncontrolled asthma. *N Engl J Med.* 2017;377(10):936–946.
101. Jackson DJ, Makrinioti H, Rana BM, et al. IL-33-dependent type 2 inflammation during rhinovirus-induced asthma exacerbations in vivo. *Am J Respir Crit Care Med.* 2014;190(12):1373–1382.
102. Londei M, Kenney B, Los G, Marino MH. A phase 1 study of ANB020, an anti-IL-33 monoclonal antibody in healthy volunteers. *J Allergy Clin Immunol.* 2017;139(2):AB73.
103. MedImmune LLC. Safety and tolerability of MEDI3506 in healthy subjects. In: Subjects With COPD and Healthy Japanese Subjects. [Clinical Trial]. 2017;

<https://clinicaltrials.gov/ct2/show/study/NCT03096795>; 2018 Accessed 13.09.18.

104. Lawrence MG, Steinke JW, Borish L. Cytokine-targeting biologics for allergic diseases. *Ann Allergy Asthma Immunol.* 2018;120(4):376–381.
105. Regeneron Pharmaceuticals, Sanofi. *Study of REGN3500 and dupilumab in patients with asthma.* <https://ClinicalTrials.gov/show/NCT03112577>; 2019.
106. Regeneron Pharmaceuticals, Sanofi. *Study of safety, tolerability, and pharmacokinetics of multiple ascending doses of REGN3500 in adults with moderate asthma.* <https://ClinicalTrials.gov/show/NCT02999711>; 2018.
107. Sanofi, Regeneron Pharmaceuticals. *Evaluation of SAR440340 and as combination therapy with dupilumab in moderate-to-severe asthma patients.* <https://ClinicalTrials.gov/show/NCT03387852>; 2019.
108. Janssen Research & Development LLC. *A single ascending dose study in healthy participants and multiple ascending dose study of CNTO 7160 in participants with asthma and participants with atopic dermatitis.* <https://ClinicalTrials.gov/show/NCT02345928>; 2017.
109. Beale J, Jayaraman A, Jackson DJ, et al. Rhinovirus-induced IL-25 in asthma exacerbation drives type 2 immunity and allergic pulmonary inflammation. *Sci Transl Med.* 2014;6(256):256ra134.
110. Ballantyne SJ, Barlow JL, Jolin HE, et al. Blocking IL-25 prevents airway hyperresponsiveness in allergic asthma. *J Allergy Clin Immunol.* 2007;120(6):1324–1331.
111. Foxman EF, Storer JA, Fitzgerald ME, et al. Temperature-dependent innate defense against the common cold virus limits viral replication at warm temperature in mouse airway cells. *Proc Natl Acad Sci USA.* 2015;112(3):827–832.
112. Singh M, Singh M, Jaiswal N, Chauhan A. Heated, humidified air for the common cold. *Cochrane Database Syst Rev.* 2017;8:CD001728.
113. Al Himdani S, Javed MU, Hughes J, et al. Home remedy or hazard?: management and costs of paediatric steam inhalation therapy burn injuries. *Br J Gen Pract.* 2016;66(644):e193–e199.
114. Baartmans M, Kerkhof E, Vloemans J, et al. Steam inhalation therapy: severe scalds as an adverse side effect. *Br J Gen Pract.* 2012;62(600):e473–e477.
115. Belmonte JA, Dominguez-Sampedro P, Perez E, Suelves JM, Collado JM. Severe burns related to steam inhalation therapy. *An Pediatr (Barc).* 2015;82(2):95–99.
116. Murphy SM, Murray D, Smith S, Orr DJ. Burns caused by steam inhalation for respiratory tract infections in children. *BMJ.* 2004;328(7442):757.
117. Wallis BA, Turner J, Pearn J, Kimble RM. Scalds as a result of vapour inhalation therapy in children. *Burns.* 2008;34(4):560–564.
118. Doherty M, Buchy P, Standaert B, Giaquinto C, Prado-Cohrs D. Vaccine impact: Benefits for human health. *Vaccine.* 2016;34(52):6707–6714.
119. Palmenberg AC, Spiro D, Kuzmickas R, et al. Sequencing and analyses of all known human rhinovirus genomes reveal structure and evolution. *Science.* 2009;324(5923):55–59.
120. McIntyre CL, Knowles NJ, Simmonds P. Proposals for the classification of human rhinovirus species A, B and C into genotypically assigned types. *J Gen Virol.* 2013;94(Pt 8):1791–1806.
121. Barclay WS, al-Nakib W, Higgins PG, Tyrrell DA. The time course of the humoral immune response to rhinovirus infection. *Epidemiol Infect.* 1989;103(3):659–669.
122. Sherry B, Mosser AG, Colonno RJ, Rueckert RR. Use of monoclonal antibodies to identify four neutralization immunogens on a common cold picornavirus, human rhinovirus 14. *J Virol.* 1986;57(1):246–257.
123. Gern JE, Dick EC, Kelly EA, Vrtis R, Klein B. Rhinovirus-specific T cells recognize both shared and serotype-restricted viral epitopes. *J Infect Dis.* 1997;175(5):1108–1114.

124. Muehling LM, Mai DT, Kwok WW, Heymann PW, Pomes A, Woodfolk JA. Circulating memory CD4 + T cells target conserved epitopes of rhinovirus capsid proteins and respond rapidly to experimental infection in humans. *J Immunol.* 2016;197(8):3214–3224.
125. Gaido CM, Stone S, Chopra A, Thomas WR, Le Souef PN, Hales BJ. Immunodominant T-cell epitopes in the VP1 capsid protein of rhinovirus species A and C. *J Virol.* 2016;90(23):10459–10471.
126. Doggett JE, Bynoe ML, Tyrrell DA. Some attempts to produce an experimental vaccine with rhinoviruses. *Br Med J.* 1963;1(5322):34–36.
127. Mitchison DA. Prevention of colds by vaccination against a rhinovirus: a report by the scientific committee on common cold vaccines. *Br Med J.* 1965;1(5446):1344–1349.
128. Perkins JC, Tucker DN, Knope HL, et al. Evidence for protective effect of an inactivated rhinovirus vaccine administered by the nasal route. *Am J Epidemiol.* 1969;90(4):319–326.
129. Buscho RF, Perkins JC, Knopf HL, Kapikian AZ, Chanock RM. Further characterization of the local respiratory tract antibody response induced by intranasal instillation of inactivated rhinovirus 13 vaccine. *J Immunol.* 1972;108(1):169–177.
130. Douglas Jr. RG, Couch RB. Parenteral inactivated rhinovirus vaccine: minimal protective effect. *Proc Soc Exp Biol Med.* 1972;139(3):899–902.
131. Hamory BH, Hamparian VV, Conant RM, Gwaltney Jr. JM. Human responses to two decavalent rhinovirus vaccines. *J Infect Dis.* 1975;132(6):623–629.
132. Hughes JH, Mitchell M, Hamparian VV. Rhinoviruses: kinetics of ultraviolet inactivation and effects of UV and heat on immunogenicity. *Arch Virol.* 1979;61(4):313–319.
133. Bachmann MF, Kundig TM, Kalberer CP, Hengartner H, Zinkernagel RM. Formalin inactivation of vesicular stomatitis virus impairs T-cell- but not T-help-independent B-cell responses. *J Virol.* 1993;67(7):3917–3922.
134. Reed SG, Orr MT, Fox CB. Key roles of adjuvants in modern vaccines. *Nat Med.* 2013;19(12):1597–1608.
135. Rappuoli R, Mandl CW, Black S, De Gregorio E. Vaccines for the twenty-first century society. *Nat Rev Immunol.* 2011;11(12):865–872.
136. Cooney MK, Wise JA, Kenny GE, Fox JP. Broad antigenic relationships among rhinovirus serotypes revealed by cross-immunization of rabbits with different serotypes. *J Immunol.* 1975;114(2 Pt 1):635–639.
137. Fox JP. Is a rhinovirus vaccine possible? *Am J Epidemiol.* 1976;103(4):345–354.
138. McCray J, Werner G. Different rhinovirus serotypes neutralized by antipeptide antibodies. *Nature.* 1987;329(6141):736–738.
139. Katpally U, Fu TM, Freed DC, Casimiro DR, Smith TJ. Antibodies to the buried N terminus of rhinovirus VP4 exhibit cross-serotypic neutralization. *J Virol.* 2009;83(14):7040–7048.
140. Edlmayr J, Niespodziana K, Popow-Kraupp T, et al. Antibodies induced with recombinant VP1 from human rhinovirus exhibit cross-neutralisation. *Eur Respir J.* 2011;37(1):44–52.
141. Bartlett NW, Walton RP, Edwards MR, et al. Mouse models of rhinovirus-induced disease and exacerbation of allergic airway inflammation. *Nat Med.* 2008;14(2):199–204.
142. Register RB, Uncapher CR, Naylor AM, Lineberger DW, Colonno RJ. Human-murine chimeras of ICAM-1 identify amino acid residues critical for rhinovirus and antibody binding. *J Virol.* 1991;65(12):6589–6596.
143. Yin FH, Lomax NB. Establishment of a mouse model for human rhinovirus infection. *J Gen Virol.* 1986;67(Pt 11):2335–2340.

144. Blanco JC, Core S, Pletneva LM, March TH, Boukhvalova MS, Kajon AE. Prophylactic antibody treatment and intramuscular immunization reduce infectious human rhinovirus 16 load in the lower respiratory tract of challenged cotton rats. *Trials Vaccinol.* 2014;3:52–60.
145. Glanville N, McLean GR, Guy B, et al. Cross-serotype immunity induced by immunization with a conserved rhinovirus capsid protein. *PLoS Pathog.* 2013;9(9): e1003669.
146. McLean GR, Walton RP, Shetty S, et al. Rhinovirus infections and immunisation induce cross-serotype reactive antibodies to VP1. *Antiviral Res.* 2012;95(3):193–201.
147. Patel MC, Pletneva LM, Boukhvalova MS, Vogel SN, Kajon AE, Blanco JCG. Immunization with live human rhinovirus (HRV) 16 induces protection in cotton rats against HRV14 infection. *Front Microbiol.* 2017;8:1646.
148. Perkins JC, Tucker DN, Knopf HL, Wenzel RP, Kapikian AZ, Chanock RM. Comparison of protective effect of neutralizing antibody in serum and nasal secretions in experimental rhinovirus type 13 illness. *Am J Epidemiol.* 1969;90 (6):519–526.
149. Lee S, Nguyen MT, Currier MG, et al. A polyvalent inactivated rhinovirus vaccine is broadly immunogenic in rhesus macaques. *Nat Commun.* 2016;7:12838.
150. Stobart CC, Nosek JM, Moore ML. Rhinovirus biology, antigenic diversity, and advancements in the design of a human rhinovirus vaccine. *Front Microbiol.* 2017;8:2412.
151. Schlingmann B, Castiglia KR, Stobart CC, Moore ML. Polyvalent vaccines: high-maintenance heroes. *PLoS Pathog.* 2018;14(4):e1006904.

CHAPTER 10

Techniques for detection and research

Brian Gregory George Oliver[1,2]
[1]Woolcock Institute of Medical Research, The University of Sydney, Glebe, NSW, Australia
[2]School of Life Sciences, University of Technology Sydney, Ultimo, NSW, Australia

10.1 OVERVIEW

There is no standard or agreed upon methodology for the detection of human respiratory viruses in general, and even when a single respiratory virus is considered, for example, rhinovirus (RV), there are numerous methodologies used in clinical practice and in research studies. The outcomes of variation in methodology are not known, and perhaps are somewhat irrelevant during active symptomatic infections. During such periods, there are high virus loads, and as such any subtleties in assay sensitivity are unlikely to be important. Furthermore, if there is positive result from a diagnostic test and concomitant symptoms of an infection (runny nose, cough, etc.), it does not take a 30-minute debate to work out what is going on. To quote Douglas Adams: "If it looks like a duck, and quacks like a duck, we have at least to consider the possibility that we have a small aquatic bird of the family anatidae on our hands." However, not all infections are symptomatic. In such instances a positive diagnostic test result in patients who are asymptomatic causes confusion. The concept that asymptomatic RV infections can occur is relatively new. This was perhaps the result of initial studies that sampled people with symptoms of an upper respiratory tract infection, or only during exacerbations. We and others have found frequent asymptomatic infections in children and adults. The frequency of asymptomatic infections in healthy people is relatively low, for example, 8.3% of university students had asymptomatic RV infection.[1] However, the detection of asymptomatic infections might be influenced by underlying respiratory disease, for example, in adult patients with bronchiectasis we found asymptomatic RV infection in 83% of patients during winter.[2] In children with asthma,

Rhinovirus Infections
DOI: https://doi.org/10.1016/B978-0-12-816417-4.00010-X

detection of asymptomatic RV infections is still relatively high (36%) compared with detection rates of 57% during exacerbations.[3] Furthermore, RV infection can often be masked by underlying respiratory disease. For example, is cough or phlegm a symptom of asthma or a symptom of infection, or a symptom of both? There is no simple answer to this question, but it is important to consider when interpreting RV test results. The common cold questionnaire is often used as a screening tool for respiratory viral infections and is based upon symptoms of infection. In people with asthma, Powell et al. addressed the question of whether the common cold questionnaire can distinguish people with asthma who were infected with respiratory viruses from those that were not infected; they found that the common cold questionnaire was not able to distinguish people with asthma who were infected with virus from those that were not infected.[4]

In the last 10 years, molecular-based testing for respiratory viruses has become commonplace, even now as point of care tests.[5] The results from the use of such testing for RV have led to a number of assumptions, arguments, and the elephants in the room, which now need to be considered.

In this chapter there are both a narrative overview of these concepts and a more detailed technical description.

10.2 NARRATIVE DESCRIPTION

10.2.1 Does polymerase chain reaction detect replication competent virus or fragments of viral RNA not directly associated with active rhinovirus infection?

The advent of polymerase chain reaction (PCR) of RV in 1993 led to greater detection of respiratory viruses in respiratory secretions,[6] but more recently also created the idea that PCR identifies fragments of nucleic acids from previous infections and not infective virions. This is a topic that is greatly contested, and so that there is no doubt about my point of view here it is: 99.99% of the time PCR detects viruses that are either trying to, or are actively infecting the host.

So, what does this mean? If we use the scenario that a colleague is interested in studying airborne transmission of viruses, the likely scenario is that they will aerosolize a pathogen that does not cause infections in humans. Bacteriophages are commonly used,[7] but if the colleague was interested in airborne transmission of RV, they might use something with greater similarity, for example, equine rhinitis A virus[8] (formerly known as equine RV). If I was exposed to the aerosol, for example, I walked through their lab, and swabs and/or washings of my respiratory tract, skin, and clothing were taken (let us just say over a 2-day period, for argument's sake), it is highly likely that as occurs with RV, equine rhinitis A virus will be present on my clothes,[9] skin,[10] and in my respiratory tract.[11]

Of course, no one would think that the presence of equine rhinitis A virus in a clothes' swab equals infection, but what happens if the virus is found in a nasal swab or lung washings? On its own this does not equate to an active infection. Let us now put this in the context of typical clinical scenario. If we consider healthcare workers who are exposed to bioaerosols from a respiratory department outpatient clinic in winter, or a study volunteer who is waiting in the waiting room for a study visit, sampling of either person's airway might detect RV. I would still argue that this virus is capable of causing an infection, but the presence of the virus does not equate to an infection occurring. However, if the amount of virus increased over time this would be good evidence of active infection, and furthermore if the rise in virus was associated with symptoms and/or elevation in biomarkers of infection then it would be reasonable to conclude that active infection was occurring.

10.2.2 Should you always carry out an assay which demonstrates virus infectivity?

If PCR detection alone is not sufficient to confirm that active infection is occurring, then why not carry out an assay that detects the presence of infective virions, also commonly referred to as replication competent virus. Examples of virus replication assay are plaque forming unit assays[12] or a viral titration assay.[13] In some ways a virus replication assay is more informative than PCR as it is empirical proof that you have a replication competent (infectious) virus in a given sample. However, continuing on from the example earlier, if samples were positive for replication competent equine rhinitis A, this does not mean that I am currently infected. If we accept that equine rhinitis A virus cannot infect humans, all this result means is that I have replication competent virus in my respiratory secretions.

From a practical point of view, the major disadvantage of assays for live virus is that no single assay is universally applicable.[14] Each virus replication assay uses specific cell lines that are permissive to certain viruses, so to detect equine rhinitis A virus, an equine cell line would be needed. This is perhaps the major reason why PCR was initially welcomed by diagnostic labs. From a research perspective, even though assays for live virus are more informative that PCR they are rarely done to identify/quantify viruses in unknown samples. For example, if you have no idea which virus(es) are present this could lead to a lot of cell culture–based assays, which when combined with a lower sensitivity of detection in comparison with PCR[15] may not necessarily convince anyone that negative results are in fact a true negative.

Furthermore, even when only considering RV there is no guarantee that the infection model (assay) is appropriate. Twenty years ago, RV research was simple: there were two groups of RV,[16] distinguished by the receptors which they used to infect cells, with detection occurring either by assays for live virus (PFU or viral titration assays typically utilizing HeLa cells) or PCR-based assays. As molecular techniques progressed, RV groups were redefined based upon their molecular sequence (phylogenetic similarity), and this led to three groups, A, B, and C, being identified.[17] A slight problem existed, however: only groups A and B were detectable by infectivity assays. In 2012 the problem was solved by researchers from Medimmune who discovered that ciliated epithelial cells were permissive to RV C, whereas undifferentiated epithelial cells (which lack cilia) were not.[18] This finding helped to explain the earlier observation of how RV C replication was occurring in organ culture.[19]

10.2.3 Can polymerase chain reaction lead to false positive results?

So far all we have established is that no single diagnostic test is definitive. But what about the possibility of false positive test through the detection of RNA fragments? My opinion in the case of RV is that this is highly unlikely to be relevant. RNA is extremely fragile and rapidly degraded by ribonucleases. Pathogenic respiratory viruses are mainly single-stranded RNA viruses that are relatively fragile, with most unable to remain infective after a couple of days at either room temperature or 37°C. There are numerous examples in the literature, with initial virus titer, dry/wet storage, temperature, and exposure to UV irradiation the most important contributors to overall virus survival.[20,21] Thomas et al. carried out an interesting study looking at the survival of influenza on Swiss bank notes. Switzerland has plastic bank notes, and therefore Swiss bank notes can be assumed to represent a relatively inert surface. Bank notes were inoculated with different strains of influenza, and infectivity assessed over a week. Of the four strains of influenza, two could not be detected 4 hours postinfection, and only one of the four remained infective at 3 days postinoculation. However, PCR detection occurred at up to 10 days postinoculation, demonstrating the difference in assay sensitivity.[22]

10.2.4 Practical ways to detect rhinovirus

The classical approach to identifying RV infections was to inoculate nasal samples onto various cell lines, testing for acid lability (to differentiate enterovirus from RV). This method of detection is still in use; however, it has been shown to be less sensitive than detection by PCR techniques, and in older studies has probably led to underestimation of the frequency of rhinoviral infections.

Unlike other respiratory viruses, the detection of RV is relatively simplistic in that only lytic cell culture–based assays or PCR are widely used, and realistically PCR is used as the assay of choice. In comparison, other respiratory viruses can be (diagnostically) identified by the use of antibody staining of infected cells or the use of antibodies in immunoassays or immunochromatographic (lateral flow) rapid tests (e.g., dipstick tests).[23,24] In recent years, commercially available antibodies against RV have become available for research purposes, but in the author's opinion it is unlikely that they would be developed into a diagnostic test without the release of a specific antirhinoviral medication. Furthermore, the structure of the virion is known to differ among different RV types, and it was this difference that allowed the initial classification into serotypes based on the use of neutralizing antibodies.[25] Therefore, depending on what the antibody recognizes it might have limited ability to detect all RV types.

There are differences between PCR assays, but for the sake of simplicity these can be divided into older PCR assays where the products need to be run on a gel, or newer quantitative PCR (qPCR) assays, where the fluorescence of the product is detected cycle by cycle, giving a real-time readout of amplification. In our lab the sensitivity of the two assays is similar, and if needed both assays can utilize preamplification approaches (e.g., nested PCR). The choice of assay is often arbitrary, with investigators' personal preference, funding, or even the age of the investigator being the deciding factor. Personally, I would always recommend that for people unfamiliar with RV PCR, or if samples have been collected differently from what the group is used to dealing with, gel-based PCR detection should initially be used. The big advantage that gel-based PCR has is that it is easy to evaluate sample quality; for example, by assessing if there are multiple products formed, or if degradation of the sample has occurred (smearing of the band).

10.2.5 How to take samples: overview

This is perhaps the area that has the greatest variability in published studies, and in my opinion incorrect sampling can radically change the outcome of a study. The exact sampling procedure is often a trade-off between what is available locally, what is permissible by ethics committees, and when the PCR is being done externally what the diagnostic/research lab expects. I would urge investigators to spend some time thinking about the best approach and evaluating data under local conditions. For example, what is appropriate for a field-based study in Central Australia where temperatures are typically in excess of 45°C, and a hospital-based study with a local diagnostic lab will be radically different. For an overview of sampling and analysis, see Fig. 10.1. There are two important components of sampling: the nature of the sample (washings,

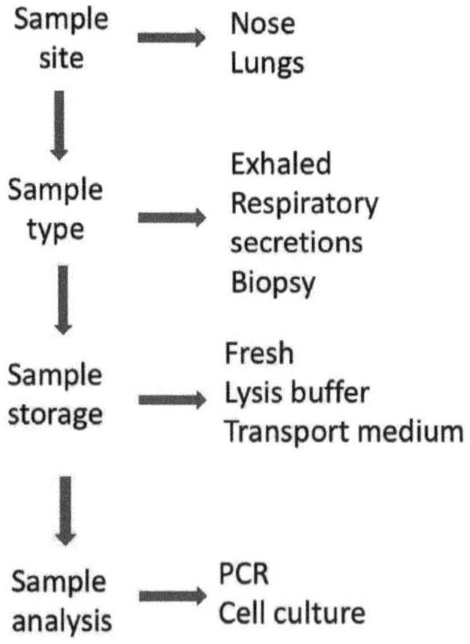

Figure 10.1 An overview of RV sampling and detection. *RV*, Rhinovirus.

swab, and biopsy), and the storage of the sample prior to analysis, which will be further discussed in the following sections.

10.2.6 Sampling site

The temperature of the upper airways is 33°C while the lower airways have a temperature of 37°C. Initially this temperature differential was thought to be sufficient to inhibit RV replication, as most viruses are highly adapted to replication within the local environment. For example, many viruses have evolved to be able to infect the stomach, which has a very acidic pH; in contrast, at such a low pH, RV is unable to replicate. As such RV infection of the epithelium in the upper airways was the initial research focus; however, RV research, especially during the last 20 years, has shown that RV also infects the lower airways.[26] As with the upper airways, the classical target of RV research in the lower airways is the airway epithelium, and as such much work has been devoted to this area.

The idea that RV would be unable to replicate at 37°C was probably derived from some of the initial RV isolation techniques carried out by Tyrrell and Parsons in 1960.[27] They found that by lowering the temperature of the tissue culture vessels to 33°C, and reducing the pH to 7 (mimicking the conditions found in the nose), a new RV serotype could be identified. Using these conditions and using different cell lines several other RV serotypes were isolated. Experiments to disprove this doctrine were carried out by Papadopoulos et al.[28] In their study, RV infection was shown to be successful at both temperatures, with slightly higher virus production obtained with infections occurring at 33°C. However, some of the serotypes tested replicated as well or even better at 37°C.[28] Thus it would appear that the preferable temperature of replication is serotype specific among RVs.

Accumulation of evidence of RV infection of the lower airways in vivo was marred by the bronchoscopic technique needed to take samples. During bronchoscopy the flexible fiberoptic bronchoscope is first introduced into either the nose or the mouth. Even in studies that have bronchoscoped patients via the oral route, contamination of the scope with virions during its passage into the lungs is difficult to rule out. However, with the use of in situ hybridization techniques, the presence of RV RNA within bronchial epithelial cells has confirmed RV as a pathogen of the lower airways.[26] In this paper the group also shows RV infection of the subepithelial cells, which were not specifically identified but could have been mesenchymal in origin.[26]

In a practical sense what this means is that if a single sampling site, that is, upper or lower airways, is used in a study it might not collect all possible RV infections. While this is a simple enough concept, it is further complicated by the type of sample collected. For example, if you want to collect samples from the lung, exhaled breath, induced or spontaneously expectorated sputum, lung washings (bronchoalveolar lavage fluid), biopsy (brush biopsy, cupped biopsy, cryobiopsy), or swabs (including bronchosorption and nasosorption[29]) are some of the choices available. These collection techniques have varying degrees of technical expertise with some needing to be carried out within a hospital setting, and vary in the potential danger to the patient/volunteer. So which sample to use? At this point we have what I would consider to be an overwhelming list of possibilities for sampling. More often than not, the choice of lower airways sample collected is based on what is available, where the study is being carried out (some methodologies listed earlier should only be done in a hospital environment), or what was done in study X that the current study is based upon. For RV research, exhaled breath, induced or spontaneously expectorated sputum, and lung washings (bronchoalveolar lavage fluid) are preferable sample types to lung biopsies. Lung biopsies sample relatively small areas, and therefore have a decreased possibility of containing RV in comparison with other methods.

In 2008 we aimed to develop a technique to noninvasively detect RV in the lungs. We based our initial research on collection of RV contained in exhaled breath samples using a material called electret. Electret is made from electrostatically charged fibers that are woven into a cloth-like material. It is commonly used in filters, or as dusting cloths due to its ability to capture particles. Using "homemade" sampling devices in our proof of concept studies we were able to demonstrate that RV was exhaled from people with naturally occurring infections.[30] However, the detection efficiency was not as good as that achieved by collecting nasal mucous. At the same time, other research groups developed different technology to detect exhaled respiratory viruses,[31] with some more successful than others.[32] We continued to refine the device design,[33] and even used our devices in longitudinal study of RV presence in school-aged children.[34] These studies yielded some interesting data and generated new research hypotheses, but overall reduced RV detection in comparison to nasal washings persisted. In 2016 we decided to evaluate a new methodology,

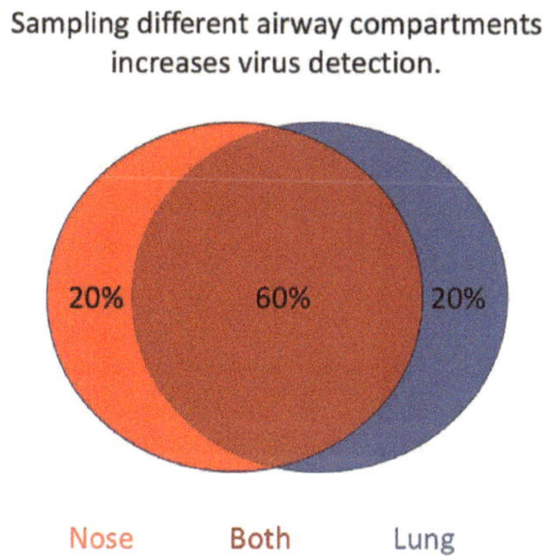

Figure 10.2 Increased RV detection occurs when multiple samples from the same patient are taken. *RV*, Rhinovirus.

but rather than make a new device we utilized spirometry filters, and collected these from patients who were undergoing routine lung function testing.[35] Using spirometry filters, we found that detection was now equivalent to that achieved from sampling BAL or nasal mucus.[2] Unexpectedly, what our study also revealed was that when sample-collection methodologies were compared (e.g., exhaled breath and BAL) 60% of patients had samples from both methodologies that were positive, and 20% of patients were positive in only BAL, and the remaining 20% positive in only exhaled samples (see Fig. 10.2).

We do not know the exact reason why detection with spirometry filters was better than our homemade filters. The most likely reason is not in fact differences in the two filters but how differences in how the samples were collected. In our previous research we collected exhaled samples from people during tidal breathing. In a lung function test, patients are asked to take a deep breath and exhale forcefully emptying their lungs. In comparison with tidal breathing, such maneuvers are known generate more particles,[36,37] and the we think that this is the likely reason why detection sensitivity was increased.

10.2.7 Sample preparation/storage

Once you have decided upon the optimal sample type for your particular cohort/study, the next consideration is how to store the samples. In general, even under optimized conditions freezing of samples reduces viral infectivity by around 1 log. As a personal anecdote, my lab spent around 1 year optimizing the freezing conditions for respiratory syncytial virus to maximize virus infectivity. RV by comparison is relatively hardy, but repetitive freeze–thaw cycles need to be avoided. At the time of collection, the downstream analysis methodology should be established, and this allows the samples to be stored optimally. Our personal experience is that if PCR is the downstream analysis methodology storing samples in RNA lysis/isolation buffer provides optimal storage conditions. However, a wide range of virus storage mediums are commercially available. These might give greater yields depending upon how quickly the samples can be frozen, or if virus infectivity assays are the downstream analysis method. As with all samples that degrade rapidly, rapid freezing and storage (−80°C or below) allows for relatively long-term storage of samples. If samples are to be stored for more than 6 months, under ideal conditions a test sample should also be stored so that a measure of degradation can be made. In the author's opinion it is worth evaluating two to three different sample storage methodologies prior to embarking upon a study for the first time.

10.2.8 The ideal study

Hopefully, the information above has been useful in refining your clinical trial design. However, it presents a number of potential options. My personal preference for a sampling strategy would be to collect exhaled samples from spirometry filters and one other sample type. Preferably, the second sample should be from the upper airway. Samples should be collected on at least two timepoints, if not more. If the study is aiming to look at the prevalence of viruses during periods of acute symptoms, it is important to collect samples in the steady state so that background infection rates can be calculated. Once collected, samples should be divided into two tubes and stored in different locations; this may sound obvious, but I know of several clinical trials that have lost samples to equipment failure. Every effort should be made to minimize losses to degradation. For example, cDNA is more robust that RNA, so why not convert samples into cDNA. As with any study the downstream analysis should be decided at the start of the study, and if possible, kits should be purchased at the start. We had one study where the manufacturer decided to discontinue a particular RNA isolation kit, and we had stored samples in RNA lysis buffer for that particular kit. Whichever analysis method you choose, it is always preferable to have some estimation of viral load. Without this, there is the potential for reviewers to ask some interesting questions, but the relationship between virus load and how it relates to symptoms is not as straightforward as it seems. In general, during a cold, peak symptoms correlate with high viral titers, and viral titers are higher in children. However, low titers should not be ignored, but can be difficult to obtain if the virus testing has been outsourced. Diagnostic labs tend to have arbitrary cutoffs for RV PCR, determining what is considered a positive test result and what is negative. It is well worth remembering the seminal study from Prof. Johnston in COPD. The idea of the study was to define the minimum infective dose of RV that would result in an increase of symptoms (an exacerbation) of COPD. For safety reasons the study was designed to start with a very low dose of infection, and gradually increase the infective dose. However, experimental infection even with two virions was sufficient to cause an exacerbation.[38]

10.3 TECHNICAL CONSIDERATIONS TO DETECTING RHINOVIRUS

10.3.1 Polymerase chain reaction

As described earlier there are different types of PCR, which can be divided into standard PCR, real-time qPCR, or the newer digital PCR. It is important to remember that the sensitivity and specificity of all these PCR methodologies is in part determined by both the quality of the initial sample, and the viral RNA extraction methodology, the reverse transcription method used, and the primer design.

10.3.2 Extraction methods

There are several commercial automated extraction technologies available, and almost universally they have better sample to sample reproducibility and faster speed of extraction; however, there are differences in extraction effeminacy across different systems.[39–41] The use of an automated extraction system is recommended if extraction and analysis will occur across different sites.

Verheyen et al. compared five different automated extraction platforms (easyMAG, bioMerieux, m2000sp, Abbott, MagNA Pure LC 2.0, Roche, QiaSymphony, Qiagen, sample preparation module of the VERSANT kPCR Molecular System, Siemens). While not using samples relevant to respiratory virology, they found marked differences across the different systems when using stool samples, and different quantities of viral RNA from serum samples.[42] Yang et al. used nasal wash samples that were spiked with respiratory pathogens and compared six automated systems and one manual extraction methodology. While they did not use RV, they found that the KingFisher mL and easyMAG produced 1- to 3-log wider linearity and around four times more RNA from samples spiked with human influenza virus and respiratory syncytial virus. For more advanced analysis methods such as genomic analysis similar differences between extraction methodologies have also been observed for respiratory viruses.[43]

10.3.3 Reverse transcription

Reverse transcription enzymes convert viral RNA into a double-stranded DNA often referred to as complementary DNA or cDNA. There is a multitude of RT enzymes available, with some incorporated into automated extraction technologies. However, not all RT enzymes perform equally. With conventional PCR (e.g., qPCR) and good quality RNA, most RT enzymes will perform adequately, but for samples with low viral copy number (a low number of virions) some evaluation of different commercial products is recommended. In cases were low copy numbers of viral RNA are expected techniques such as linear amplification are possible,[44] but because RNA is amplified prior to PCR enumeration of viral RNA copy number is not possible.

10.3.4 Primer design

The art of designing PCR primers has almost been forgotten with PCR assays available as preoptimized of the shelf commodities. However, good primer design can increase PCR sensitivity by 100–1000-fold.[14] Initial PCR primers for RV[15] are capable of detecting several picornaviruses (e.g., enteroviruses), while some qPCR primers detect only RV,[16] and even differentiate the three RV species from each other.[45]

10.3.5 Polymerase chain reaction sensitivity

A seminal study from the Bardin Lab established the superior sensitivity of qPCR over conventional PCR (sensitivity of 72% vs 39%),[46] but with both qPCR and conventional PCR it is possible to carry out nested PCR[47] to increase assay sensitivity. In a population-based study of around 1000 people nested PCR has changed RV detection rates from 6% to

24%.[48] Older strategies to increase sensitivity with conventional PCR to incorporate a degenerate internal probe, which can used in a ELISA-based assay to increase sensitivity by a factor of 10,[49] are by and large not used frequently but may have still utility. Increasing sensitivity of a PCR reaction comes at a cost of decreasing specificity especially if a low fidelity polymerase is used, and care needs to be taken if this approach is going to be used for sequencing.

10.3.6 Digital polymerase chain reaction

Digital PCR is not a new technique—the term was coined in 1999[50]—but had in fact been used over the preceding decade.[51] The main difference between conventional or qPCR and digital PCR is that in digital PCR the reaction is divided into thousands or millions of reactions such that each reaction contains a single molecule of interest or not as the case might be, and a digital (all or none) signal is obtained. One advantage that digital PCR has over other forms of PCR is that it is less affected by PCR efficiency (the ability to logarithmically increase PCR product),[52,53] which translates to decreased PCR variation,[54] that is, there is increased precision in digital PCR. This increased precision was elegantly demonstrated by Sedlak et al.,[45] who compared the ability of qPCR and digital PCR to distinguish RV genotypes.

10.3.7 Molecular assays to detect different rhinovirus genotypes

There are at least 168 different genotypes of RV,[55] which at the moment might be more important from a research point of view than as a clinical diagnostic. RV can be genetically divided into three groups: RV A, B, and C. RV C is the more recently discovered group. It was discovered in 2006[56] using molecular techniques, and initially caused great interest as it appeared that the symptoms of infection[57,58] were more severe in comparison to RV A or B. The increased symptoms might be specific to children as this does not occur in adults.[59] Furthermore, several more recent studies suggest that RV A causes more severe infections. For example, in a study of 1003 children with severe acute respiratory infections RV A accounted for 14% of infections while RV B was 2.1% and RV C 5.6%.[60] Nonetheless understanding which genotype of RV is associated with severe symptoms or exacerbations of respiratory diseases has important implications for future antiviral drug development. As mentioned earlier it is possible to distinguish different genotypes of RV using digital PCR, but from a research perspective this yields insufficient information. The current gold standard to define genotype of RV is sequencing.

10.3.8 Standard sequencing to determine rhinovirus genotype

Any PCR product can be sequenced to define its exact molecular sequence. As sequencing costs have dropped dramatically, with typical costs of around US$10 per sample for a nonpurified PCR product, in some labs sequencing is now routine. Sequencing not only discovered RV C, it allows study of the phylogenetic similarity of RV infections in a person over time; it is also an incredibly powerful tool to understand viral churn within the community.[61] Some thought, however, needs to be given to the primer design used in sequencing, as if the primer targets a conserved region of the RV genome it will not be possible to distinguish genotypes. Fortunately, RV sequences in the viral protein 4–2 region are highly variable and when combined with the 5′UTR regions can be used to distinguish genotypes.[62]

10.3.9 Alternative approaches to polymerase chain reaction sequencing

PCR-based sequencing is a robust technique; however, it is predicated by the choice of PCR primers used, as sequencing is only possible if there is a PCR product. Next-generation sequencing, which is now often referred to as RNA sequencing (RNAseq), offers an alternative approach.[63] Simplistically RNAseq is unbiased sequencing of all cDNA, with identification made possible by aligning sequences to reference genomes or assembling novel transcripts. RNAseq has been used for well over a decade[64,65] and now a robust well-validated approach to examining the transcriptome. As RNAseq utilizes cDNA libraries, if, for example, bronchial biopsies are used and viruses are present they would also be reverse transcribed and sequenced. Typically, the presence of viruses in reparatory samples is not reported in transcriptomic studies, but this creates an opportunity given the large amount of publicly available respiratory RNAseq data files available (e.g., in the Gene Expression Omnibus[66]). A recent study has used RNAseq to identify RV. Wesolowska-Andersen et al. utilized nasal brushings in a dual qPCR and RNAseq analysis. RNAseq had a sensitivity of 86% compared with qPCR, but detected viruses in qPCR negative samples. Perhaps unsurprisingly, the major advantage of RNAseq approaches is obtaining information around the host response simultaneously, and Wesolowska-Andersen et al. were able to demonstrate evidence of immune cell airway infiltration, downregulation of cilia genes, and type 2 inflammation in patients with high viral copy number.[67] A slight twist on this approach using RNAseq has also been used to define an infected transcriptome signature, and the signature used to identify patients with respiratory virial infections.[68]

Sequencing approaches are now routinely carried out in microbiome research, and we and others routinely use both metagenomic[69] and 16s rRNA gene sequencing.[70] However, neither of these approaches will detect RNA viruses as they both sequence DNA.

10.3.10 Future perspectives

It is highly likely that RV infections do not occur in isolation and are not always associated with symptoms, perhaps best exemplified by the concept of the respiratory virome.[71] What composition of, or perturbations of, the respiratory virome means for long-term patient outcomes is currently unknown. Furthermore, we should not consider RV infections in isolation, as the interaction between RV infection and other respiratory pathogens including bacteria and viruses infections is an area of clinical importance,[72] and is related to clinical outcomes.[69,73] Interestingly experimental RV infection has been shown to change the composition of the lung microbiome[74] but not the nasal or gut microbiome,[75] but more studies are needed to fully investigate this.

In the author's opinion it is highly likely that future studies incorporate RNAseq, bacterial microbiome through metagenomic and other omics platforms such as proteomics and metabolomics to fully understand the host's response to infection.

REFERENCES

1. Granados A, Goodall EC, Luinstra K, Smieja M, Mahony J. Comparison of asymptomatic and symptomatic rhinovirus infections in university students: incidence, species diversity, and viral load. *Diagn Microbiol Infect Dis*. 2015;82(4):292–296.
2. Mitchell AB, Mourad B, Buddle L, Peters MJ, Oliver BGG, Morgan LC. Viruses in bronchiectasis: a pilot study to explore the presence of community acquired respiratory viruses in stable patients and during acute exacerbations. *BMC Pulm Med*. 2018;18 (1):84.

3. Esquivel A, Busse WW, Calatroni A, et al. Effects of omalizumab on rhinovirus infections, illnesses, and exacerbations of asthma. *Am J Respir Crit Care Med.* 2017;196 (8):985–992.
4. Powell H, Smart J, Wood LG, et al. Validity of the common cold questionnaire (CCQ) in asthma exacerbations. *PLoS One.* 2008;3(3):e1802.
5. Basile K, Kok J, Dwyer DE. Point-of-care diagnostics for respiratory viral infections. *Expert Rev Mol Diagn.* 2018;18(1):75–83.
6. Ireland DC, Kent J, Nicholson KG. Improved detection of rhinoviruses in nasal and throat swabs by seminested RT-PCR. *J Med Virol.* 1993;40(2):96–101.
7. Turgeon N, Toulouse MJ, Martel B, Moineau S, Duchaine C. Comparison of five bacteriophages as models for viral aerosol studies. *Appl Environ Microbiol.* 2014;80 (14):4242–4250.
8. Gilkerson JR, Bailey KE, Diaz-Mendez A, Hartley CA. Update on viral diseases of the equine respiratory tract. *Vet Clin North Am Equine Pract.* 2015;31 (1):91–104.
9. Gralton J, McLaws ML, Rawlinson WD. Personal clothing as a potential vector of respiratory virus transmission in childcare settings. *J Med Virol.* 2015;87 (6):925–930.
10. L'Huillier AG, Tapparel C, Turin L, Boquete-Suter P, Thomas Y, Kaiser L. Survival of rhinoviruses on human fingers. *Clin Microbiol Infect.* 2015;21(4):381–385.
11. Gwaltney Jr. JM, Jordan Jr. WS. Rhinoviruses and respiratory disease. *Bacteriol Rev.* 1964;28:409–422.
12. Arruda E, Crump CE, Hayden FG. In vitro selection of human rhinovirus relatively resistant to soluble intercellular adhesion molecule-1. *Antimicrob Agents Chemother.* 1994;38(1):66–70.
13. Sachs LA, Schnurr D, Yagi S, Lachowicz-Scroggins ME, Widdicombe JH. Quantitative real-time PCR for rhinovirus, and its use in determining the relationship between TCID50 and the number of viral particles. *J Virol Methods.* 2011;171 (1):212–218.
14. Halonen P, Herholzer J, Ziegler T. Advances in the diagnosis of respiratory virus infections. *Clin Diagn Virol.* 1996;5(2-3):91–100.
15. Johnston SL, Pattemore PK, Sanderson G, et al. Community study of role of viral infections in exacerbations of asthma in 9-11 year old children. *Br Med J.* 1995;310 (6989):1225–1229.
16. Lewis-Rogers N, Seger J, Adler FR. Human rhinovirus diversity and evolution: how strange the change from major to minor. *J Virol.* 2017;91(7). pii: e01659-16.
17. McErlean P, Shackelton LA, Andrews E, et al. Distinguishing molecular features and clinical characteristics of a putative new rhinovirus species, human rhinovirus C (HRV C). *PLoS One.* 2008;3(4):e1847.
18. Hao W, Bernard K, Patel N, et al. Infection and propagation of human rhinovirus C in human airway epithelial cells. *J Virol.* 2012;86(24):13524–13532.
19. Bochkov YA, Palmenberg AC, Lee WM, et al. Molecular modeling, organ culture and reverse genetics for a newly identified human rhinovirus C. *Nat Med.* 2011;17 (5):627–632.
20. Sattar SA, Karim YG, Springthorpe VS, Johnson-Lussenburg CM. Survival of human rhinovirus type 14 dried onto nonporous inanimate surfaces: effect of relative humidity and suspending medium. *Can J Microbiol.* 1987;33(9):802–806.
21. Ansari SA, Springthorpe VS, Sattar SA, Rivard S, Rahman M. Potential role of hands in the spread of respiratory viral infections: studies with human parainfluenza virus 3 and rhinovirus 14. *J Clin Microbiol.* 1991;29(10):2115–2119.
22. Thomas Y, Vogel G, Wunderli W, et al. Survival of influenza virus on banknotes. *Appl Environ Microbiol.* 2008;74(10):3002–3007.

23. Yu ST, Thi Bui C, Kim DTH, VT Nguyen A, Thi Trinh TT, Yeo SJ. Clinical evaluation of rapid fluorescent diagnostic immunochromatographic test for influenza A virus (H1N1). *Sci Rep*. 2018;8(1):13468.
24. Yang M, Clavijo A, Graham J, Pasick J, Neufeld J, Berhane Y. Evaluation of diagnostic applications of monoclonal antibodies against avian influenza H7 viruses. *Clin Vaccine Immunol*. 2010;17(9):1398−1406.
25. Oberste MS, Maher K, Flemister MR, Marchetti G, Kilpatrick DR, Pallansch MA. Comparison of classic and molecular approaches for the identification of untypeable enteroviruses. *J Clin Microbiol*. 2000;38(3):1170−1174.
26. Papadopoulos NG, Bates PJ, Bardin PG, et al. Rhinoviruses infect the lower airways. *J Infect Dis*. 2000;181(6):1875−1884.
27. Tyrrell DA, Parsons R. Some virus isolations from common colds. III. Cytopathic effects in tissue cultures. *Lancet*. 1960;1(7118):239−242.
28. Papadopoulos NG, Sanderson G, Hunter J, Johnston SL. Rhinoviruses replicate effectively at lower airway temperatures. *J Med Virol*. 1999;58(1):100−104.
29. Hansel TT, Tunstall T, Trujillo-Torralbo MB, et al. A comprehensive evaluation of nasal and bronchial cytokines and chemokines following experimental rhinovirus infection in allergic asthma: increased interferons (IFN-gamma and IFN-lambda) and type 2 inflammation (IL-5 and IL-13). *EBioMedicine*. 2017;19:128−138.
30. Huynh KN, Oliver BG, Stelzer S, Rawlinson WD, Tovey ER. A new method for sampling and detection of exhaled respiratory virus aerosols. *Clin Infect Dis*. 2008;46(1):93−95.
31. Fabian P, McDevitt JJ, Lee WM, Houseman EA, Milton DK. An optimized method to detect influenza virus and human rhinovirus from exhaled breath and the airborne environment. *J Environ Monit*. 2009;11(2):314−317.
32. Houspie L, De Coster S, Keyaerts E, et al. Exhaled breath condensate sampling is not a new method for detection of respiratory viruses. *Virol J*. 2011;8:98.
33. Stelzer-Braid S, Oliver BG, Blazey AJ, et al. Exhalation of respiratory viruses by breathing, coughing, and talking. *J Med Virol*. 2009;81(9):1674−1679.
34. Tovey ER, Stelzer-Braid S, Toelle BG, et al. Rhinoviruses significantly affect day-to-day respiratory symptoms of children with asthma. *J Allergy Clin Immunol*. 2015;135(3):663−669.e12.
35. Mitchell AB, Mourad B, Tovey E, et al. Spirometry filters can be used to detect exhaled respiratory viruses. *J Breath Res*. 2016;10(4):046002.
36. Larsson P, Bake B, Wallin A, et al. The effect of exhalation flow on endogenous particle emission and phospholipid composition. *Respir Physiol Neurobiol*. 2017;243:39−46.
37. Schwarz K, Biller H, Windt H, Koch W, Hohlfeld JM. Characterization of exhaled particles from the healthy human lung—a systematic analysis in relation to pulmonary function variables. *J Aerosol Med Pulmonary Drug Deliv*. 2010;23(6):371−379.
38. Mallia P, Message SD, Kebadze T, Parker HL, Kon OM, Johnston SL. An experimental model of rhinovirus induced chronic obstructive pulmonary disease exacerbations: a pilot study. *Respir Res*. 2006;7:116.
39. Aebischer A, Beer M, Hoffmann B. Development and validation of rapid magnetic particle based extraction protocols. *Virol J*. 2014;11:137.
40. Kim Y, Han MS, Kim J, Kwon A, Lee KA. Evaluation of three automated nucleic acid extraction systems for identification of respiratory viruses in clinical specimens by multiplex real-time PCR. *Biomed Res Int*. 2014;2014:430650.
41. Perandin F, Pollara PC, Gargiulo F, Bonfanti C, Manca N. Performance evaluation of the automated NucliSens easyMAG nucleic acid extraction platform in comparison with QIAamp Mini kit from clinical specimens. *Diagn Microbiol Infect Dis*. 2009;64(2):158−165.

42. Verheyen J, Kaiser R, Bozic M, Timmen-Wego M, Maier BK, Kessler HH. Extraction of viral nucleic acids: comparison of five automated nucleic acid extraction platforms. *J Clin Virol*. 2012;54(3):255–259.
43. Zhang D, Lou X, Yan H, et al. Metagenomic analysis of viral nucleic acid extraction methods in respiratory clinical samples. *BMC Genom*. 2018;19(1):773.
44. Van Gelder RN, von Zastrow ME, Yool A, Dement WC, Barchas JD, Eberwine JH. Amplified RNA synthesized from limited quantities of heterogeneous cDNA. *Proc Natl Acad Sci USA*. 1990;87(5):1663–1667.
45. Sedlak RH, Nguyen T, Palileo I, Jerome KR, Kuypers J. Superiority of digital reverse transcription-PCR (RT-PCR) over real-time RT-PCR for quantitation of highly divergent human rhinoviruses. *J Clin Microbiol*. 2017;55(2):442–449.
46. Dagher H, Donninger H, Hutchinson P, Ghildyal R, Bardin P. Rhinovirus detection: comparison of real-time and conventional PCR. *J Virol Methods*. 2004;117 (2):113–121.
47. Arruda E, Hayden FG. Detection of human rhinovirus RNA in nasal washings by PCR. *Mol Cell Probes*. 1993;7(5):373–379.
48. Andeweg AC, Bestebroer TM, Huybreghs M, Kimman TG, de Jong JC. Improved detection of rhinoviruses in clinical samples by using a newly developed nested reverse transcription-PCR assay. *J Clin Microbiol*. 1999;37(3):524–530.
49. Rawlinson WD, Waliuzzaman Z, Carter IW, Belessis YC, Gilbert KM, Morton JR. Asthma exacerbations in children associated with rhinovirus but not human metapneumovirus infection. *J Infect Dis*. 2003;187(8):1314–1318.
50. Vogelstein B, Kinzler KW. Digital PCR. *Proc Natl Acad Sci USA*. 1999;96 (16):9236–9241.
51. Saiki RK, Gelfand DH, Stoffel S, et al. Primer-directed enzymatic amplification of DNA with a thermostable DNA polymerase. *Science*. 1988;239(4839):487–491.
52. Hall Sedlak R, Jerome KR. The potential advantages of digital PCR for clinical virology diagnostics. *Expert Rev Mol Diagn*. 2014;14(4):501–507.
53. Hindson BJ, Ness KD, Masquelier DA, et al. High-throughput droplet digital PCR system for absolute quantitation of DNA copy number. *Anal Chem*. 2011;83 (22):8604–8610.
54. Hayden RT, Gu Z, Ingersoll J, et al. Comparison of droplet digital PCR to real-time PCR for quantitative detection of cytomegalovirus. *J Clin Microbiol*. 2013;51 (2):540–546.
55. Kuroda M, Niwa S, Sekizuka T, et al. Molecular evolution of the VP1, VP2, and VP3 genes in human rhinovirus species C. *Sci Rep*. 2015;5:8185.
56. Lamson D, Renwick N, Kapoor V, et al. MassTag polymerase-chain-reaction detection of respiratory pathogens, including a new rhinovirus genotype, that caused influenza-like illness in New York State during 2004–2005. *J Infect Dis*. 2006;194 (10):1398–1402.
57. Calvo C, Garcia ML, Pozo F, Reyes N, Perez-Brena P, Casas I. Role of rhinovirus C in apparently life-threatening events in infants, Spain. *Emerg Infect Dis*. 2009;15 (9):1506–1508.
58. Xiang Z, Gonzalez R, Xie Z, et al. Human rhinovirus group C infection in children with lower respiratory tract infection. *Emerg Infect Dis*. 2008;14(10):1665–1667.
59. Wark PA, Tooze M, Powell H, Parsons K. Viral and bacterial infection in acute asthma and chronic obstructive pulmonary disease increases the risk of readmission. *Respirology*. 2013;18(6):996–1002.
60. Zhao Y, Shen J, Wu B, Liu G, Lu R, Tan W. Genotypic diversity and epidemiology of human rhinovirus among children with severe acute respiratory tract infection in Shanghai, 2013-2015. *Front Microbiol*. 2018;9:1836.

61. Wisdom A, Leitch EC, Gaunt E, Harvala H, Simmonds P. Screening respiratory samples for detection of human rhinoviruses (HRVs) and enteroviruses: comprehensive VP4-VP2 typing reveals high incidence and genetic diversity of HRV species C. *J Clin Microbiol.* 2009;47(12):3958–3967.
62. Palmenberg AC, Spiro D, Kuzmickas R, et al. Sequencing and analyses of all known human rhinovirus genomes reveal structure and evolution. *Science.* 2009;324 (5923):55–59.
63. Mardis ER. Next-generation DNA sequencing methods. *Annu Rev Genomics Hum Genet.* 2008;9:387–402.
64. Cheung F, Haas BJ, Goldberg SM, May GD, Xiao Y, Town CD. Sequencing Medicago truncatula expressed sequenced tags using 454 Life Sciences technology. *BMC Genom.* 2006;7:272.
65. Bainbridge MN, Warren RL, Hirst M, et al. Analysis of the prostate cancer cell line LNCaP transcriptome using a sequencing-by-synthesis approach. *BMC Genom.* 2006;7:246.
66. Yick CY, Zwinderman AH, Kunst PW, et al. Transcriptome sequencing (RNA-Seq) of human endobronchial biopsies: asthma versus controls. *Eur Respir J.* 2013;42 (3):662–670.
67. Wesolowska-Andersen A, Everman JL, Davidson R, et al. Dual RNA-seq reveals viral infections in asthmatic children without respiratory illness which are associated with changes in the airway transcriptome. *Genome Biol.* 2017;18(1):12.
68. Landry ML, Foxman EF. Antiviral response in the nasopharynx identifies patients with respiratory virus infection. *J Infect Dis.* 2018;217(6):897–905.
69. Turturice BA, McGee HS, Oliver B, et al. Atopic asthmatic immune phenotypes associated with airway microbiota and airway obstruction. *PLoS One.* 2017;12(10): e0184566.
70. Copeland E, Leonard K, Carney R, et al. Chronic rhinosinusitis: potential role of microbial dysbiosis and recommendations for sampling sites. *Front Cell Infect Microbiol.* 2018;8:57.
71. Mitchell AB, Oliver BGG, Glanville AR. Translational aspects of the human respiratory virome. *Am J Respir Crit Care Med.* 2016;194(12):1458–1464.
72. Oliver BG, Lim S, Wark P, et al. Rhinovirus exposure impairs immune responses to bacterial products in human alveolar macrophages. *Thorax.* 2008;63(6):519–525.
73. Wang Z, Bafadhel M, Haldar K, et al. Lung microbiome dynamics in COPD exacerbations. *Eur Respir J.* 2016;47(4):1082–1092.
74. Molyneaux PL, Mallia P, Cox MJ, et al. Outgrowth of the bacterial airway microbiome after rhinovirus exacerbation of chronic obstructive pulmonary disease. *Am J Respir Crit Care Med.* 2013;188(10):1224–1231.
75. Lehtinen MJ, Hibberd AA, Mannikko S, et al. Nasal microbiota clusters associate with inflammatory response, viral load, and symptom severity in experimental rhinovirus challenge. *Sci Rep.* 2018;8(1):11411.

INDEX

Note: Page numbers followed by "*f*" and "*t*" refer to figures and tables, respectively.

K

L

M

N

O

P

FSC
www.fsc.org
MIX
Papier | Fördert gute Waldnutzung
FSC® C083411

Kolibri 360 GmbH
Ferdinand-Jühlke-Straße 7
99095 Erfurt, Deutschland
produktsicherheit@kolibri360.de